Towards a Just and Ecologically Sustainable Peace

Joseph Camilleri · Deborah Guess
Editors

Towards a Just and Ecologically Sustainable Peace

Navigating the Great Transition

palgrave
macmillan

Editors
Joseph Camilleri
La Trobe University
Melbourne, VIC, Australia

Deborah Guess
University of Divinity
Melbourne, VIC, Australia

ISBN 978-981-15-5020-1 ISBN 978-981-15-5021-8 (eBook)
https://doi.org/10.1007/978-981-15-5021-8

This Palgrave Macmillan imprint is published by the registered company Springer Nature Singapore Pte Ltd.
The registered company address is: 152 Beach Road, #21-01/04 Gateway East, Singapore 189721, Singapore

Preface

This volume is part of a wider project which has gathered momentum over a number of years. It began with a Symposium *Ecological Aspects of War* held on September 2015 at Trinity College, Melbourne, in which scholars from Australia considered critically the intersections between war and ecology. From this symposium emerged two scholarly books: *Ecological Aspects of War: Engagements with Biblical Texts*, edited by Keith Dyer and Anne Elvey, with Deborah Guess, (Bloomsbury T&T Clark, 2017), and *Ecological Aspects of War: Religious and Theological Perspectives from Australia*, edited by Anne Elvey, Deborah Guess and Keith Dyer. (Adelaide, ATF Press: A Forum for Theology in the World, 2017), and a journal article: Anne Elvey, Keith Dyer and Deborah Guess, 'Ecological Aspects of War—Imagining Creaturely Mission', *Australian Journal of Mission Studies* 10, no. 2 (December 2016): 40–48.

Subsequently, the project more intentionally incorporated the theme 'Towards a Just and Ecologically Sustainable Peace' with Professor Joseph Camilleri as Chair. The project gave rise to several events, the first of which was a reflective public forum held at the Islamic Council of Victoria in Melbourne on Remembrance Day 2018 titled *A World at Peace with Itself: Elusive Dream or Achievable Goal?* attended by 100 people. It was followed by the centrepiece of the project, a major two-day conference *Earth at Peace* held at Pilgrim Theological College, Parkville, Melbourne, in April 2019 attended by 140 people.

The 2019 conference was unique and significant. It intentionally sought to make connections between scholarly and practical dimensions, with many of its presenters encompassing the worlds of academia and civic engagement. It had the usual components of a conference program (with keynote presentations and a number of panels) but also more dynamic aspects, including seven working groups tasked with recommending specific courses of action, a plenary interactive session with international speakers, and a hypothetical scenario on how Australia might handle a mass influx of climate change refugees from neighbouring Pacific Islands in line with the principles of a just and ecologically sustainable peace. Another highlight of the conference was a public forum (which attracted an audience of 200 people) with interactive involvement from keynote speakers including Behrouz Boochani, a refugee then held in detention by the Australian Government, an Iranian-Kurdish journalist and writer of the award-winning book *No Friend But the Mountains: Writing from Manus Prison*.

Another notable dimension of the Earth at Peace project was the emphasis on linking thought, action, poetry and art. The conference venue hosted a major six-week exhibition titled *Just Art* by the noted peace artist William Kelly as well as a cultural evening of live music and poetry performances which responded both to the conference theme and to Kelly's artwork.

This volume is an integral part of the Earth at Peace project and a direct outcome of the conference. It includes contributions from keynote speaker Heather Eaton and others involved in the conference. A subsidiary volume focusing on the project's art and poetry component is currently in preparation.

Two organisations were integral in planning and funding the conference: the Centre for Research in Religion and Social Policy at the University of Divinity, and Pax Christi Australia. In addition, we are indebted to a number of organisations which either financially sponsored or in some other way supported the conference. These include: Australian Association of the Study of Religions; Australian Research Theology Foundation Inc.; Borderlands Co-operative; Christian Brothers; Ethos—Centre for Christianity and Society; Faith Communities Council of Victoria; Global Reconciliation; Islamic Council of Victoria; Journeys for Climate Justice; Loreto Sisters; Medical Association for the Prevention of War; Melbourne Unitarian Peace Memorial Church; Peace and Conflict Studies Institute Australia (PaCSIA); Pilgrim Theological College, University of Divinity; Social Policy Connections; Society of Friends, Victorian Regional Meeting; Redemptorist Congregation; Religions for Peace; Trinity College Theological School, University of Divinity; Victorian Council of Churches. We express our appreciation to the many individual donors, with special thanks to the generous contribution by Elizabeth Proust and Brian Lawrence. We are also grateful to this volume's two anonymous peer reviewers whose insights provided valuable feedback to our contributors.

Melbourne, Australia
January 2020

Joseph Camilleri
Deborah Guess

Contents

List of Contributors

Tony Birch Victoria University, Carlton, VIC, Australia

Mark G. Brett Whitley College, University of Divinity, Kew, VIC, Australia

Joseph Camilleri La Trobe University, Melbourne, VIC, Australia

Bruce Duncan University of Divinity, Melbourne, VIC, Australia

Heather Eaton Saint Paul University, Ottawa, ON, Canada

Anne Elvey Monash University, Melbourne, VIC, Australia; University of Divinity, Melbourne, VIC, Australia

Salim Farrar Sydney Law School, University of Sydney, Camperdown, NSW, Australia

Deborah Guess Pilgrim Theological College, University of Divinity, Melbourne, VIC, Australia

Norman Habel Flinders University, Bedford Park, SA, Australia

Shelini Harris Australian National University, ACT, Canberra, Australia

Zuleyha Keskin Charles Sturt University, Melbourne, VIC, Australia

Freya Mathews La Trobe University, Melbourne, VIC, Australia

Mehmet Ozalp Charles Sturt University, Melbourne, VIC, Australia

Allan Patience School of Social and Political Sciences, University of Melbourne, Melbourne, VIC, Australia

Ariel Salleh School of Social and Political Sciences, University of Sydney, Sydney, NSW, Australia

Chaiwat Satha-Anand Thammasat University, Bangkok, Thailand

CHAPTER 1

Introduction

Deborah Guess

This is not the first point in history when people have recognised that they live at a time of existential crisis and transition: in some periods human beings have believed that something like the 'end times' might be approaching. To look at the apprehensions of former times through the lens of historical hindsight can bring a comforting sense of proportion lacking in some current dystopian projections. Yet we cannot ignore the record of many societies that suffered massive trauma, decline and ultimate destruction.[1] It seems undeniable that we are, at the very least, living at a time of significant transition. This is evident in rapidly accelerated social change, increasing violence, the continuing risk posed by nuclear weapons, COVID-19, and, in particular, the phenomenon of climate change and the tangible evidence it provides that the systems which support life on Earth are becoming increasingly unstable. For some

[1] For example, see: Jared Diamond, *Collapse: How Societies Choose to Fail or Survive* (London: Allen Lane, 2005); Ronald Wright, *A Short History of Progress* (Toronto: Anansi, 2004).

D. Guess (✉)
Pilgrim Theological College, University of Divinity, Melbourne, VIC, Australia

J. Camilleri and D. Guess (eds.), *Towards a Just and Ecologically Sustainable Peace*, https://doi.org/10.1007/978-981-15-5021-8_1

thinkers there is increasing consensus that we are at the very least facing crisis, probably collapse, perhaps even extinction.[2]

The unprecedented attention given in recent years to a range of ecological questions that coalesce around climate disruption has sanctioned the largely unheeded warnings of environmentalists over the past half-century and more. At the present time, protests and calls for action serve to reflect, and advance, a considerable amount of social ferment which, among other things, is challenging some of the central tenets of Western modernity such as the growth imperative.

The idea that a sharp distinction exists between nature and culture is being repudiated on two sides: by the evidence of increasingly frequent and destructive events such as drought, fire and flood that palpably manifest the reality of human-made climate change; and in the renaming of our present epoch as the Anthropocene (the epoch of the human), an idea that premises an intersection—perhaps it is better described as a collision—between human history and culture on the one hand and earth history and geology on the other. Much about the concept of the Anthropocene is open for further discussion,[3] but its basic thesis is now broadly accepted that humankind is a geological agent who has exceeded planetary limits and has irreversibly altered the planet's geological make-up: the transition into this new geological era is said to be 'unquestionably the deepest and most profound event in recent history'.[4]

Given the prominence of the issue of climate disruption, the problem of other forms of violence might seem more marginal. Yet this is far from the case. In recent times military spending has continued to expand, global conflicts have escalated, and technological development has made weapons increasingly lethal. When the overarching questions of ecology, violence and peace are held together, a clearer understanding emerges of their interrelationship. Military activity not only has a destructive impact on the other-than-human world, it also magnifies social inequality and intensifies an ethos of violence. Similarly, environmental factors such as

[2] Collapse, catastrophe and extinction are three possible future scenarios discussed by Jem Bendell, 'Deep Adaptation: A Map for Navigating Climate Tragedy', *IFLAS Occasional Paper*, no. 2 (2018): 19–21.

[3] For discussion of the concept of the 'Anthropocene', and competing understandings of when the transition to the Anthropocene began, see Chris Otter et al., 'Roundtable: The Anthropocene in British History', *Journal of British Studies* 57 (2018): 568–96.

[4] Ibid., 568.

changes in land use or an altered climate can cause famine, poverty and dislocation which in turn lead to social unrest, violence and war: even in times of peace, it can be said that: 'the production, acquisition and transfer of arms represent a pernicious misallocation of human and financial resources, a profound disregard for the cries of the poor and the cries of the Earth'.[5]

In light of this, it is unsurprising that there is increasing critique of the centuries-old and already weakened 'just war' theory.[6] The dominant strand in the Christian tradition, and until recently in international law, the intention of the just war doctrine was to place conditions before the use of force can be deemed just, but instead it has often been used to justify rather than prevent or condemn war. Conscious of the fact that the 'just war' concept has little to say about social, economic or environmental justice, and therefore about the present global predicament, several voices in international law, in the UN system, in labour and social movements, in intellectual circles, in the Christian churches and other religious and ethical traditions are calling for a shift from 'just war' to 'just peace', and more broadly to the concept of a just and ecologically sustainable peace which seeks a confluence of peace, justice and environment. In the opinion of one of the editors of this book, this confluence will require a fundamental intellectual and ethical shift towards 'an inclusive ethic which transcends the five maladies of our time: parochialism, populism, militarism, extremism and anthropocentrism'.[7]

The question of how to make the transition to this inclusive ethos forms the bedrock of this book. Given the increasing globalisation of violence against humans and the wider Earth community, a defining question of our time, and the central focus of this volume, is: Where can humanity turn for guidance and inspiration, and what conditions might be required for a shift to a peace which is just and ecologically sustainable? This collection of essays focuses on some of the theoretical principles—philosophical, ecological and spiritual/religious—that can guide this shift,

[5] Joseph Camilleri, 'Australia Is Sleepwalking into an Orgy of Militarism', *New Matilda*, 13 March 2019.

[6] Joseph A. Camilleri to John Menadue—Pearls and Irritations, 11 February 2019, https://johnmenadue.com/joseph-a-camilleri-just-peace-the-only-antidote-to-the-age-of-violence-part-1/?fbclid=IwAR1h-KhbAwIBLjfXX8YYxaggJyV9HbBjGtel_DteDcq-Mk0BE9F67BICbTU.

[7] Ibid.

and considers some of the many ways that it might be expressed in practical and active ways.

This volume necessarily adopts a multifaceted approach. The often rigid division between fields of study that underlay Enlightenment thinking has today become less tenable as the decline of that vision has acknowledged the complexity and diversity of human societies, and as recognition has grown of the interdependence of life forms and their reliance on the good functioning of Earth systems. The issues currently facing humankind can most thoroughly be interrogated by using a number of different lenses and a plurality of narratives. No single approach which focuses on a particular human tendency, a single political-economic approach or one philosophical or religious idea is likely to adequately encompass the scale or complexity of the interrelationship between violence, ecological catastrophe and social justice. A multifaceted approach befits the present time when the validity of the 'grand narrative' has been increasingly contested and dismantled. There is therefore a broad thematic scope to this book which gathers Indigenous scholars, peace studies and non-violence specialists, scholars of international relations, philosophers, ecological feminists, ecological theologians and religious scholars, both Australian and international, to address these questions from their combined professional and activist expertise. This multidisciplinary approach reveals and reinforces the integral connections between the key topics of justice, peace and ecology.

Although in many ways international in scope, this volume has taken shape in the Asia-Pacific region and specifically in Australia and its unique situation. Under-acknowledged colonial contact wars, Indigenous relationship to and cultivation of land (Country) and continuing Indigenous calls for social and ecological justice comprise an important part of the book's framework. Since at least the early 1990s, Australia has been the site of pioneering work in ecological philosophy, ecological feminism and ecological biblical hermeneutics (as evidenced in the internationally acclaimed work of the late thinkers Val Plumwood and Denis Edwards, and our contributors Ariel Salleh, Freya Mathews and Norman Habel). In Australia, issues of imperial power, land use (notably mining),[8] water mismanagement and deforestation, inadequate responses to climate

[8] Australia is one of the world's largest coal exporters yet the present government supports the opening of a number of new coal mines. For further discussion, see Chapter 12 of this volume by Tony Birch.

change, military alliances, treatment of refugees and asylum seekers, and Islamophobia, such as enacted by an Australian in the tragic shootings at two mosques in Christchurch on 15 March 2019, call for mature reflection, sustained dialogue and appropriate political responses.

The primary thesis of this volume is a fundamentally hopeful one, that a just and ecologically sustainable peace is urgently needed. The chapters of this book reflect the rich variety of its authors' areas of expertise and practical engagement. Although each contribution is unique, dominant threads from one contributor at times intersect with those of other contributors. This makes for a dense yet holistic tapestry, rich in diversity yet coalescing at distinct points into comprehensive intellectual discourse and suggestions for action that, in the final two chapters and in the two appendices, are drawn together to form principles and actions that begin to articulate some of the foundations of a just and ecologically sustainable peace.

Structurally, the book is divided into three main parts: (I) *In Search of a Holistic Approach*; (II) *Cosmological and Religious Perspectives*; and (III) *Questioning the Colonial Mindset*. Acknowledging the interconnectedness of an increasingly complex world, the book's tripartite structure enables it both to encompass the international scope of broader intellectual, philosophical, religious and political questions, and address or exemplify those questions in localised and tangible ways.

Part 1 (*In Search of a Holistic Approach*) sets the scene for the volume as a whole, naming various types of violence and exploring the underlying concept of just peace. Its deeply rooted philosophical and political themes are international in vision even while they also hone in on local or specific questions. All four chapters share a concern with the problem of violence towards the 'other' and call, each in its particular way, for a nonviolent ethic that entails dialogue, reciprocity, listening and inclusiveness.

The section begins with Chapter 2 where Joseph Camilleri examines different forms of violence (physical, structural/social and ecological) and notes the premise that violence 'has become a defining feature of life in the twentieth and twenty-first centuries'. He goes on to draw out the connections between the types of violence and suggests that violence and the indifference which accompanies it are the outcome of exclusionist mindsets, policies and institutions, and in response identifies the key to the holistic notion of a just and ecologically sustainable peace in what he call an 'inclusivist ethic'. Camilleri revisits just war theory, summarising its development and discussing its numerous weaknesses, not least its inability to address the ecological turn. As a reaction to the impasse to which just war thinking has led us, he traces the emergence of the concept of just peace, and the way it is beginning to intrude with increasing force in both secular and religious circles—in the Christian and Islamic traditions, UN pronouncements and initiatives, and the Earth Charter. Rich possibilities, Camilleri suggests, are offered by dialogue, at the core of which lies the practice of truly listening to the other. Here, he references Tu Weiming's neo-Confucian concept of 'anthropocosmic unity' which rejects anthropocentrism and suggests that social and ecological harmony are achieved by a form of self-cultivation, adopted both individually and communally, and summarised in terms of four pillars: self, community, Earth and Heaven. Camilleri applies Tu Weiming's idea to three related fields of engagement: citizenship (which engages not just a dedicated few but a sizeable and reasonable cross-section of society); governance (which integrates inclusive decision-making processes and institutions); and radically inclusive dialogue (which encompasses diverse cultures and civilisations). Finally, he envisages a 'dialogical citizenship' grounded in community education, reworked educational curricula, 'an imagination that is cross-national, cross-cultural and planetary in scope and inspiration' and communal dialogical encounters.

In Chapter 3, Heather Eaton continues the discussion of the nature of violence, acknowledging that humanity is at a crossroads where many pathways intersect. Her compelling argument for a strategy of nonviolence that might lead to a just and ecologically sustainable peace focuses on three interrelated themes. First, various forms of violence have become normalised both physically and culturally/socially and are exemplified in violence against women, expressed in physical and sexual violence, as well as in a lack of human rights and gender injustice, and in violence towards the entire Earth community—a moral and spiritual offence which calls for the counter-response of ecological integrity. Second, she argues for developing 'nonviolent resistance, and social imaginaries, or visions, that provide direction, inspiration and energy'. Nonviolence involves disengaging, through education and training, with domination, and deploying active resistance so that violence is transformed 'into relations of reciprocity, respect and interdependence'. Eaton insists that although there is a place for analysis and critique in social transformation, the development of worldviews and social imaginaries is more important: ethical positions, prophetic voices and the quality of courage are better able to address the danger and despair that ecological activism entails than analysis alone. Bringing about peace for the entire Earth community, she says, requires becoming ecologically literate—learning how ecosystems operate and fully accepting the extent to which human beings are embedded in Earth systems. Third, in common with other contributors to this volume, Eaton argues that there is a need for greater capacity for wonder or awe (aspects not only of religion and philosophy but also of science) which is able to bypass analysis and critique and be in themselves educational, persuasive and transformative.

The essay by Ariel Salleh (Chapter 4) continues Eaton's focus on violence towards women and Earth by providing a cogent and insightful eco-feminist perspective on the current crisis. She discusses how women are particularly oppressed by wars and other forms of patriarchal violence and argues that a key to justice and peace involves critiquing more stringently the idea and practice of 'othering' or objectifying. She notes that treating nature as a resource 'is foundational to the operation of patriarchal, colonial, and capitalist power today' and that power of all kinds involves some form of cultural 'othering': patriarchy 'others' women and capitalism 'others' workers. To this end, Salleh creatively reinterprets the 'Great Chain of Being' model as the 'Great Chain of Othering', arguing that violent othering reflects an inability to cope with difference and leads to ecological and social degradation. The way to counter the system of 'othering', she suggests, is to adopt 'holding' activities (exemplified in Aldo Leopold's land ethic and in global social movements endorsing simpler ways of living) which instead of leading to conflict and entropy 'are integrative and energise life-affirming exchanges between human and natural processes'.

Freya Mathews' chapter concludes the first part of the book, offering a carefully argued and imaginative analysis that effectively draws on both new ecological perspectives and older theories, notably historical materialism. She begins by asking the question: Why have ecological philosophy and associated disciplines failed to bring about a sufficient change in consciousness and action? Mathews identifies dualism as the philosophical principle which underpins both the objectification of nature and the domination of human others—whether the other be women or the working class. Using a particular understanding of historical materialism, she argues that social and economic systems deeply influence the dominant ideology and prevent new values from being put into practice; therefore, ideological structures are unlikely to be changed by critical argument or moral persuasion. On the other hand, Mathews suggests that the human urge for affiliation may be more powerful even than the limitations of a society's social-economic system. Foreshadowing Part II of the book, Matthews argues for a change of consciousness and a new social imaginary, perhaps through a new non-anthropocentric religion/cosmology based around a community which: gives priority to science; is fundamentally Earth-friendly; and has a strong praxis component, actively participating in on-the-ground conservation and observation, as exemplified by the Australian Aboriginal practice of 'walking the land'.

Part II: Cosmological and Religious Perspectives

Much of the broad philosophical scope of Part I is reflected, either explicitly or implicitly, in Part II which explores ideas and practices from some of the world's major religious traditions. Suggestions made in Part I such as a revised or new religion or cosmology (Camilleri, Mathews) or a renewed sense of awe and wonder (Eaton), are taken up in this section by engaging the overall theme of a just and ecologically sustainable peace with specific religious traditions, texts and practices. Essays on two representative major faith traditions—Islam and Christianity—are followed by two broader essays that reference a wider range of spiritual and religious ideas and practices.

Zuleyha Keskin and Mehmet Ozalp (Chapter 6) open this section by expounding an Islamic approach to ecology and peace which opens up rich possibilities for dialogue with other faith traditions. Drawing on the Qur'an and on various exegetes and theologians including twentieth-century Muslim scholar Said Nursi, three main ideas are identified. First, human beings have the potential to be something more than whatever they currently are: a potential that can be extended to apply to all of creation, making it possible to bring about a closer relationship between humankind and its environment. Second, knowledge of God and of the natural world is central to the Islamic tradition: nature mirrors God's names and attributes and contains signs of God; therefore, it is instrumental in facilitating knowledge of God and can be said to have value. Third, all of creation has a purpose 'which it yearns to fulfil' and which involves interdependence between communities of creatures, a phenomenon that is also understood to be fundamental to the way ecosystems operate. Keskin and Ozalp argue that these concepts, linked as they are by the overarching notions of harmony, balance and the universal principle of justice in the Islamic worldview, contribute to the understanding of a just and ecologically sustainable peace.

A practical and specific insight from the Islamic tradition which draws on a legal background is found in a detailed study by Salim Farrar in Chapter 7 which makes a strong case for the importance of truth and reconciliation commissions (TRCs) in Islamic countries. Drawing on Islamic scholarship, Farrar describes the way TRCs operate and shows how they are consistent with central tenets of Islamic thought. His two excellent case studies show how TRCs in Islamic settings (the Aceh TRC and the Moroccan ERC) have functioned, followed by an overview of the wider role such bodies are able to provide. Importantly, the chapter seeks to extend the relevance of such commissions beyond human rights to encompass the wider topic of environmental destruction, thereby emphasizing the value of the concept of stewardship in the Islamic tradition.

In Chapter 8, biblical scholar Norman Habel uses methodology developed in the *Earth Bible* series to reinterpret key concepts from the Wisdom scriptures. He notes that the Christian tradition has at times rather readily accepted as normative various biblical traditions which have not promoted peace in or with Earth, and it has tended to ignore the tradition of innate Wisdom, which Habel understands to be the inherent cosmic driving force in nature. He argues that in the Wisdom tradition (especially in the text of Job 28), God both observes and fixes innate wisdom in the planet. The notion of God as observer of nature, he argues, provides a model for human beings to develop a far greater habit of observing nature, a quality which might lead to the ecologically important attitude of awe and wonder. The book of Job, Habel argues, can be understood as rejecting the idea of a God of retributive justice in favour of the idea of a God of wisdom. Finally, it is human discernment of innate wisdom (equivalent to eco-Wisdom) that leads to a peace consciousness. Habel's reflections in this chapter and at other interactive sessions in the 2019 conference (mentioned in the Preface to this volume) are given further expression in his two appendices in this book: Appendix A gives the background and principles for Earth Consciousness, and Appendix B offers six principles for making peace with the land of Australia.

Also from the Christian tradition, Bruce Duncan (Chapter 9) discusses the statements of Pope Francis in relation to social and ecological justice, referencing the encyclical *Laudato Si'* as well as other writings and actions. Duncan expounds Francis's extensive background in Latin American social justice, noting the way this has led, especially in *Laudato Si'*, to a concern for the well-being of all forms of life on earth and to Francis's unflinching critique of the current economic system and the ideology that underpins it. He also notes some of the strong reactions to the Pope's thought from commercial and business interests and from Catholic conservatives. In this exposition of the Pope's moral agenda, Duncan demonstrates the practical ways in which religious authority can promote change for the common good and points to possible future action by the Christian churches. Importantly, the chapter also indicates the way that interfaith dialogue and collaboration have resulted from Pope Francis's initiatives in the areas of ecology, peace and social justice.

In Chapter 10, Shelini Harris takes a wider philosophical, religious and spiritual perspective, articulating a critique (following Val Plumwood and others) of the frameworks of narrow rationality and economic calculus that have endorsed ecological destruction by supporting the idea of a separation between humankind and nature. Theoretically speaking, she argues, the major obstacle to the whole endeavour of peace with Earth is the Euro-centric mechanistic notion that 'what cannot be measured is considered not to exist', a notion that is also expressed in practices of Western-trained environmental managers. Instancing a number of religious conceptions and practices drawn mainly from the Buddhist and Hindu traditions, Harris suggests that religion and spirituality are often (though not always or necessarily) able to create a changed mindset which resists dualism and rationality in favour of an ecologically positive understanding of the interconnectedness of reality. The argument is given tangible expression in the example of a Buddhist soil/water conservation system used in Sri Lanka. Harris considers that a just and ecologically sustainable peace can be attained by giving a greater priority to the non-dualistic epistemologies that reside in the spiritual values and goals that are found within some Eastern approaches as well as in Western thinkers such as Val Plumwood.

In the final chapter of this section, Chaiwat Satha-Anand continues Harris' topic of the epistemology of rationalism to ask whether there is an alternative mode of being human that could be considered as a possible philosophical replacement of the dominant one. Again resonating with the volume's prior theme of how we are to respond to the 'other', Satha-Anand begins by framing his question in relation to some of the highly pragmatic ways that Western popular culture has responded to ecological questions. He then contests the dominant epistemic belief, originating with Descartes, that Satha-Anand considers to have colonised the way in which we think. He goes on to suggest an alternative epistemology based on 'breathing', a concept whose resonances he locates in Buddhist and Islamic spiritual teachings, and which he argues are exemplified in the practical action taken by the women of the Chipko movement in India, where villagers successfully employed non-violent action to protect their forest from logging.

Part III: Questioning the Colonial Mindset from Australian Perspectives

The third part of the book continues to engage with a number of previous themes, fleshing them out in relation to the Australian situation. The volume's international scope is maintained in the theme of colonisation and its effects. The first two chapters discuss colonisation directly in relation to Indigenous Australia; the last chapter deals with the question of present-day Australia's international relations.

As an indigenous Australian, Tony Birch (Chapter 12) begins by discussing the way that ecological problems are caused primarily by richer and northern/western nations yet disproportionately affect poorer and Indigenous peoples. The deeper issues associated with climate change, he says, have been well known to Indigenous people since white colonisation: 'Within the Australian context climate injustice is a two-hundred-year (and counting) project'; therefore, climate change cannot be tackled without taking climate justice seriously. In a way that intersects with themes from the first part of this book, Birch points out that the legacy of colonialism is a dual violence against nature and against Indigenous peoples, and he recounts the agricultural, extractive and manufacturing industries that put pressure on both Indigenous people and the Australian environment, these being sheep/wool, gold and—particularly—coal. Understanding Indigenous opposition to the currently proposed Adani coal mine can, argues Birch, help non-Indigenous Australians to realise 'that country is an autonomous entity with inherent rights rather than a resource to be endlessly plundered'. Birch concludes by asking a crucial question: Whether the Australian nation is 'mature enough to accept an invitation to stand with Indigenous people to protect country and, in doing so, break the shackles of the destructive logic of colonialism?'

Chapter 13 addresses some of the fundamental philosophical questions previously raised and also tracks and expounds Birch's thought. Anne Elvey's focus is on indigenous dispossession in Australia and she argues for a renewed engagement with Indigenous sovereignty. Given the extent to which ecological knowledge is deeply indebted to First Nations epistemologies, she asks whether there is any authority on which this knowledge may legitimately be employed by non-Indigenous people especially in the light of the many intersecting types of violence employed by colonialism. Echoing Eaton, Salleh and Mathews, Elvey critiques the 'othering' that accompanies a dualistic understanding of the world. In particular, she notes the tendency for 'others' to be cross-identified 'so that Indigenous people for example are both naturalised and feminised, and in the process both people and Country are subject to appropriative violence'. Elvey argues that a decolonising praxis is necessary for both colonised and coloniser, something that will entail: truth-telling (not only bringing to consciousness the violence and theft that undergird Australian history but acknowledging the ecological devastation caused nationally and globally by Australian agriculture, mining and manufacturing); contextualising ecological action in relation to Country (respecting and giving precedence to Indigenous agency and leadership); recognising and respecting Indigenous activist ancestries (recognising Indigenous protest and activism as undergirding non-Indigenous struggles in relation to justice, peace and ecology); and submitting to Indigenous sovereignty in a way that goes beyond mere acknowledgement and moves towards changed structural and constitutional governance.

The question of what it means to live in a country which has undergone colonisation forms the implicit background for Allan Patience (Chapter 14) who focuses on Australia's self-understanding and its international political engagement. Patience calls into question the prevailing foreign policy assumption that Australia is, and should present itself to the world as, an independent 'middle power' an argument that has thus far undergirded the ANZUS treaty between Australia and the United States. As a consequence of this policy, Patience suggests that Australia is highly dependent on, and subservient to, the United States and has therefore become a 'mendicant nation'. Australia, he argues, has become a supplicant to the United States for its security: it participates in unjustifiable regional wars, and its sovereignty is potentially compromised by housing US communication facilities. Further, Patience considers that Australia has become a 'laggard state' in relation to climate change and in its treatment of refugees. He argues that it is opportune for Australia to re-think its global position, perhaps recalling the post-World War II Non-Aligned Movement and adopting the role modelled by some Nordic states in terms of niche multilateral diplomacy. Such options might, he argues, contribute to peace-making and to ecologically sustainable ways of operating.

Mark Brett accepted an invitation to make a response to these highly diverse but often complementary contributions, and his reflections comprise Chapter 15. From an essentially theological perspective, Brett considers that the contributions offer 'prophetic hope'. Making significant connections between their core meanings and wider social, environmental and political questions, he especially reflects on some of these meanings in relation to the Australian Indigenous context.

Acknowledging that humankind faces an unprecedented level of existential threat, this holistic and diverse collection ranges from the academic and analytical to the contextual and practical, from intellectual questions to courses of action, from philosophy to religion and spirituality, and from the global to the local. It offers substantial pointers for navigating the great transition with the aim of establishing a just and ecologically sustainable peace.

Bibliography

Bendell, Jem. 'Deep Adaptation: A Map for Navigating Climate Tragedy'. *IFLAS Occasional Paper*, no. 2 (2018).

Camilleri, Joseph. 'Australia Is Sleepwalking into an Orgy of Militarism'. *New Matilda*, 13 March 2019.

Camilleri, Joseph A. 'Just Peace: The Only Antidote to the Age of Violence—Part 1'. In *John Menadue—Pearls and Irritations*, edited by John Menadue, 2019. https://johnmenadue.com/joseph-a-camilleri-just-peace-the-only-antidote-to-the-age-of-violence-part-1/. Accessed 1 June 2020.

Davison, Nicola. 'The Anthropocene Epoch: Have We Entered a New Phase of Planetary History?' *The Guardian*, 30 May 2019.

Diamond, Jared. *Collapse: How Societies Choose to Fail or Survive*. London: Allen Lane, 2005.

Otter, Chris, Alison Bashford, John L. Brooke, Frederik Albritton Jonsson, and Jason M. Kelly. 'Roundtable: The Anthropocene in British History'. *Journal of British Studies* 57 (2018): 568–96.

Wright, Ronald. *A Short History of Progress*. Toronto: Anansi, 2004.

Deborah Guess is an Honorary Research Associate and Adjunct Lecturer at Pilgrim Theological College, University of Divinity, Melbourne. Her primary research areas are eco-theology and Christology. She is currently writing a monograph exploring the ecotheological meaning/s of place. Recently published book chapters include: 'Earth as Home-Place: Eco-Theology and the Incarnation', in *Reimagining Home: Understanding, Reconciling and Engaging with God's Stories Together*, Darrell Jackson et al. (eds.), Macquarie Park: Morling Press, 2019; and 'Oil Beyond War and Peace: Rethinking the Meaning of Matter', in *Ecological Aspects of War: Religious and Theological Perspectives from Australia*, Anne Elvey, Deborah Guess and Keith Dyer (eds.), Adelaide: ATF Press, 2017.

PART I

In Search of a Holistic Approach

CHAPTER 2

A Just and Ecologically Sustainable Peace: The Policy Imperative of Our Time

Joseph Camilleri

Ours is a unique moment in the human story. This is not a novel observation. For over a century representatives of diverse intellectual traditions have been foreshadowing the end of an era, whether it be 'the decline of the West',[1] the 'crisis of civilization'[2] or 'the end of history'.[3] A number of labels have entered the lexicon of the last several decades, including post-war, post-colonial, post-industrial, post-Westphalian, post-modern and now post-secular, all suggesting that we are at the end of an era, yet conceding that the shape of things to come remains at best blurred and uncertain.

[1]Oswald Spengler, *The Decline of the West*, trans. and notes Charles Francis (London: G. Allen & Unwin, 1926).

[2]Joseph A. Camilleri, *Civilization in Crisis: Human Prospects in a Changing World* (Cambridge: Cambridge University Press, 1976). See also José Ortega y Gasset, *Man and Crisis*, trans. Mildred Adams (London: Allen & Unwin, 1959).

[3]Francis Fukuyama, *The End of History and the Last Man* (New York: Free Press, 1992).

J. Camilleri (✉)
La Trobe University, Melbourne, VIC, Australia
e-mail: j.camilleri@latrobe.edu.au

J. Camilleri and D. Guess (eds.), *Towards a Just and Ecologically Sustainable Peace*, https://doi.org/10.1007/978-981-15-5021-8_2

Two world wars, the Great Depression, the Holocaust, the advent of nuclear weapons, the Cold War, numerous armed conflicts many of which are still raging, genocides, mass displacement of peoples, periodic financial crises and now climate change—itself emblematic of a more pervasive ecological crisis—leave little doubt that we are going through a period of profound upheaval.

What are we to make of all this? And how well equipped are we to navigate the turbulent seas ahead? Simply put, the contemporary human predicament is symptomatic of a transition of epochal proportions. It points to nothing less than the end of modernity itself. The foundational intellectual and institutional pillars of modernity are well known. They include state sovereignty, nationalism, empire building (premised on western dominance of the non-Western world), and the ever-widening subordination of science and technology to the economics of growth and capital accumulation. A good deal of evidence suggests that the modern epoch may have reached its limits. In *Worlds in Transition*, five limits are identified as critical to an understanding of the current transition: *limits to sovereignty, to empire, to national identity as a legitimising norm of governance, to growth and even to science and technology*.[4]

There is perhaps a sixth limit, which subsumes and amplifies the other five, namely the *limit to violence*. As these limits reach their climax, we are faced with an inescapable question: How might we retreat from the brink and move in more promising directions? How might we navigate the 'great transition'? This is the context in which this chapter explores the possibilities of a paradigm shift in the way we understand, connect and institutionalise justice, peace and care of the Earth.

Violence and Its Victims

Violence, by virtue of its scale, intensity and pervasiveness, has become a defining feature of life in the twentieth and twenty-first centuries.[5] The destructiveness unleashed by the structures and instruments of violence is without precedent, and can now be said to threaten not just the human future, but the functioning of the planet itself. This is precisely what

[4] *Worlds in Transition*, 149–158.

[5] Eric Hobsbawm, 'War and Peace in the 20th Century', *London Review of Books*, 24(4), 21 February 2002, 16–18, https://www.lrb.co.uk/v24/n04/eric-hobsbawm/war-and-peace-in-the-20th-century, accessed 9 September 2018.

we mean when we say that nuclear weapons and climate change are the two existential threats of our time, though there is more to catastrophic violence than just these two threats.[6]

To understand the far-reaching implications of the violence epidemic unleashed since the onset of industrialisation, we need to consider violence in its three forms: physical, structural and ecological. Johan Galtung, regarded as the father of modern peace research, argued as far back as the 1960s that violence had both direct and indirect manifestations.[7] He distinguishes between direct or physical violence and social or structural violence. Direct violence, which we associate with the battlefield and other forms of organised conflict, involves the piercing, crushing, tearing, poisoning, burning, exploding, evaporating, starving, torturing of human bodies. Invariably, the physical injury inflicted on a person also carries an important mental or emotional dimension. Indirect or social violence, which we associate with the slum or ghetto, preventable diseases and hunger, involves inequality, poverty, discrimination, social constraints and lifelong division of labour. These distinctions are important for both analysis and remedial action, but equally important is the commonality which lies at the core of all violence. Both direct violence and indirect violence entail human intervention, the consequence of which is to hinder, frustrate or negate human fulfilment and so *violate* what it is to be human.

Galtung subsequently extended his typology of violence by introducing the notion of 'cultural violence', which he associates with those aspects of culture—the symbolic sphere of human existence—exemplified by religion and ideology, language and art, empirical science and formal science that are used to justify physical or structural violence. The violent act is rendered acceptable in society by placing it in a cultural context which downplays or makes invisible its injurious effects, focusing instead on what are said to be its virtuous qualities. And so, the killing of another person, normally viewed as murder, takes on an entirely different complexion when it occurs on the battlefield. In this case, criminal conduct is rebadged as heroic in inspiration and execution, as the ultimate self-sacrifice in the pursuit of a higher good, the defence of the

[6] See Nick Bostrom and Milan M. Cirkovic (eds), *Global Catastrophic Risks* (Oxford: Oxford University Press, 2008).

[7] Johan Galtung, 'Violence, Peace and Peace Research', *Journal of Peace Research*, 6(3), 1969, 167–191.

nation, its values and its freedom. As Hedges aptly puts it, 'war gives us resolve, a cause. It allows us to be noble', and in the process 'it dominates culture, distorts memory, corrupts language, and infects everything around it, even humor, which becomes preoccupied with the grim perversities of smut and death'.[8]

The potency of mythmaking, in particular its effectiveness in legitimising the worst human excesses and sustaining the highest levels of indifference to human suffering, is undeniable. But this may not be sufficient reason to treat it as a distinct mode or form of violence. In many respects, cultural violence is a subset of structural or social violence in that it provides a set of institutional mechanisms that help legitimise the acquisition, use and threatened use of force on the one hand and inequality and subjugation on the other. There is, however, an altogether different form of violence, namely ecological violence,[9] that is in a category of its own. Physical violence and social violence are understood primarily as inflicting injury or death on human beings, whereas ecological violence is directed primarily at other forms of life. This is a critically important form of violence for it points to the accelerating decline of the Earth's natural life-support systems, which in turn places human society in jeopardy. The UN report on biodiversity and ecosystem services released in May 2019 is at pains to convey the unprecedented scale and impact of the human assault on 'nature'. Though nature can be conceptualised in quite different ways, as indicated by such notions as biodiversity, ecosystems, Mother Earth, Gaia, systems of life, the fact remains that nature's gifts are vital to human existence and well-being. To quote the report:

> Nature, through its ecological and evolutionary processes, sustains the quality of the air, fresh water and soils on which humanity depends, distributes fresh water, regulates the climate, provides pollination and pest control and reduces the impact of natural hazards.[10]

[8] Chris Hedges, *War Is a Force That Gives Us Meaning* (New York: Anchor Books, 2003).

[9] See Horacio de la Cueva Salcedo, 'Environmental Violence and Its Consequences', *Latin American Perspectives*, 42(5), September 2015, 19–26; Brandon Absher, 'Toward a Concept of Ecological Violence: Martin Heidegger and Mountain Justice', *Radical Philosophy Review*, 15(1), 2012, 89–101.

[10] IPBES, 2019, *Global Assessment Report on Biodiversity and Ecosystem Services of the Intergovernmental Science-Policy Platform on Biodiversity and Ecosystem Services*, E.S. Brondizio, J. Settele, S. Díaz, and H.T. Ngo (eds) (Bonn: IPBES Secretariat, 2019), https://www.ipbes.net/global-assessment-biodiversity-ecosystem-services#report, accessed 22 May 2019.

The report, compiled over three years by more than 450 scientists and diplomats, has been described as 'the most thorough planetary health check ever undertaken'.[11] Its main finding is that nature across most of the globe has been 'significantly altered by multiple human drivers, with the great majority of indicators of ecosystems and biodiversity showing rapid decline'.[12] Biodiversity—the diversity within species, between species and of ecosystems—is said to be declining faster than at any time in human history, with some one million species facing extinction within a matter of decades. This loss of diversity, including genetic diversity, represents one of the most serious threats to global food security, for it undermines the resilience of farming, making it increasingly vulnerable to pests, pathogens and climate change.

It goes without saying that direct violence and indirect violence, distinct though they are in their symptoms and underlying logic, are nonetheless closely interconnected in both their causes and consequences.[13] Exploitation of the poor and discrimination against minorities sooner or later create conditions conducive to communal and international tensions. Similarly, protracted armed conflict imposes a heavy psychological and emotional burden on both combatants and civilians—not only for the duration of the conflict but often for decades to come. As a succession of recent conflicts have shown—in Afghanistan, Iraq, Syria, Libya and many parts of Africa—the cost to public health, education, social and economic infrastructure, and political institutions is immense and prolonged.

The same mutually reinforcing dynamic connects physical and structural violence with ecological violence. War and peacetime military activity can have a hugely detrimental impact on the natural environment. Water pollution resulting from the use of depleted uranium, release of toxic dust

[11] Jonathan Watts, 'Human Society Under Urgent Threat from Loss of Earth's Natural Life', *The Guardian*, 6 May 2019, www.theguardian.com/environment/2019/may/06/human-society-under-urgent-threat-loss-earth-natural-life-un-report, accessed 22 May 2019.

[12] Summary for policymakers of the global assessment report on biodiversity and ecosystem services of the Intergovernmental Science-Policy Platform on Biodiversity and Ecosystem Services (Advance unedited version), 6 May 2019, 3.

[13] 'The Global Burden of Armed Violence', Report published by the Geneva Declaration Secretariat, September 2008, http://www.genevadeclaration.org/fileadmin/docs/Global-Burden-of-Armed-Violence-full-report.pdf, accessed 18 October 2018.

and carbon emissions associated with heavy military operations, extensive damage done to natural habitats by bombing campaigns, the risks posed by unexploded landmines and fragmentation bombs, the long-lasting effects of radioactive contamination from nuclear weapons testing and the likely catastrophic effects of a nuclear war on climate change and ecosystems are a few obvious examples.[14] The converse is equally true. We now understand more clearly how environmental degradation can provoke armed conflict.[15] Soil erosion, desertification, air and water pollution often lead to sudden and inevitably destabilising mass migrations, as we have already seen in Darfur and Syria, and are likely to see in the South Pacific as Island nations prepare for humanitarian catastrophes borne by increased extreme weather events and rising sea levels.[16]

Physical violence, structural violence and ecological violence are connected not just by virtue of the cause and effect, action-reaction feedback loop. They are also connected, perhaps in more fundamental ways, by the underlying logic of exploitation, of which they are the outward expression. As liberation theologian, Leonardo Boff, puts it:

> The same logic of the ruling system, based on profit and social manipulation, that leads to the exploitation of workers also leads to the spoliation of entire nations and eventually to the depredation of nature itself.[17]

Exploitation, whether of people or the environment, inevitably rests on the apparatus of exclusion.[18] Weapons of mass destruction and

[14] Jürgen Scheffran, Michael Brzoska, Jasmin Kominek, P. Michael Link, and Janpeter Schilling, 'Climate Change and Violent Conflict', *Science*, 336(6083), 18 May 2012, 869–871.

[15] Urmilla Bob and Salomé Bronkhorst, 'Environmental Conflicts: Key Issues and Management Implications', *African Journal on Conflict Resolution*, 10(2), 2010, 9–30.

[16] John R. Campbell, 'Climate Change, Migration and Land in Oceania', Toda Peace Institute, Policy Brief, No. 37, April 2019, https://toda.org/policy-briefs-and-resources/policy-briefs/climate-change-migration-and-land-in-oceania.html, accessed 10 May 2019.

[17] Leonardo Boff, 'Liberation Theology and Ecology: Alternative, Confrontation or Complementarity', in Leonardo Boff and Virgil Elizondo (eds), *Ecology and Poverty: Cry of the Earth, Cry of the Poor* (Maryknoll, NY: Orbis Books, 1995), 73.

[18] John Braithwaite and Bina D'Costa, 'Identifying Social Inclusion and Exclusion', in *Leaving No One Behind: The Imperative of Inclusive Development*, Report on the World Social Situation 2016 (New York: United Nations, 2016), 17–31; Danielle Allen, 'Invisible Citizens: Political Exclusion and Domination in Arendt and Ellison', in Melissa S. Williams and Stephen Macedo (eds), *Political Exclusion and Domination* (New York: New York University Press, 2005), 29–76.

human-driven deforestation are best understood as the outcome of exclusionist attitudes, policies and institutions. People and nature are routinely excluded from the decisions likely to adversely affect them either now or in the future. Potentially critical voices able to give expression to these concerns are marginalised, for fear that giving them due consideration would threaten the rewards of exploitation, whether it be wealth, power or privilege. Indefinite exclusion serves to entrench continued domination of the political, economic and cultural levers of society, but to function at all effectively, exclusion has to institutionalise indifference to the harm and suffering such domination inevitably leaves in its train. This has prompted Pope Francis to label the process as the 'globalisation of indifference'. His incisive portrayal of the phenomenon bears repeating:

> The culture of comfort, which makes us think only of ourselves, makes us insensitive to the cries of other people, makes us live in soap bubbles which, however lovely, are insubstantial; they offer a fleeting and empty illusion which results in indifference to others; indeed, it even leads to the globalization of indifference. In this globalized world, we have fallen into globalized indifference. We have become used to the suffering of others: it doesn't affect me; it doesn't concern me; it's none of my business![19]

Once economy, politics and culture, and with them the patterns of consumption and production, are governed by the dual mechanism of exclusion and indifference, violence, be it physical, structural or ecological, or some combination of the three, is the inevitable outcome.

The Futility of 'Just War' Theory

The idea of the just war in the West has its origins in classical Greek and Roman civilisation and biblical Hebraic culture.[20] This body of thought, which has developed as a blend of religious and secular influences, seeks

[19] The globalisation of indifference, which has been a recurring theme of Pope Francis's pontificate, was first enunciated in a homily delivered during his visit to Lampedusa on 8 July 2013. http://w2.vatican.va/content/francesco/en/homilies/2013/documents/papa-francesco_20130708_omelia-lampedusa.html, accessed 2 August 2018.

[20] See Michael Walzer, 'The Ethics of Warfare in the Jewish Tradition', *Philosophia*, 40(4), December 2012, 633–641; Cian O'Driscoll, 'Rewriting the Just War Tradition: Just War in Classical Greek Political Thought and Practice', *International Studies Quarterly*, 59(1), March 2015, 1–10. Cicero is widely regarded as the foremost Roman exponent

to place limits on the use of force, notably in the context of armed conflict. Though it has been subjected to mounting criticism, especially since the advent of the nuclear age, it still features prominently in intellectual and political discourse. Yet, as we shall see, its inability to offer anything resembling ethically useful and policy-relevant guidance to the multidimensional challenge of violence has never been starker.

In the Christian tradition, just war thinking emerged slowly and did not gain widespread currency until the medieval period.[21] The early Church resisted participation in wars waged by the Roman Empire. Some practised a passive shunning of physical violence (pacifism), while others sought to actively resist evil and in particular injustice (nonviolence).[22]

However, by the end of the second century condemnation of war featured less and less in the writings of the church fathers. The idea of the just war was introduced somewhat tentatively by Clement of Alexandria (c.150–215) and was more explicitly considered by Ambrose (c.339–397). But it was not until Augustine (354–430) that the problem of violence was given detailed attention.[23] At first, he argued that it was right to use force against heretics in the interests of their own spiritual health and the safety of the state. In later years, Augustine developed a more nuanced position. Fighting, he now argued, could proceed only for a just purpose, namely to ensure the common peace and safety when these were threatened by the aggressor. In other words, the recourse to war was justified as the lesser of two evils, when responding to the aggressor's cruelty and destructiveness. Moreover, war could be waged only on the authority of God, though in practice this meant the authority of the ruler.[24]

of the just war concept. See Cicero, *On Duties*, edited by M.T. Griffin and E.M. Atkins (Cambridge: Cambridge University Press, 1991), bk. 1, §34.

[21] For an illuminating account tracing the development of just war thinking from the Greco-Roman intellectual traditions through to the early Church and then medieval Church, see Rory Cox, 'The Ethics of War up to Thomas Aquinas', in Seth Lazar and Helen Frowe (eds), *The Oxford Handbook of Ethics of War* (Oxford: Oxford University Press, 2018), 99–121.

[22] Kirk R. MacGregor, 'Nonviolence in the Ancient Church and Christian Obedience', *Themelios*, 33(1), 2008, 16–28.

[23] John Langan, 'The Elements of St. Augustine's Just War Theory', *The Journal of Religious Ethics*, 12(1), Spring, 1984, 19–38.

[24] Augustine, *Answer to Faustus, a Manichean*, Vol. I/20 (translation and notes by Roland Teske, S.J., edited by Boniface Ramsey) (Hyde Park, NY: New City Press, 2007),

It was only centuries later that Thomas Aquinas (1224–1275) refined the concept of just war in his *Summa Theologica*.[25] For war to be just, Aquinas reasoned, it must satisfy three conditions. It had to be waged: (a) under the authority of the prince (or ruler); (b) for the purpose of protecting the state against aggression, and even then only if peaceful means to remedy the situation have been tried and failed; (c) with the right intention, and even then the intended action had to take account of all foreseeable consequences.

The just war doctrine in its recognisable modern form was not developed until the early seventeenth century when Hugo Grotius (1583–1645), widely regarded as the father of modern international law, stripped away a good deal of the trappings of theological discourse and grounded just war theory firmly in natural law.[26] War, he argued, could be considered just if it is waged for the right purpose, for example, defence, recovery of property, or punishment of aggression. This came to be known as *jus ad bellum*.[27]

Grotius went on to establish rules to govern what is legally permissible in the conduct of war—now commonly referred to as *jus in bello*.[28] Under these rules, the means used were strictly limited to what was necessary to achieve victory. These were meant to be humanitarian constraints on war, from which has gradually emerged the body of international law we now know as international humanitarian law (IHL). Designed to tame the brutality of war, these laws include the Geneva Conventions, the Genocide Convention and the Rome Statute that established the International Criminal Court.

352; Augustine, *Letters 156–210*. Vol. II/3 (translation and notes by Roland Teske, edited by Boniface Ramsey) (Hyde Park, NY: New City Press, 2004), 262.

[25] Thomas Aquinas, *The Summa Theologica*. (New York: Benziger Bros, 1947) (translated by Fathers of the English Dominican Province), Second Part of the Second Part, Question 40, https://dhspriory.org/thomas/summa/SS/SS040.html#SSQ40OUTP1, accessed 12 February 2018.

[26] Hugo Grotius, *On the Law of War and Peace* [*De jure belli ac pacis*] (edited and annotated by Stephen C. Nett) (Cambridge: Cambridge University Press, 2012).

[27] For a through exposition of the concept and its origins, see Larry May '*Jus ad Bellum*', in Larry May (ed.), *The Cambridge Handbook of the Just War* (Cambridge: Cambridge University Press, 2018), 219–233.

[28] Okimoto Keichiro, 'The Relationship between *Jus Ad Bellum* and *Jus In Bello*', in Marc Weller (ed.), *The Oxford Handbook of the Use of Force in International Law* (Oxford: Oxford University Press, 2015), 1209–1223.

Notwithstanding good intentions, just war prescriptions have singularly failed to stem the tide of war in the twentieth and twenty-first centuries. Since the two world wars, the Holocaust and the atomic bombings of Hiroshima and Nagasaki, we have seen the spread of nuclear weapons and a succession of bloody conflicts and genocides. The death toll resulting from war in the twentieth century is estimated by Eric Hobsbawm at 187 million,[29] but a more accurate figure maybe 231 million,[30] while the number of armed conflicts since 1946 has steadily risen and now stands at 50 or more in any one year.

Simply put, just war has proved a remarkably poor guide to prudent, let alone ethical decision making.[31] Whatever relevance they may have once had, just war principles have been overwhelmed by the dramatic changes ushered in by the industrial revolution, the rise of global empires, the relentless global expansion of production, trade and finance, and the increasingly prominent role of non-state actors in armed conflict.

As a consequence, the notion of 'right authority', one of the bedrocks of just war theory, has proven to be its Achilles' heel. More often than not, it is unclear who exercises legitimate authority, and can therefore pronounce on what is or is not just.[32] When oppressive states resort to force to protect their hold on power, who can legitimately exercise authority on behalf of that state? When powerful military and business

[29] Eric Hobsbawm, *The Age of Extremes: A History of the World, 1914–1991* (New York: Vintage Books, 1994), 12.

[30] This figure is arrived at in the most exhaustive analysis yet of the available evidence by Milton Leitenberg, 'Deaths in War and Conflicts in the 20th Century', 3rd edition, Peace Studies Program, Cornell University, Occasional Paper 29, 2006, 14, https://www.clingendael.org/sites/default/files/pdfs/20060800_cdsp_occ_leitenberg.pdf, accessed 25 October 2018.

[31] See Mary Kaldor, 'From Just War to Just Peace', in C. Reed, and D. Ryal (eds), *The Price of Peace: Just War in the Twenty First Century* (Cambridge, UK: Cambridge University Press, 2007), 255–274; Richard T. McSorley, *Kill? For Peace?*, revised edition (Washington, DC: Georgetown University, 1982). Some writers accept that the traditional concept of just war is under challenge in light of contemporary political and military realities, but in seeking to redeem the concept their logic, often tortuous, is less than compelling. See James Turner Johnson, 'Just War as It Was and Is', *First Things*, January 2005, https://www.firstthings.com/article/2005/01/just-war-as-it-was-and-is, accessed 14 June, 2017.

[32] What constitutes legitimate authority has been and remains a highly contentious question. See Joseph Raz, *The Authority of Law: Essays on Law and Morality*, 2nd edition (Oxford: Oxford University Press, 2009), 3–26.

elites, whose influence is often hidden from public view, impose their will on governments regarding the actual or threatened use of force, where does 'right authority' lie? When a state claims that it is using force as a last resort, that fighting has a reasonable chance of success, or that the overriding aim is to establish the conditions for peace, who can sit in judgement on such claims? The questions abound, but the answers are few and far between.

Remarkably, just war theory has little to say about the production, acquisition and transfer of arms regardless of how lethal their use may be, or how pernicious the investment in war machines may be. Just war thinking seems unaware that arms kill simply by their very existence.[33] Just war has even less to say about other forms of structural violence, including gross inequalities of wealth and income, discriminatory practices or repression of minorities. It has literally nothing to say about ecological violence that degrades our natural environment and endangers the plant and animal species that inhabit it.

The Shift to Just Peace and the Ecological Turn

Even when well intentioned, the attempts of just war exponents to link security and justice seem increasingly ill-equipped to address our contemporary predicament. Not surprisingly, a growing number of scholars and practitioners, some from a secular, others from a religious standpoint, are exploring different ways of understanding the relationship between peace and justice, and what this might mean for reimagining the human future.

The notion of 'just peace'[34] represents a significant step in this direction, though linking peace and justice is hardly a new idea. Over the

[33] Sam Perlo-Freeman, 'The Opportunity Cost of World Military Spending', SIPRI, 5 April 2016, https://www.sipri.org/commentary/blog/2016/opportunity-cost-world-military-spending, accessed 14 September, 2018.

[34] The concept is still at an embryonic stage and is open to varying interpretations. See Pierre Allanand Alexis Keller (eds), *What Is a Just Peace?* (Oxford: Oxford University Press, 2006); Robert E. Williams Jr. and Dan Caldwell, '*Jus Post Bellum*: Just War Theory and the Principles of Just Peace', *International Studies Perspectives*, 7(4), November 2006, 309–320; Gerald W. Schlabach, 'What Is "Just Peace"? A Collation and Synthesis Based on Catholic and Ecumenical Christian Sources', February 2018, http://www.geraldschlabach.net/documents/justpeacecollation.pdf, accessed 5 May 2019.

last two hundred years—from Kant's essay *Perpetual Peace: A Philosophical Sketch* to recent cosmopolitan theories of global governance,[35] much effort has gone into thinking through the relationship between peace and justice in both national and international contexts, in relation to conflict prevention, conflict resolution and post-conflict peacebuilding.[36]

Over the last hundred years lawyers, diplomats, NGO advocates and others have sought to develop a legal and organisational framework that recognises the interests of different parties and enshrines a wide range of civil, political, social, economic and cultural rights, and importantly the principle of non-discrimination on the basis of race, religion, ethnicity, gender or sexual orientation.[37] The tentative efforts of the United Nations system to combine human rights and economic and social development with conflict prevention, peacemaking and peacebuilding are an embryonic attempt to forge a just peace agenda.[38] In 2015, UN member states adopted unanimously the *2030 Agenda for Sustainable Development*, including the seventeen Sustainable Development Goals (SDGs). The initiative, officially described as a 'shared blueprint for peace and prosperity for people and the planet, now and into the future'[39] may not live up to expectations. It is nevertheless a useful milestone along this journey.

Other noteworthy contributions to the just peace agenda have come from within the Christian tradition. The World Council of Churches (WCC) has played a particularly useful advocacy role. Building on its work for peace, justice and human rights in the Middle East and South

[35] See D. Archibugi and D. Held (eds), *Cosmopolitan Democracy: An Agenda for a New World Order* (Cambridge: Polity, 1995); Gillian Brock and Darrel Moellendorf (eds), *Current Debates on Global Justice* (Dordrecht: Springer, 2004); Gerard Delanty, *The Cosmopolitan Imagination: The Renewal of Critical Social Theory* (Cambridge: Cambridge University Press, 2009).

[36] See, for example, Elisabeth Porter, *Connecting Peace, Justice, and Reconciliation* (Boulder, CO: Lynne Rienner, 2015).

[37] Neil Hibbert, 'Human Rights and Social Justice', *Laws*, 6(2), 2017, 1–16, https://doi.org/10.3390/laws6020007, accessed 13 May 2019.

[38] Thomas G. Weiss, 'The United Nations: Before, During and After 1945', *International Affairs*, 91(6), November 2015, 1221–1235.

[39] UN DESA, *The Sustainable Development Goals Report 2018*, https://unstats.un.org/sdgs/report/2018/.

Africa and its Restorative Justice programme in the 1990s, the WCC initiated the Decade to Overcome Violence (2001–2010).[40] This was quickly followed by the International Ecumenical Peace Convocation (IEPC) in 2011 which issued *An Ecumenical Call to Just Peace*, and two years later the *Statement on the Way of Just Peace* which called for

> ... a collective and dynamic yet grounded process of freeing human beings from fear and want, of overcoming enmity, discrimination and oppression, and of establishing conditions for just relationships that privilege the experience of the most vulnerable and respect the integrity of creation.[41]

In April 2016, a conference convened by the Pontifical Council for Justice and Peace and other Catholic organisations, notably Pax Christi International, called on the Church to abandon 'just war theory', initiate a global conversation on nonviolence and just peace, call 'unjust world powers' to account, and support those engaged in nonviolence. These sentiments were echoed a few months later in Pope Francis's message for the 2017 World Day of Peace *Nonviolence: A Style of Politics for Peace*.[42]

Scholars and religious leaders in other faith traditions have voiced similar sentiments. In Islam, leading thinkers have begun to articulate new ways of responding to the defining issues of the current period of transition. Many see the process of intellectual and spiritual renewal as inextricably linked to *ijtihad* (best translated as independent reasoning), the key elements of which are a thorough knowledge of theology, revealed texts and legal theory (*usul al-fiqh*), a highly developed capacity for legal reasoning and command of classical Arabic.[43] Others have called for the reformulation of Islamic concepts, notably those that have

[40] Guillermo Kerber, '"Peace with the Earth" in the Context of the Decade to Overcome Violence', *The Ecumenical Review*, 63(1), March 2011, 5–15.

[41] 'Statement on the Way of Just Peace', adopted by the WCC 10th Assembly on 8 November 2013. www.oikoumene.org/en/resources/documents/assembly/2013-busan/adopted-documents-statements/the-way-of-just-peace, accessed 13 November 2018.

[42] Message of His Holiness Pope Francis for the Celebration of the Fiftieth World Day of Peace, 'Nonviolence: A Style of Politics for Peace', 1 January 2017, http://w2.vatican.va/content/francesco/en/messages/peace/documents/papa-francesco_20161208_messaggio-l-giornata-mondiale-pace-2017.html, accessed 7 May 2019.

[43] Abdulaziz Othman Altwajiri, 'Towards a Renewal of Islamic Thought', *Islam Today*, 31, 2015, 13–34.

universal applicability, not least equality of human beings and rejection of racism:

> Umma is easily 'universalisable'. In a period of globalization where people are becoming citizens of the global village they are in reality members of the same community. Thus the Umma of Islam (Ummat al-Islamiyya) may be extended to the Umma of humankind (Ummat al-Insaniyya).[44]

Islam's focus on justice (*adala*) and the virtues of compassion (*sabr*) and solidarity (*ta'awun*) are seen as having particular resonance in an increasingly interdependent world. These scholars have drawn on the Quran and the Prophet's teachings and practice, not least the Medina Charter, which he negotiated and promulgated for the multi-religious ten thousand-strong citizens of the city-state of Medina in 622 CE.[45] They point to a rich reservoir of resources which can be brought to bear on the peaceful resolution of conflicts, including a range of mediation, peacemaking and consultative strategies and practices.[46]

A series of interfaith initiatives has enhanced, at least at the declaratory level, the religious commitment to an integrated vision of justice and peace. Two initiatives are especially instructive in this context. In October 2007, 137 Muslim scholars and religious leaders issued a path-breaking declaration *A Common Word between Us and You*, in which they identified through detailed reference to both Christian and Islamic sacred texts the common ground between Islam and Christianity. The declaration, arguably the most forthright statement of common ground since the days of the Prophet, begins by affirming that '[w]ithout peace and justice between these two religious communities, there can be no meaningful peace in the world'.[47] An equally revealing initiative followed a decade later. In the first-ever papal visit to the Arabian peninsula in March 2019,

[44] Mehdi Mozaffari, *Globalization and Civilizations* (London: Routledge, 2002), 212.

[45] See Muhammad Hamidullah, *The First Written Constitution of the World*, 3rd edition (Lahore: Sh. Muhammad Ashraf, 1975).

[46] Mohammed Abu-Nimer, 'A Framework for Nonviolence and Peacebuilding in Islam', *Journal of Law and Religion*, 15(1–2), 217–265; Mohammed Abu-Nimer and S. Ayse Kadayifci-Orellana, 'Muslim Peace-Building Actors in Africa and the Balkan Context: Challenges and Needs', *Peace & Change*, 33(4), October 2008, 549–581.

[47] 'A Common Word between Us and You', an open letter issued on 13 October 2007, https://www.acommonword.com/the-acw-document/ accessed 11 May 2019.

Pope Francis jointly signed with the Grand Imam of al-Azhar, Ahmad Al-Tayyeb, head of Sunni Islam's most prestigious seat of learning, a historic declaration, calling for peace between nations, religions and races based on 'full citizenship' understood as respect of the rights of all, in particular the victims of war, injustice and persecution.[48]

This all too brief review of current trends makes it clear that we are in the early stages of a transformative reassessment of the centrality and interconnectedness of peace and justice. This development mirrors and builds upon a sharper discernment of the complex ways in which the institutions and instruments of physical and structural violence connect with and reinforce each other. This is not to say that we have as yet a fully developed holistic conception of the relationship between justice and peace, which is the underlying rationale of the 'just peace' paradigm. Much less do we have a blueprint that can translate diagnosis into effective remedial action. In any case, the just peace formulation remains deficient if it does not take full cognisance of our deepening ecological predicament. In the words of Rob Nixon:

> For most of our planet's people there are more immediate terrors than a terrorist attack: creeping deserts that reduce farms to sand; the incremental assaults of climate change compounded by deforestation; not knowing where tonight's meal will come from; unsafe drinking water ... Such quotidian terrors haunt the lives of millions immiserated, abandoned, and humiliated by authoritarian rule and by a neoliberal world order. Under such circumstances, slow violence (often coupled with direct repression) can ignite tensions, creating flashpoints of desperation and explosive rage.[49]

Simply put, the relationship between justice and peace on the one hand and care of the Earth on the other is close and pervasive, reflecting the close interconnections between ecological and other forms of violence. It follows, for example, that stewardship of the Earth's ecosystems is not

[48] 'A Document on Human Fraternity for World Peace and Living Together', Abu Dhabi, 4 February 2019, http://w2.vatican.va/content/francesco/en/travels/2019/outside/documents/papa-francesco_20190204_documento-fratellanza-umana.html, accessed 28 May 2019.

[49] Rob Nixon, 'Planting the Seeds of Peace', *Counterpunch*, 27 September 2011, https://www.counterpunch.org/2011/09/27/planting-the-seeds-of-peace/, accessed 28 May 2019. See also Randall Amster, *Peace Ecology* (New York: Routledge, 2016).

just sensible environmental policy, but an indispensable pillar of conflict prevention and post-conflict peacebuilding. But there is more to the triad of justice, peace and ecology than converging policy applications. What fundamentally links the three constitutive principles of a just and ecologically sustainable peace is an inclusive global ethic, for only such an ethic offers a viable antidote to the logic of exclusion. Only such an ethic can provide the stimulus and inspiration needed to transcend the violence unleashed by the exploitation of people and nature.

The *Earth Charter*, issued in 2000 after a decade-long multicultural and multi-sectoral global dialogue,[50] and Pope Francis's encyclical *Laudato Si'*[51] are two milestones in the development of such an ethic. What is especially striking are the similarities and synergies between the two documents. The ethic of inclusiveness is clearly spelt out in the preamble of the Charter:

> As the world becomes increasingly interdependent and fragile, the future at once holds great peril and great promise. To move forward we must recognize that in the midst of a magnificent diversity of cultures and life forms we are one human family and one Earth community with a common destiny. We must join together to bring forth a sustainable global society founded on respect for nature, universal human rights, economic justice, and a culture of peace.

It is articulated with comparable clarity and force in paragraph 92 of *Laudato Si'*:

> … when our hearts are authentically open to universal communion, this sense of fraternity excludes nothing and no one. It follows that our indifference or cruelty towards fellow creatures of this world sooner or later affects the treatment we mete out to other human beings… We can hardly consider ourselves to be fully loving if we disregard any aspect of reality.

[50] 'The Earth Charter', officially launched on 29 June 2000 in a ceremony at The Peace Palace in The Hague, Netherlands, http://earthcharter.org/discover/the-earth-charter/, accessed 8 April 2018.

[51] *Encyclical Letter Laudato Si' of the Holy Father Francis On Care for Our Common Home*, 24 May 2015, http://w2.vatican.va/content/francesco/en/encyclicals/documents/papa-francesco_20150524_enciclica-laudato-si.html, accessed 29 May 2015.

It is, of course, one thing to affirm that all living things are inextricably linked and interdependent as are the principles of justice, peace and respect for nature, and quite another to explain how this interconnectedness finds expression in the self, in culture and in the institutional fabric of society.

Self-Cultivation

How, then, might we conceptualise the relationship between self, culture and social and political institutions on the one hand and the ethical imperative of a just and ecologically sustainable peace? To this complex and pivotal question, there can be no simple or single answer, but the notion of 'anthropocosmic unity' systematically developed by Tu Weiming offers an insightful framework within which to locate the question. Tu Weiming, regarded by many as the pre-eminent neo-Confucian scholar of our time, is at pains to distinguish 'the anthropocosmic worldview, in which the human is embedded in the cosmic order' from 'an anthropocentric worldview, in which the human is alienated, either by choice or by default, from the natural world'.[52] Though in itself not an original idea, Tu Weiming's treatment of it benefits from a rich and prolonged encounter between Occident and Orient, notably between modern Confucian thought and Western social theory and religious studies. Here, we find fascinating similarities and points of connection with Thomas Berry's cosmology which has as its overarching theme humanity's broken relationship with nature.[53] Like Tu Weiming, Berry ascribes this break to modernity, in particular to the inability of contemporary scientific, religious and philosophical narratives to establish a meaningful relationship between humanity and the Earth's ecosystems.

In the light of modernity's failure, Tu Weiming turns to the civilisations of the axial age to see 'what wisdom they can offer to reorient the human developmental trajectory of the modern world in light of the growing environmental crisis'.[54] He acknowledges the enormity of the task—dysfunctional families, failed states and world disorder, all of which he

[52] Tu Weiming, 'The Ecological Turn in New Confucian Humanism: Implications for China and the World', *Daedalus*, 130(4), Fall 2001, 244.

[53] See Thomas Berry, *Evening Thoughts: Reflecting on Earth as Sacred Community* (San Francisco, CA: Sierra Club Books, 2006), 18–22.

[54] Tu Weiming, 'The Ecological Turn in New Confucian Humanism', 243.

characterises as different symptoms of the same underlying ailment. For Tu Weiming, the key idea in neo-Confucianism, or as he calls it holistic Confucian humanism, is the 'unity of Humanity and Heaven'—a unity of which the Earth is an integral part. It is this unity that now lies in tatters. To repair that which is broken requires that we rethink 'the human'. Tu Weiming calls for 'a comprehensive spiritual humanism' capable of integrating the four pillars of human flourishing: self, community, Earth and Heaven. What distinguishes the neo-Confucian holistic vision are the four indispensable relationships that are said to connect the four pillars: (1) fruitful interaction between self and community; (2) harmonisation of community, which encompasses regulation of family, governance of the state and maintenance of world order; (3) a sustainable harmonious relationship between the human species and nature; and (4) mutual responsiveness between the 'human heart-and-mind' and 'the way of heaven'.[55]

If these four relationships are to achieve the desired level of harmony and fully realise their potential, much will depend on self-cultivation. It is worth stressing here that for Tu Weiming cultivation of self is as much a communal as an individual effort, a dynamic process rather than a static structure.[56] In the neo-Confucian schema, the communal dimension of life is itself a complex one, encompassing family, village, country, world and cosmos. In Tu Weiming's words:

> Self-realization as a communal act presupposes a personal commitment for harmonizing the family, governing the state, and bringing peace to the world. The full realization of personhood entails the real possibility of transcending selfishness, nepotism, parochialism, nationalism, and anthropocentrism.[57]

To this list of afflictions could have been added patriarchy, racism and despotism. The nub of Tu Weiming's argument is nevertheless clear. The realisation of personhood, or cultivation of self, holds the key to all four relationships, not least the relationship between humans and the Earth, which is why the self must free itself from the chains of anthropocentrism and cultivate an anthropocosmic appreciation of life's present travails and future possibilities. Tu Weiming describes the Earth as 'our prophet', for it

[55] Tu Weiming, 'The Ecological Turn in New Confucian Humanism', 253–255.

[56] Tu Weiming, 'The Ecological Turn in New Confucian Humanism', 256–257.

[57] Tu Weiming, 'The Ecological Turn in New Confucian Humanism', 244.

is 'the sound of the Earth' that can 'guide us on to a new path of survival'. In pursuing this path, Tu Weiming invites us to learn from Indigenous traditions precisely because they have cultivated over millennia a unique life-sustaining relationship with the land. For him, understanding and internalising this deeper relationship with Mother Earth will make it possible to cultivate the self in ways that can redirect the modern trajectory. Understood in its communal and cosmic sense, cultivation of self offers a powerful lever and an incentive to think through the practical implications of a just and ecologically sustainable peace. Here, we can do no more than attempt a preliminary exploration of these possibilities by reference to three related fields of engagement: citizenship, governance and dialogue.

Towards a Just and Ecologically Sustainable Peace: A Strategic Approach

It is fair to say that much of the literature concerned with identifying pathways for advancing justice, peace and care of the Earth falls into two main baskets: one focuses on the theory and practice of nonviolence,[58] and the other on alternative policy prescriptions and visions of the future.[59] This body of thought and the advocacy and activism it has inspired have greatly contributed to a sharper understanding of the dangers posed by current

[58] This has become quite a large field attracting both scholarly and practical guidance, of which the following are notable examples. Heather Eaton and Laurel Michelle Levesque (eds), *Advancing Nonviolence and Social Transformation: New Perspectives on Nonviolent Theories* (Sheffield: Equinox, 2016); Michael N. Nagler, *The Nonviolence Handbook: A Guide for Practical Action* (San Francisco, CA: Berrett-Koehler, 2014); Marty Branagan, *Global Warming, Militarism and Nonviolence: The Act of Active Resistance* (New York: Palgrave Macmillan, 2013); Erica Chenoweth and Maria J. Stephan, *Why Civil Resistance Works: The Strategic Logic of Nonviolence* (New York: Columbia University Press, 2011); Gene Sharp, *Waging Nonviolent Struggle: 20th Century Practice and 21st Century Potential* (Boston, MA: Porter Sargent, 2005); Peter Ackerman and Jack Duvall, *A Force More Powerful: A Century of Nonviolent Conflict* (New York: Palgrave Macmillan, 2000).

[59] Tessa Finley, 'Future Peace: Breaking Cycles of Violence through Futures Thinking', *Journal of Futures Studies*, 16(3), March 2012, 47–62. For an insightful perspective on the future which combines aspiration with a degree of strategic realism and foreshadows the analysis offered in this essay, see Richard Falk, 'Changing the Political Climate: A Transitional Imperative', *Great Transition Initiative*, September 2014, https://www.greattransition.org/publication/changing-the-political-climate-a-transitional-imperative, accessed 12 September 2014.

policies and practices, an enhanced willingness and capacity to reimagine the future, and creative forms of active resistance. But there is more to advancing the cause of a just and ecologically sustainable peace. A strategic approach is needed to give the concept its full meaning and content, and provide it with intellectual and organisational coherence.

The preceding analysis points to several requirements. First, the pathways identified and pursued need to engage not just a minuscule minority, but at least a sizeable and reasonably representative cross-section of society. They must call forth not just the heroic deeds of the few who are willing to sacrifice much for this or that cause, but the energies of a wider public as it goes about the business of everyday life. The adoption of nonviolent approaches by current climate emergency activist groups, notably Extinction Rebellion, and the passionate global movement ignited by the sixteen-year-old Swedish climate activist Greta Thunberg,[60] may well constitute a transformative moment in the application of nonviolence to the current ecological crisis.

Secondly, a strategy that is premised on self-cultivation and the four relationships described above must somehow be integrated into the cultural and institutional fabric of the community, be it local, national or international. Importantly, the strategy must connect justice, peace and ecological flourishing not just in its conception but in all aspects of its execution. Last but not least is the identity of the interlocutors. With some notable exceptions, much of the literature features Western voices explicitly or implicitly addressing Western audiences, which greatly limits the efficacy, some might say authenticity, of the enterprise. It is difficult to see how any strategy that is informed primarily if not exclusively, by Western currents of thought and geared to the needs and circumstances of Western societies can realistically expect to have global relevance or resonance. A strategy seeking to advance a just and ecologically sustainable peace cannot but be cognisant of and sensitive to the priorities and perspectives of diverse cultures and civilisations.

A tall order indeed, but one which does not preclude promising points of entry. The present transitional period faces the task of devising, perhaps for the first time in human evolution, a set of responses that can reconcile

[60] 'Global Climate Strike: Greta Thunberg And Huge Crowds Protest', *Forbes*, 20 September 2019, https://www.forbes.com/sites/oliviagieger/2019/09/20/global-climate-strike-greta-thunberg-and-huge-crowds-protest/#4ee4d1dc36fb, accessed 22 September 2019.

the needs of the human family (universalism) with those of its constituent parts on the one hand (pluralism) and those of the other members of the Earth community on the other (ecologism).[61] Such reconciliation is unlikely to materialise unless it is grounded in the intellectual, emotional and experiential dimensions of life. Only then is cultivation of self likely to acquire its full meaning and potency, and pave the way for a renewed understanding of citizenship.

A citizen has generally been defined as a member of a political community who enjoys the rights and assumes the obligations of membership.[62] Over time a minimalist version of citizenship, sometimes referred to as 'private' or 'passive' citizenship, has gained ascendancy in much of the Western world. It asserts that the affairs of government are best left in the hands of a periodically elected elite, leaving citizens to concentrate on their individual rights and private interests. This has proven to be a recipe for unprecedented levels of public cynicism, periodic outbursts of discontent, more often than not the politics of withdrawal, and now the troubling rise of populist extremism.[63]

The question is: Can the ethos of citizenship be renewed in ways that begin to address the multiple and mutually reinforcing crises of contemporary life? Can citizenship have as its defining purpose cultivation of self? Both questions—the two are closely interlinked—can be answered in the affirmative once territorially bound notions of citizenship give way to a multi-spatial, multidimensional frame of reference where the emphasis is on acceptance of social and cultural diversity within and across societies, constructive coexistence between states and an expanding ecological consciousness. This conception of citizenship is not just 'global' or 'postnational' but 'dialogical' in scope and inspiration. It is informed by the deepest insights that science can offer and the wisdom of the world's enduring ethical traditions. It is premised on the proposition that dialogue

[61] A. Dobson, *Green Political Thought*, 4th edition (London: Routledge, 2007); Brian Baxter, *Ecologism: An Introduction* (Edinburgh: Edinburgh University Press, 1999); Mark J. Smith, *Ecologism: Towards Ecological Citizenship* (Minneapolis, MN: University of Minnesota Press, 1998).

[62] T.H. Marshall, *Citizenship and Social Class* (Cambridge: Cambridge University Press, 1950), 28–29.

[63] I review the scope and drivers of the crisis of liberal democracy in an Op-Ed 'Democracy in Crisis', *The Age*, 29 December 2014, http://www.theage.com.au/comment/democracy-in-crisis-20141228-126quk.html, accessed 29 December 2014.

offers the most promising basis for fruitful relationships between humans and their respective cultural and political formations as well as between humanity and other members of the Earth community.

At the heart of the dialogical project is the ethic of inclusion, which stands in sharp contrast to the logic of exclusion that underlies much of contemporary violence. Those to be included are not individuals understood as atomised globules of interest and desire but cultivated selves, social persons who bring to the communal table a range of identities and solidarities. In this context, citizenship has a dual function. It gives expression to these identities and solidarities and enables the negotiation of differences in ways that are mutually beneficial and capable of enhancing not just the human common good but the 'Commonwealth of life'.[64]

Dialogue is best understood as an encounter across cultural, religious, philosophical, civilisational boundaries, in which the citizen listens to the other, becomes open, sensitive, even vulnerable to the needs, concerns and grievances of the other.[65] Equally, it involves an encounter and deepening intimacy with the other biological species and their respective habitats. In the process, the citizen can be said to embark on a journey as much of 'self-discovery' as of 'discovery of the other'. In the process of dialogue, every entity that is actually or potentially affected by a project, decision or policy thereby becomes a stakeholder whose voice must be heard. The stakeholders, it cannot be stressed enough, are not just the living but those yet to be born, not just humans but other inhabitants of the biosphere and their respective life-support systems.

If dialogue is as much about listening as it is about speaking, it follows that those holding public office (whether elected or appointed), industrialists, financiers, lawyers, doctors, architects, town planners, religious and

[64] Peter G. Brown and Geoffrey Garver, 'Humans and Nature: The Right Relationship', *Minding Nature,* 2(1), April 2009, 8–16, https://www.humansandnature.org/filebin/pdf/minding_nature/April_2009_Humans_and_Nature.pdf, accessed 17 June 2016.

[65] Leading contributors to the philosophy of dialogue include: Martin Buber, *Between Man and Man*, trans. Ronald Gregor-Smith (New York: Routledge, 2002); Hans-Georg Gadamer *Truth and Method*, revised edition, trans. Joel Weinsheimer and Donald G. Marshall (New York: Crossroads, 1989); Jürgen Habermas, *The Inclusion of the Other: Studies in Political Theory* (Cambridge, MA: MIT Press, 1998); Tu Weiming, 'Confucianism and Civilization', in Majid Tehranian and David W. Chappell (eds), *Dialogue of Civilizations: A New Peace Agenda for a New Millennium* (London: I.B Tauris, 2002), 83–90; Fred Dallmayr, *Dialogue Among Civilizations: Some Exemplary Voices* (New York: Palgrave/St. Martin's Press, 2002).

community leaders have an obligation to listen and give voice to those whose voices are struggling to be heard. An equally heavy but distinctive responsibility rests with philosophers, religious scholars, intellectuals of one kind or another, but also with poets, artists and publicists to shed light on the nature and magnitude of the contemporary predicament but also on the rich and still largely untapped resources that can enrich the dialogical project and thereby enhance the prospects of a just and ecologically sustainable peace.

Dialogical citizenship also requires congruent forms and institutions of governance. Here, the signs are less than reassuring. True enough, the post-1945 period has seen a number of encouraging developments, in particular increased codification of inter-state co-operation; the internationalisation of the legal responsibility to protect individual persons and minorities; a growing consensus that customary and treaty law are meant to advance such norms as peace, development, human rights, and ecological balance; and the emergence of a multi-spatial system of governance in which local, provincial, regional and global institutions complement and often compete with nation-states suffering from a debilitating legitimacy deficit. But these innovations are still a long way from bearing fruit. For a while at least dialogical citizenship may have to make strides in what is a decidedly less than ideal institutional environment.

In any case, citizenship cannot confine its attention to the institutions of governance. Given the pre-eminent role of the market in resource allocation and the ensuing inequalities of wealth and income, citizenship requires engagement with corporate centres of decision making. This in turn requires jurisdictions at various levels to establish the regulatory frameworks conducive to dialogical interaction between corporate institutions and stakeholders affected by their decisions. The 'environmental impact statement' mechanism now widely used by provincial and national jurisdictions,[66] though more honoured in the breach than in the observance, embryonically at least, points in the right direction. The consultative status accorded by the UN system to a range of civil society organisations and their inclusion, at least at the periphery of international negotiations—more centrally when it comes to responding to pandemics and humanitarian emergencies—is suggestive of a trend that is worth

[66] Anne N. Glucker, Peter P.J. Driessen, Arend Kolhoff, and Hens A.C. Runhaar, 'Public Participation in Environmental Impact Assessment: Why, Who and How?', *Environmental Impact Assessment Review*, 43, November 2013, 104–111.

pursuing with increased vigour. The time has come to press for more effective consultative processes and to this end work towards an internationally coordinated approach that links thematically and organisationally the different strands of the justice/peace/ecology agenda.

The potency of dialogical engagement lies precisely in the ability to connect that which otherwise would remain fragmented. It follows that in the coming decades much thought and innovation will need to go into making the citizenship experience whole. This means equipping the citizenry to connect, as befits the paradigm of a just and ecologically sustainable peace, the multiple issues, sites and participants of dialogue not just in a technical and organisational sense, but intellectually and experientially.

The new model of citizenship canvassed here will not materialise overnight. No technical fix or social engineering can bring it to fruition. Even if it is to engage only a sizeable minority—let's say fifteen or even ten per cent of the world's population—a major cultural shift will be necessary, and education and training will be integral to this transformative process.[67] This is too large a subject to be addressed within the confines of this essay. A few observations may nevertheless help to convey something of the magnitude of the task ahead.

First, while formal educational institutions—from kindergartens to universities—will play a critical role, they will need to be complemented by less formal programs in community education, for the needs of dialogical citizenship will continue to evolve and grow in line with the far-reaching technological, socio-economic, political and environmental changes still to come.[68]

Second, the purpose of formal education will have to be rethought. Education for dialogical citizenship will need to occupy centre stage,

[67] Elemer Hankiss, 'Paideia in an Age of Uncertainty', in Alan M. Olson, David M. Steiner, and Irian S. Tuuli (eds), *Educating for Democracy* (Lanham, MD: Rowman and Littlefield, 2004), 143–161.

[68] Neil Hopkins, *Citizenship and Democracy in Further and Adult Education* (Dordrecht: Springer, 2014); Sobhi Tawi, 'Education for "Global Citizenship": A Framework For Discussion', UNESCO, Education Foresight and Research, Working Papers, August 2013, https://unesdoc.unesco.org/ark:/48223/pf0000223784, accessed 9 May 2016.

and not be seen as a mere appendage to the serious business of vocational training dictated largely by the needs of industry.[69] The challenges posed by the 'great transition' and the opportunities offered by dialogical discourse and practice will need to inform the reworking of curriculum and learning materials across all levels of education. To this end, the reform of teacher training content and method should be seen as a high priority. Effective implementation will require sustained policy support from all tiers of governance, pedagogical guidance from educationalists, intellectuals, artists and practitioners, and innovative use of both traditional and social media. Thirdly, the pedagogical enterprise, including all aspects of teaching and learning, would greatly benefit from nurturing an imagination that is cross-national, cross-cultural and planetary in scope and inspiration. Finally, a priority for the new pedagogy will be the fostering of formal and informal dialogical encounters among educators and students and between them, parents and the wider community.

As the challenges posed by the current transitional period reach a climactic stage in coming decades, new forms of civic engagement will crucially bear on the outcome. The task is dauntingly complex, the terrain will not always be fertile, and the level of inertia and at times outright resistance in political, bureaucratic and corporate circles will be far from negligible. Yet, it is difficult to imagine us navigating the 'great transition' at all safely in the absence of a transnational citizenry conscious of the tension between competing interests, between cultural plurality and human universality, and between humanity and planet Earth. Grounding cultivated selves in the ethic of inclusion holds the key to overcoming these polarities and advancing the prospect of a just and ecologically sustainable peace.

[69] See Greg Mannion, Gert Biesta, Mark Priestley, and Hamish Ross, 'The Global Dimension in Education and Education for Global Citizenship: Genealogy and Critique', *Globalisation, Societies and Education*, 9(3–4), September–November 2011, 443–456; Mike Bottery, *The Challenges of Educational Leadership: Values in a Globalized Age* (London: Paul Chapman Publishing, 2004).

Bibliography

Absher, Brandon. 'Toward a Concept of Ecological Violence: Martin Heidegger and Mountain Justice'. *Radical Philosophy Review*, 15(1), 2012, 89–101.

Abu-Nimer, Mohammed. 'A Framework for Nonviolence and Peacebuilding in Islam'. *Journal of Law and Religion*, 15(1–2), 217–265.

Abu-Nimer, Mohammed and S. Ayse Kadayifci-Orellana. 'Muslim Peace-Building Actors in Africa and the Balkan Context: Challenges and Needs'. *Peace & Change*, 33(4), October 2008, 549–581.

Ackerman, Peter and Jack Duvall. *A Force More Powerful: A Century of Nonviolent Conflict*. New York: Palgrave Macmillan, 2000.

Allan, Pierre and Alexis Keller (eds). *What Is a Just Peace?* Oxford: Oxford University Press, 2006.

Allen, Danielle. 'Invisible Citizens: Political Exclusion and Domination in Arendt and Ellison'. In Melissa S. Williams and Stephen Macedo (eds). *Political Exclusion and Domination*. New York: New York University Press, 2005, 29–76.

Altwajiri, Abdulaziz Othman. 'Towards a Renewal of Islamic Thought'. *Islam Today*, 31, 2015, 13–34.

Amster, Randall. *Peace Ecology*. New York: Routledge, 2016.

Aquinas, Thomas. *The Summa Theologica. 1947* (translated by Fathers of the English Dominican Province). New York: Benziger Bros, 1947–48.

Archibugi, D. and D. Held (eds). *Cosmopolitan Democracy: An Agenda for a New World Order*. Cambridge: Polity, 1995.

Augustine. *Answer to Faustus, a Manichean*, Vol. I/20 (translation and notes by Roland Teske, S.J., edited by Boniface Ramsey). Hyde Park, NY: New City Press, 2007.

Augustine. *Letters 156–210*. Vol. II/3 (translation and notes by Roland Teske, edited by Boniface Ramsey). Hyde Park, NY: New City Press, 2004.

Baxter, Brian. *Ecologism: An Introduction*. Edinburgh: Edinburgh University Press, 1999.

Berry, Thomas. *Evening Thoughts: Reflecting on Earth as Sacred Community*. San Francisco, CA: Sierra Club Books, 2006.

Bob, Urmilla and Salomé Bronkhorst. 'Environmental Conflicts: Key Issues and Management Implications'. *African Journal on Conflict Resolution*, 10(2), 2010, 9–30.

Boff, Leonardo. 'Liberation Theology and Ecology: Alternative, Confrontation or Complementarity'. In Leonardo Boff and Virgil Elizondo (eds). *Ecology and Poverty: Cry of the Earth, Cry of the Poor*. Maryknoll, NY: Orbis Books, 1995.

Bostrom, Nick and Milan M. Cirkovic (eds). *Global Catastrophic Risks*. Oxford: Oxford University Press, 2008.

Bottery, Mike. *The Challenges of Educational Leadership: Values in a Globalized Age*. London: Paul Chapman Publishing, 2004.

Braithwaite, John and Bina D'Costa. 'Identifying Social Inclusion and Exclusion'. In *Leaving No One Behind: The Imperative of Inclusive Development*. Report on the World Social Situation 2016. New York: United Nations, 2016, 17–31.

Branagan, Marty. *Global Warming, Militarism and Nonviolence: The Act of Active Resistance*. New York: Palgrave Macmillan, 2013.

Brock, Gillian and Darrel Moellendorf (eds). *Current Debates on Global Justice*. Dordrecht: Springer, 2004.

Brown, Peter G. and Geoffrey Garver. 'Humans and Nature: The Right Relationship'. *Minding Nature*, 2(1), April 2009, 8–16. https://www.humansandnature.org/filebin/pdf/minding_nature/April_2009_Humans_and_Nature.pdf. Accessed 17 June 2016.

Buber, Martin. *Between Man and Man*, translated Ronald Gregor-Smith. New York: Routledge, 2002.

Camilleri, Joseph A. *Civilization in Crisis: Human Prospects in a Changing World*. Cambridge: Cambridge University Press, 1976.

Camilleri, Joseph A. 'Democracy in Crisis'. *The Age*, 29 December 2014. http://www.theage.com.au/comment/democracy-in-crisis-20141228-126quk.html. Accessed 29 December 2014.

Camilleri, Joseph A. and Jim Falk. *Worlds in Transition: Evolving Governance across a Stressed Planet*. Cheltenham, UK: Edward Elgar, 2009.

Campbell, John R. 'Climate Change, Migration and Land in Oceania'. Toda Peace Institute, Policy Brief, No. 37, April 2019. https://toda.org/policy-briefs-and-resources/policy-briefs/climate-change-migration-and-land-in-oceania.html. Accessed 10 May 2019.

Chenoweth, Erica and Maria J. Stephan. *Why Civil Resistance Works: The Strategic Logic of Nonviolence*. New York: Columbia University Press, 2011.

Cicero. *On Duties*. M.T. Griffin and E.M. Atkins (eds). Cambridge: Cambridge University Press, 1991.

'A Common Word between Us and You', an open letter issued on 13 October 2007. https://www.acommonword.com/the-acw-document/. Accessed 11 May 2019.

Cox, Rory. 'The Ethics of War up to Thomas Aquinas'. In Seth Lazar and Helen Frowe (eds). *The Oxford Handbook of Ethics of War*. Oxford: Oxford University Press, 2018, 99–121.

Dallmayr, Fred. *Dialogue among Civilizations: Some Exemplary Voices*. New York: Palgrave/St. Martin's Press, 2002.

Delanty, Gerard. *The Cosmopolitan Imagination: The Renewal of Critical Social Theory*. Cambridge: Cambridge University Press, 2009.

Dobson, A. *Green Political Thought*. 4th edition. London: Routledge, 2007.

'A Document on Human Fraternity for World Peace and Living Together'. Abu Dhabi, 4 February 2019. http://w2.vatican.va/content/francesco/en/travels/2019/outside/documents/papa-francesco_20190204_documento-fratellanza-umana.html. Accessed 28 May 2019.
'The Earth Charter', officially launched on 29 June 2000 in a ceremony at The Peace Palace in The Hague, Netherlands. http://earthcharter.org/discover/the-earth-charter/. Accessed 8 April 2018.
Eaton, Heather and Laurel Michelle Levesque (eds). *Advancing Nonviolence and Social Transformation: New Perspectives on nonviolent Theories*. Sheffield: Equinox, 2016.
Falk, Richard. 'Changing the Political Climate: A Transitional Imperative'. *Great Transition Initiative*. September 2014. https://www.greattransition.org/publication/changing-the-political-climate-a-transitional-imperative. Accessed 12 September 2014.
Finley, Tessa. 'Future Peace: Breaking Cycles of Violence through Futures Thinking'. *Journal of Futures Studies*, 16(3), March 2012, 47–62.
Fukuyama, Francis. *The End of History and the Last Man*. New York: Free Press, 1992.
Gadamer, Hans-Georg. *Truth and Method*. Revised edition, trans. Joel Weinsheimer and Donald G. Marshall. New York: Crossroads, 1989.
Galtung, Johan, 'Violence, Peace and Peace Research'. *Journal of Peace Research*, 6(3), 1969, 167–191.
Geneva Declaration Secretariat. 'The Global Burden of Armed Violence'. Geneva, Geneva Declaration Secretariat, 2008. http://www.genevadeclaration.org/fileadmin/docs/Global-Burden-of-Armed-Violence-full-report.pdf. Accessed 18 October 2018.
Gieger, Olivia. 'Global Climate Strike: Greta Thunberg and Huge Crowds Protest'. *Forbes*, 20 September 2019. https://www.forbes.com/sites/oliviagieger/2019/09/20/global-climate-strike-greta-thunberg-and-huge-crowds-protest/#4ee4d1dc36fb. Accessed 22 September 2019.
Glucker, Anne N., Peter P.J. Driessen, ArendKolhoff, and Hens A.C. Runhaar. 'Public Participation in Environmental Impact Assessment: Why, Who and How?'. *Environmental Impact Assessment Review*, 43, November 2013, 104–111.
Grotius, Hugo. *On the Law of War and Peace* [*De jure belli ac pacis*] (edited and annotated by Stephen C. Nett). Cambridge: Cambridge University Press, 2012.
Habermas, Jürgen. *The Inclusion of the Other: Studies in Political Theory*. Cambridge, MA: MIT Press, 1998.
Hamidullah, Muhammad. *The First Written Constitution of the World*. 3rd edition. Lahore: Sh. Muhammad Ashraf, 1975.

Hankiss, Elemer. 'Paideia in an Age of Uncertainty'. In Alan M. Olson, David M. Steiner, and Irian S. Tuuli (eds). *Educating for Democracy*. Lanham, MD, Rowman and Littlefield, 2004.

Hedges, Chris. *War Is a Force That Gives Us Meaning*. New York: Anchor Books, 2003.

Hibbert, Neil. 'Human Rights and Social Justice', *Laws*, 2017, 6(2), 1–16. https://doi.org/10.3390/laws6020007. Accessed 13 May 2019.

Hobsbawm, Eric. *The Age of Extremes: A History of the World, 1914–1991*. New York: Vintage Books, 1994.

Hobsbawm, Eric. 'War and Peace in the 20th Century'. *London Review of Books* 24(4), 21 February 2002, 16–18. https://www.lrb.co.uk/v24/n04/eric-hobsbawm/war-and-peace-in-the-20th-century. Accessed 9 September 2018.

Hopkins, Neil. *Citizenship and Democracy in Further and Adult Education*. Dordrecht: Springer, 2014.

IPBES. *Global Assessment Report on Biodiversity and Ecosystem Services of the Intergovernmental Science-Policy Platform on Biodiversity and Ecosystem* Services. E.S. Brondizio, J. Settele, S. Díaz, and H.T. Ngo (eds). Bonn: IPBES, 2019. https://www.ipbes.net/global-assessment-biodiversity-ecosystem-services#report. Accessed 22 May 2019.

Johnson, James Turner. 'Just War as It Was and Is'. *First Things*, January 2005.https://www.firstthings.com/article/2005/01/just-war-as-it-was-and-is. Accessed 14 June 2017.

Kaldor, Mary. 'From Just War to Just Peace'. In C. Reed and D. Ryal (eds). *The Price of Peace: Just War in the Twenty First Century*. Cambridge, UK: Cambridge University Press, 2007, 255–274.

Keichiro, Okimoto. 'The Relationship between *Jus Ad Bellum* and *Jus In Bello*'. In Marc Weller (ed), *The Oxford Handbook of the Use of Force in International Law*. Oxford: Oxford University Press, 2015, 1209–1223.

Kerber, Guillermo, '"Peace with the Earth" in the Context of the Decade to Overcome Violence'. *The Ecumenical Review*, 63(1), March 2011, 5–15.

Langan, John, 'The Elements of St. Augustine's Just War Theory'. *The Journal of Religious Ethics*, 12(1), Spring, 1984, 19–38.

Leitenberg, Milton. 'Deaths in War and Conflicts in the 20th Century'. 3rd edition, Peace Studies Program, Cornell University, Occasional Paper 29, 2006.https://www.clingendael.org/sites/default/files/pdfs/20060800_cdsp_occ_leitenberg.pdf. Accessed 25 October 2018.

MacGregor, Kirk R. 'Nonviolence in the Ancient Church and Christian Obedience'. *Themelios*, 33(1), 2008, 16–28.

Mannion, Greg, Gert Biesta, Mark Priestley and Hamish Ross. 'The Global Dimension in Education and Education for Global Citizenship: Genealogy and Critique'. *Globalisation, Societies and Education*, 9(3–4), September–November 2011, 443–456.

Marshall, T.H. *Citizenship and Social Class*. Cambridge: Cambridge University Press, 1950.

May, Larry. '*Jus ad Bellum*'. In Larry May (ed). *The Cambridge Handbook of the Just War*. Cambridge: Cambridge University Press, 2018, 219–233.

McSorley, Richard, T. *Kill? For Peace*? Revised edition. Washington, DC: Georgetown University, 1982.

Mozaffari, Mehdi. *Globalization and Civilizations*. London: Routledge, 2002.

Nagler, Michael N. *The Nonviolence Handbook: A Guide for Practical Action*. San Francisco, CA: Berrett-Koehler, 2014.

Narci, Nemer E. and Horacio de la Cueva Salcedo. 'Environmental Violence and Its Consequences'. *Latin American Perspectives*, 42(5), September 2015, 19–26.

Nixon, Rob. 'Planting the Seeds of Peace'. *Counterpunch*, 27 September 2011. https://www.counterpunch.org/2011/09/27/planting-the-seeds-of-peace/. Accessed 28 May 2019.

O'Driscoll, Cian. 'Rewriting the Just War Tradition: Just War in Classical Greek Political Thought and Practice'. *International Studies Quarterly*, 59(1), March 2015, 1–10.

Ortega y Gasset, Jose. *Man and Crisis*. Trans. Mildred Adams. London: Allen & Unwin, 1959.

Perlo-Freeman, Sam. 'The Opportunity Cost of World Military Spending'. SIPRI, 5 April 2016. https://www.sipri.org/commentary/blog/2016/opportunity-cost-world-military-spending. Accessed 14 September 2018.

Pope Francis. *Encyclical Letter Laudato Si' of the Holy Father Francis on Care for Our Common Home*. 24 May 2015. http://w2.vatican.va/content/francesco/en/encyclicals/documents/papa-francesco_20150524_enciclica-laudato-si.html. Accessed 29 May 2015.

Pope Francis. 'Nonviolence: A Style of Politics for Peace'. Message for the Celebration of the Fiftieth World Day of Peace, 1 January 2017. http://w2.vatican.va/content/francesco/en/messages/peace/documents/papa-francesco_20161208_messaggio-l-giornata-mondiale-pace-2017.html. Accessed 7 May 2019.

Porter, Elisabeth. *Connecting Peace, Justice, and Reconciliation*. Boulder, CO: Lynne Rienner, 2015.

Raz, Joseph. *The Authority of Law: Essays on Law and Morality*. 2nd edition. Oxford: Oxford University Press, 2009.

Scheffran, Jürgen, Michael Brzoska, Jasmin Kominek, P. Michael Link and Janpeter Schilling. 'Climate Change and Violent Conflict'. *Science*, 336(6083), 18 May 2012, 869–871.

Schlabach, Gerald W. 'What Is "Just Peace"? A Collation and Synthesis based on Catholic and Ecumenical Christian Sources'. February 2018. http://www.geraldschlabach.net/documents/justpeacecollation.pdf. Accessed 5 May 2019.

Sharp, Gene. *Waging Nonviolent Struggle: 20th Century Practice and 21st Century Potential*. Boston, MA: Porter Sargent, 2005.

Smith, Mark J. *Ecologism: Towards Ecological Citizenship*. Minneapolis, MN: University of Minnesota Press, 1998.

Spengler, Oswald. *The Decline of the West*. Trans. and notes Charles Francis. London: G. Allen & Unwin, 1926.

Tawi, Sobhi. 'Education for "Global Citizenship": A Framework for Discussion'. UNESCO, Education Foresight and Research, Working Papers, August 2013. https://unesdoc.unesco.org/ark:/48223/pf0000223784. Accessed 9 May 2016.

Tu Weiming. 'Confucianism and Civilization'. In Majid Tehranian and David W. Chappell (eds). *Dialogue of Civilizations: A New Peace Agenda for a New Millennium*. London: I.B Tauris, 2002.

Tu Weiming. 'The Ecological Turn in New Confucian Humanism: Implications for China and the World'. *Daedalus*, Fall 2001, 130(4), 244.

UN DESA. *The Sustainable Development Goals Report 2018*. New York: United Nations, 2018. https://unstats.un.org/sdgs/report/2018/.

Walzer, Michael. 'The Ethics of Warfare in the Jewish Tradition'. *Philosophia*, 40(4), December 2012, 633–641.

Watts, Jonathan. 'Human Society under Urgent Threat from Loss of Earth's Natural Life'. *The Guardian*, 6 May 2019. www.theguardian.com/environment/2019/may/06/human-society-under-urgent-threat-loss-earth-natural-life-un-report. Accessed 22 May 2019.

Weiss, Thomas G. 'The United Nations: Before, During and After 1945'. *International Affairs*, 91(6), November 2015, 1221–1235.

Williams, Robert E. Jr. and Dan Caldwell. '*Jus Post Bellum*: Just War Theory and the Principles of Just Peace'. *International Studies Perspectives*, 7(4), November 2006, 309–320.

World Council of Churches. 'Statement on the Way of Just Peace'. Adopted by the WCC 10th Assembly on 8 November 2013. www.oikoumene.org/en/resources/documents/assembly/2013-busan/adopted-documents-statements/the-way-of-just-peace. Accessed 13 November 2018.

Joseph Camilleri is Emeritus Professor, La Trobe University, Melbourne, and a Fellow of the Academy of Social Sciences in Australia. He chairs the Academic Board of La Trobe College Australia and is executive director of Alexandria Agenda, a venture in ethical consulting. In 2005, he founded the La Trobe Centre for Dialogue. Under his leadership, the Centre quickly established a national and international reputation for research, training, policy development and community engagement. Joseph Camilleri has authored or edited some thirty major books and over 100 book chapters and journal articles. His research has centred on six key areas: the contemporary role of culture and religion; peace and security; international politics of the Asia-Pacific region; Australian foreign policy; regional and global governance; and the theory and practice of dialogue, reconciliation and conflict resolution.

He has given lectures and keynote addresses around Australia and internationally, including n the United States, United Kingdom, Canada, Sweden, Norway, France, Italy, China, Japan, Indonesia, Malaysia and India.

CHAPTER 3

Crossroads and Crosshairs: Violence, Nonviolence, Critique, Vision and Wonder

Heather Eaton

> We humans have come to a crossroads in our history: we can either destroy the world or create a good future. Even climatically, the balance is shifting to dramatically change the face of the earth. Our ecosystem is in a precarious and fragile state, and our future depends on our actions as a species.
> *Sakyong Mipham Rinpoche*[1]

We are at a crossroads, with many difficult issues facing much of humanity and most of the Earth community of life. For those paying attention, it is unconscionable, heartbreaking, overwhelming and, for many, disempowering. But it should not be surprising. It is from a trajectory of many pathways: ecological disregard, social inequities, economic greed and arms proliferation. For some, this is the legacy of patriarchy: social systems built on relations of domination, exploitation and persistent violence, placing these squarely in the crosshairs.

[1] Sakyong Mipham Rinpoche, *The Shambhala Principle: Discovering Humanity's Hidden Treasure* (Harmony Books: New York, 2013), ix.

H. Eaton (✉)
Saint Paul University, Ottawa, ON, Canada
e-mail: heaton@ustpaul.ca

J. Camilleri and D. Guess (eds.), *Towards a Just and Ecologically Sustainable Peace*, https://doi.org/10.1007/978-981-15-5021-8_3

Many researchers provide facts and data, expose empirically what is going on, analyse the complex reasons and indicate what the consequences are, and will be. Solutions are presented from local to planetary level and from the relatively easy to those that require restructuring human societies. People resist, protest, critique and say change is possible, then mourn, and dare to hope. As theologian Daniel Maguire says: 'If present trends continue, we won't'.[2] Jürgen Moltmann wrote in 1985:

> Our situation today is determined by the ecological crisis of our whole scientific and technological civilization, and by the exhaustion of nature through human beings. This crisis is deadly. ... Unless there is a radical reversal in the fundamental orientation of our human societies, and unless we succeed in finding an alternative way of living and dealing with other living things and with nature, this crisis is going to end in a wholesale catastrophe.[3]

And yet, the trends have continued, with small changes. There are no radical reversals. Fundamental orientations are intact. Countless ongoing conflicts and warmongering occur in many places around the world. Mobile devices, media, entertainment and celebrity cultures provide distractions. The biosphere declines.

Researchers have analysed how we arrived at these crossroads, and what should be put in our crosshairs to strengthen responsiveness. The following focuses on three distinct yet interrelated themes that are implicated in the pathways to a just and ecologically sustainable peace. The first is a general discussion of different forms of violence and domination. This includes direct, cultural, structural and ecological forms of violence. This section concludes with a discussion on the relevance of nonviolence.

The second theme considers the importance and limits of understanding and exposing the issues. The role of analysis and critique is fundamental in revealing power structures, cultural mechanisms and actors involved in what is occurring, how and why. However, critique has its limits. Denouncing problems, especially when solutions are not evident, demobilizes, and furthers despondency. More is needed. Two

[2] Daniel Maguire, *The Moral Core of Christianity and Judaism: Reclaiming the Revolution* (Philadelphia: Fortress, 1993), 13.

[3] Jürgen Moltmann, *God in Creation*, trans. M. Kohl (New York: Harper and Row, 1985), 20.

options will be mentioned: nonviolent resistance, and social imaginaries, or visions, that provide direction, inspiration and energy. This section concludes with an outline of some efforts offering a way forward, with an emphasis on ecological literacy.

The third section focuses on the need for wonder as one aspect that might inspire, nurture a vision and guide a road ahead. Old wisdom and recent research on wonder is discussed, followed by a brief conclusion. Confronting blatant and subtle forms of violence and domination, utilizing critique to probe problems while recognizing that active resistance and vision are further steps, and considering the capacity of wonder are seen here as three elements that could contribute to a sustainable peace.

Forms of Violence and Domination

There is no future without a cessation of the endless violence that saturates most cultures and social imaginaries. The Heidelberg Institute for International Conflict Research reports annually on active conflicts.[4] Violence, terrorism, weapons, injury and the need for security dominate global news reporting. In turn, this reinforces popularist views about the necessity of weapons and wars. National security agendas are accepted without resistance, with ideologies and indoctrination claiming that war, conflict and violence are necessary to put an end to war, conflict and violence. Military preparedness consumes national budgets. The Stockholm International Peace Research Institute found that the world's military budget increased to $1.8 trillion in 2018.[5] This occurs in countries with extensive poverty, malnutrition, poor education and limited health care. The incalculable costs of war are immense. Trauma spreads and sprawls from violence. Families, communities and cultures are disrupted, displaced and damaged, sometimes for generations. Violence begets suffering and more violence.

Violence is always about domination. Johan Galtung, a founder of contemporary peace studies, considers violence to be differentiated

[4] Heidelberg Institute for International Conflict Research, https://hiik.de/conflict-barometer/current-version/?lang=en, accessed 5 March 2019.

[5] Stockholm International Peace Research Institute, 'World Military Expenditure Grows to $1.8 Trillion in 2018', 29 April 2019, https://www.sipri.org/media/press-release/2019/world-military-expenditure-grows-18-trillion-2018, accessed 3 May 2019.

into direct, structural, institutional and cultural forms.[6] These forms of violence are dynamic, and function by strengthening and maintaining each other.

Some forms of violence are sudden and terrifying: guns, mass shootings, missiles and suicide bombings. These forms of violence, for which one cannot predict or prepare, are meant to terrorize, and to keep people afraid and insecure. Even the threats of such forms of violence foster insecurity and anxiety. Such realities are now part of the cultural landscape in many regions of the world. Violence has become normalized and omnipresent.

Violence can also be slow, as is seen in the weight of social injustices, poverty, malnutrition, neglect and domination. Ecological degradation is slow at times, and irreversible deterioration can occur before there are adequate responses, which may exacerbate poverty and social injustices. Climate change can be understood as a form of slow violence. Another example, one that is resistant to change, is violence against women. Progress has been made, unevenly, and globally, yet violence against women persists. It is difficult to prevent as it is connected to the sexualizing of women's bodies and patriarchal beliefs of male rights, power and control.[7]

The United Nations deems that the gamut of violence against women is the most serious and pervasive human rights violation anywhere, and everywhere, in the world.[8] The World Health Organization considers violence to be the most significant health risk to women: a global public health problem that affects between one and two-thirds of all women.[9]

[6] Johan Galtung, 'Violence, Peace, and Peace Research', *Journal of Peace Research*, 6 (3), 1969, 167–191. Galtung and others have developed these dimensions of violence over decades of research. For Galtung's work on peace research see the organization he founded, TRANSCEND International, a global network for Peace, Development and Environment, https://www.transcend.org/galtung/ and the Galtung Institut for Peace Theory and Peace Practice, https://www.galtung-institut.de/en/home/johan-galtung/. Countless others have developed research into the many forms of structural violence.

[7] Heather Eaton, 'Gender Injustices', in Willis J. Jenkins and Mary Evelyn Tucker (eds), *Routledge Handbook of Religion and Ecology* (New York: Routledge, 2016), 348–357.

[8] UN Women does extensive research on violence against women, and how it can be transformed, http://www.unwomen.org/en/what-we-do/ending-violence-against-women/facts-and-figures, accessed 1 March 2019.

[9] World Health Organization. See various reports, https://www.who.int/reproductive health/publications/violence/en/, accessed 1 March 2019.

Violence against women must be seen within the context of direct, structural and cultural violence. Physical and sexual violence are common and global. Other issues that UN Women tackle are the lack of basic human rights: food, water, safety, shelter, education and health care for women and girls. Gender injustices are further exacerbated by intersections of class, ethnicity, gender, sexuality, age, ability and nationality. When a sexual division of labour, poverty, illiteracy, food insecurity, modesty codes, limited reproductive choices, lack of education and commanding religious or political patriarchal governance combine, there is considerable gender injustice. All of this perpetuates the continuum, from private to public, of violence against women.

Although we use the terms domestic or intimate partner violence, or intimate terrorism, over ninety per cent of such violence entails men hurting women: this is *male* violence against women. Women are most at risk from their male partners. Yet rarely do we confront directly that the perpetrators are largely men. It is becoming clear that male violence and toxic masculinities must be confronted: something many ignore, and most avoid. Male violence is seldom deliberated in gender research and programs.[10] My position is anti-essentialist, meaning that men are not *naturally* violent. Women can be violent. Women participate in sustaining patriarchy. What requires meticulous attention is the cultural gender training that condones or disregards male violence, teaches and reinforces toxic masculinities, and blatantly or subtly subjugates women. There will not be a significant shift in violence against women without confronting male violence, which extends into ideas of defence, security, protection, male honour and privilege. Overall, many (post)-industrialized cultures accept and normalize male violence as necessary and natural rather than problematic and learned.

Ecological degradation is best seen as domination of, and violence towards, the Earth. Poisoning waterways, soil and air is a brutality done towards the Earth. The repercussions on the biosphere—this intensely interconnected and integrated living Earth community—are extensive, from macro planetary systems of wind, oceans, climates and biodiversity to the micro damaging of DNA in countless life forms. The 2019 Report from the Global Assessment on Biodiversity and Ecosystem Services of

[10]Because of the #MeToo movement there is new research on patriarchy, men and masculinities. See Jared Yates Sexton, *The Man They Wanted Me to Be: Toxic Masculinity and a Crisis of Our Own Making* (Berkeley: Counterpoint, 2019).

the United Nations[11] warns: 'The health of ecosystems on which we and all other species depend is deteriorating more rapidly than ever. We are eroding the very foundations of our economies, livelihoods, food security, health and quality of life worldwide.'[12] At least a million species are threatened with extinction. Every report on the biosphere—oceans, freshwater, forests, climate change, species stability and biodiversity—such as the comprehensive 2018 Living Earth Report,[13] and the myriad global and local climate change reports reveal that the decline of the biosphere is severe. Violence towards the Earth community needs to be seen not only as a pragmatic problem, but also as a moral and spiritual offence. There will be no peace without ecological integrity.

More generally, violence saturates cultural contexts, permeating social imaginaries with news, entertainment, images and day-to-day language. Life is perceived as a series of fights. Violence saturates the Eurowestern operative worldviews and language patterns, even when these occur in constructive efforts for social transformation. We fight against tyranny, terrorism, poverty, malnutrition and climate change. We fight for peace, justice, equality and freedom. We fight against disease, and for human rights. This adversarial stance has seeped far into the social imaginary.

Forms of violence, only touched upon here, are manifold. Nonviolence is a different worldview. It is not simply an absence of violence. Nor is it passive or naïve. Nonviolence means disengaging with domination, advancing active resistance and developing conflict transformation processes and constructive cultural programs. Nonviolence has power to transform violence—in direct, structural, institutional and cultural forms—into relations of reciprocity, respect and interdependence. However, education and training are required for nonviolence. If we trained as much for nonviolence and peace as for violence and war, perhaps there would be more levers for social change.

[11] E. S. Brondizio, J. Settele, S. Díaz, and H. T. Ngo (eds), *Global Assessment Report on Biodiversity and Ecosystem Services of the Intergovernmental Science-Policy Platform on Biodiversity and Ecosystem Services* (Bonn, Germany: IPBES Secretariat, 2019).

[12] Comment made by Robert Watson, Chair of Intergovernmental Science-Policy Platform on Biodiversity and Ecosystem Services in Paris when the report was released in April and May 2019, https://www.un.org/sustainabledevelopment/blog/2019/05/nature-decline-unprecedented-report/, accessed 10 May 2019.

[13] Living Planet Report, World Wildlife Fund, 2018, https://www.worldwildlife.org/pages/living-planet-report-2018, accessed 9 December 2019.

Fortunately, the interest in, research on and training for nonviolence are growing exponentially.[14] Research is delving into what Mohandas Gandhi called ***constructive programs***, in addition to the tools for nonviolent resistance.[15] These are community actions that build structures, systems or resources that are positive alternatives to relations of domination. It is a hopeful sign.

When considering nonviolence and peace, it is important to be knowledgeable, intelligent and convincing. While many may say peace is desirable, it is not well studied. What are the conditions for building and sustaining peace? How do we educate for peace? Here, the work of Mohandas Gandhi, Martin Luther King Jr. and Johan Galtung, among others, is crucial. Contemporary research into cultures of peace and positive peace is most promising.[16] The Institute of Economics and Peace researches and promotes education, policies and practices for cultures of peace. They developed a positive peace index and report annually on country performance for positive peace. In the 2018 Report, they write: Positive Peace is systemic and requires new thinking to be properly understood.[17] This further supports Gandhi's claim: 'If intellect plays a large part on the field of violence, it plays a larger part on the field of nonviolence.'[18]

[14] For example see Waging Nonviolence, https://wagingnonviolence.org, accessed 9 December 2019, or Voices for Creative Nonviolence, http://vcnv.org, accessed 9 December 2019. See also Heather Eaton and Lauren Levesque (eds), *Advancing Nonviolence and Social Transformation: New Perspectives on Nonviolent Theories* (London: Equinox, 2016).

[15] The Metta Centre for Nonviolence provides a synopsis of this term, https://mettacenter.org/definitions/constructive-program/, accessed 1 March 2019.

[16] Hilary Cremin and Terence Bevington, *Positive Peace in Schools: Tackling Conflict and Creating a Culture of Peace in the Classroom* (Abingdon: Routledge, Taylor & Francis Group, 2017); Andrew Fitz-Gibbon, *Positive Peace: Reflections on Peace Education, Nonviolence, and Social Change*, Vol. v. 217 (Amsterdam; New York, NY: Rodopi, 2010).

[17] Institute for Economics & Peace, *Positive Peace Report 2018: Analysing the Factors That Sustain Peace*, Sydney, October 2018, 3. Available from: http://visionofhumanity.org/reports, accessed 1 March 2019.

[18] M. Gandhi (Thomas Merton, ed.), *Gandhi on Nonviolence: Selected Texts from Mohandas K. Gandhi on Nonviolence, Peace and War* (New York: New Directions, 1965), 60.

Analysis, Critique, Nonviolent Resistance and Vision

What is the role of analysis and critique at these difficult crossroads? Most efforts confronting violence, gender and ecological issues focus on exposing the dimensions within each area. Academics, intellectuals and social commentators are trained to critique: they are the critical avant-garde of society. With excellent tools for debunking beliefs, powers and illusions, these social players put poor arguments and weak intellectual reasoning in their crosshairs. Justice activists also join in with social critiques, and expose injustices, denounce crimes and criminals, and speak truth to power.

Yet these efforts, while crucial, can also limit a way forward. Such voices may become no more than the routine patrol of the critical barbarians.[19] As Bruno Latour notes in 'Why Has Critique Run out of Steam?', the critical landscape has become preoccupied with denunciation, which is necessary but insufficient to address the values, visions and emotions that load the current complex matters of concern.

From another angle, the drivers of ecological degradation, for example, are not always exposed, thus are not transformed. Veteran environmental expert, lawyer and activist Gus Speth summarizes the point:

> I used to think that top environmental problems were biodiversity loss, ecosystem collapse and climate change. I thought that thirty years of good science could address these problems. I was wrong. The top environmental problems are selfishness, greed and apathy, and to deal with these we need a cultural and spiritual transformation.[20]

The current crossroads are not transformed by the customary academic tools of scepticism, discernment and well-reasoned arguments that only expose, analyse and critique the immense problems of our era. To make matters worse, Eurowestern realms of the world are in a postmodern muddle of multiple narratives and truths. While much of this is liberating, it has also opened a space for post-factual politics, post-truth and

[19] Bruno Latour, 'Why Has Critique Run Out of Steam? From Matters of Fact to Matters of Concern', *Critical Inquiry*, Winter, 2004, 242.

[20] A quote from a conversation with Steve Curwood for *Living on Earth*, February 2015, https://loe.org/shows/segments.html?programID=15-P13-00007&segmentID=6, accessed 25 February 2019.

fake news tropes, which in turn foster a potent anti-intellectual shift and the rise of popularism. Gaps exist between understanding and transforming, and among knowledge of issues, general societal awareness, public concern and preparedness, and political leadership and policies. These gaps are hindering effective action.

Obviously, analysis and critique are essential for social transformation. Understanding the issues is a first step, but more is needed to develop ethical positions, strengthen moral outrage and supply courage for effective resistance. This is where nonviolent resistance is decisive for effective change. As mentioned above, nonviolence, among its myriad theories and practices, represents a wide range of actions from constructive social change on numerous issues to levels of resistance. At times, however, social transformations require courageous and prophetic voices to expose injustices and become an opposing social force in order to construct justice.

For example, protests can induce social transformation through education, public presence and political participation. Climate change protests, demonstrations and political activism are forms of ecological democracy. They are an antidote to unresponsive governments and militarized nation-states. Actions such as those initiated by 350.org, Extinction Rebellion and the climate strikes inspired by Swedish environmental activist Greta Thunberg are forces to be reckoned with and respected. They nourish hope.

There can be a cost. The changes needed for a sustainable future are difficult and dangerous tasks and are far more treacherous than analysis and critique alone. *Global Witness* records the deaths of environmental defenders—one of the most dangerous jobs.[21] Most environmental defenders put their lives on the line and are dedicated in a manner akin to a vocation. In the crosshairs of ecology versus economy, four environmental activists are killed each week. Even environmental journalists are threatened or murdered.[22] *Forbidden Stories* publishes reports from environmental journalists. Their *Green Blood* series tracked the investigations

[21] 'Exposing Corruption and Environmental Abuse', *Global Witness*. Operative since 1993, their theme is 'Find the Facts, Expose the Story, Change the System', https://www.globalwitness.org/en/, accessed 20 February 2019.

[22] Juliette Garside and Anthony Watts, 'Environment Reporters Facing Harassment and Murder, Study Finds', https://www.globalwitness.org/en/blog/journalism-battleground-environment-just-tip-iceberg/, accessed 9 December 2019.

of several reporters who were threatened while exposing environmental damage and other abuses by mining companies.[23]

There is ecological grief. Glenn Albrecht, expert in environmental and sustainability studies, known for his brilliant wordsmithing, coined the term *solastalgia*: the lived experience of negative ecological change, a kind of ecological melancholia.[24] Albrecht, considering the debates on climate change and the geological force of humans, suggests that the idiom 'the Anthropocene' is debilitating and futile. He writes:

> In order to counter all these negative trends within the Anthropocene we clearly need, within popular politics and culture, visions and memes of a different future. We also will need more novel conceptual development, since the foundation on which we are building right now is seriously flawed and conducive of nothing but great waves of ennui, grief, dread, solastalgia, mourning, and melancholia.[25]

He suggests that the image of the Anthropocene be replaced with the Symbiocene, which compels ecological literacy and affirms the interconnectedness of life and all living things. The biosphere is, in fact, a symbiosis of life communities living together for mutual benefit. The Symbiocene should be our ecological and cultural image. He considers that we need to consciously move from solastalgia to solaphilia, the love of place.[26]

It is a spiritual crossroads for some.[27] Eco-theologian Norman Habel writes that, 'To violate the Earth is to tear God's masks, to scar God's physical face, to desecrate God's Earthly dwelling'.[28] In a similar vein, Thomas Berry wrote: 'The natural world is the larger sacred community to which we belong. To become alienated from this community is to

[23] Forbidden Stories, https://forbiddenstories.org, accessed 25 June 2019.

[24] Glenn Albrecht, 'Solastalgia, a New Concept in Human Health and Identity', *Philosophy Activism Nature*, no. 3 (2005), 41–44.

[25] Glenn Albrecht, 'Exiting the Anthropocene and Entering the Symbiocene', *Minding Nature*, 9 (2), Spring 2016, 12.

[26] Albrecht, 'Solastalgia, a New Concept in Human Health and Identity'.

[27] Heather Eaton, 'A Spirituality of the Earth', in Norman Habel and Graham Buxton (eds), *The Nature of Things* (Eugene, Oregon: Wipf and Stock Publishers, 2017), 229–243.

[28] Norman Habel, 1998, 119.

become destitute in all that makes us human. To damage this community is to diminish our own existence'.[29]

This insight requires much reflection. Berry continues:

> A diminished Earth will dull our spiritual sensitivities and awareness. Our horizons will shrink; inner lives wither, and spirituality will expire. So integral is our inner world with the outer world that if this outer world is damaged, then the inner life of our souls is diminished proportionately…When we so ruthlessly extinguish the life forms of our period, we threaten, along with those planetary beings, the inner life of the human.[30]

This joins with the axiom of Geswanouth Slahoot, a Salish Elder from Vancouver Island, known as Chief Dan George: 'We are as much alive as we keep the Earth alive'.[31]

There is the question of how to manage our ecological grief, our solastalgia. In a beautiful, sad and inspiring book, *Mourning Nature: Hope at the Heart of Ecological Loss and Grief*, the authors narrate their fears, anger, sadness, despair and hope. There is no hope, no energy and no clarity of vision unless we grieve.[32] What we are doing, intentionally or not, is tragic. We must grieve. Only then can hope (more sophisticated than optimism) surface.

These areas of protest, threats, grief and loss are not well presented with analysis and critique alone. They require different forms to be expressed. This leads to the next topic of alternative or new visions (worldviews, social imaginaries): ways forward that hold power and promise, as evidenced in a 2019 headline in The Guardian entitled 'Enough of the climate nightmare. It's time to paint the dream'.[33] Efforts

[29] Thomas Berry, *The Dream of the Earth* (San Francisco: Sierra Club, 1988), 81.

[30] Thomas Berry, *The Christian Future and the Fate of Earth*, Mary Evelyn Tucker and John Grim (eds) (Maryknoll, NY: Orbis Books, 2009), 60.

[31] Frances Sanderson and Mark Hathaway, 'Golden Rule Interfaith Commentaries', 2019, https://www.scarboromissions.ca/golden-rule/golden-rule-interfaith-commentaries/3, accessed 7 December 2019.

[32] Ashlee Cunsolo and Karen Landman (eds), *Mourning Nature: Hope at the Heart of Ecological Loss and Grief* (Montreal: McGill-Queen's University Press, 2017).

[33] Ed Miliband, 'Enough of the Climate Nightmare: It's Time to Paint the Dream', *The Guardian*, 4 July 2019, International Edition, https://www.theguardian.com/commentisfree/2019/jul/04/enough-climate-nightmare-paint-dream-inequality, accessed 10 July 2019.

such The Earth Charter, The Universal Declaration of Human Responsibilities, Earth Democracy, People's Agreement on Climate Change and the Declaration of the Rights of Mother Earth are generating ecological and planetary charters.[34] They signal a change in vision from the nation-state to a planetary horizon. They share an ethical appeal for the common good of humanity within an Earth community. The Green New Deals, renewable energies, carbon taxes, plastic prohibitions, food security and other countless initiatives are pushing for constructive changes in economics and cultural practices.

Within these new directions, it is fundamental to develop ecological literacy: a requirement for sustainable peaceful communities. Ecological literacy means learning how ecosystems function dynamically and interactively. For example, if we address climate change, then it is important to know something of climate sciences and systems, and not conflate climate change with weather patterns. We need to learn how water works (the ingenious hydrologic cycle) and about plant-insect linkages (how trees and bees communicate) and how food grows. Other animals are often ignored, and yet many are very intelligent, complex and sentient. Mammals are social; have language, families and friends; and suffer in similar ways to humans. It is deeply troubling that new ecological knowledge of the dazzling, ingenious and intensely complex biosphere is rapidly increasing in tandem with its destruction.

At these crossroads, we can choose to become ecologically literate. This changes the kinds of analyses we do, and how we approach social and ecological issues. Social justice and human rights are expanded to incorporate ecojustice and all animal rights. The vision changes. The notion of a common good enlarges. Just peace also expands because peace must include the entire Earth community: we must not engage in wanton domination of Earth life.

From a different angle, ecological literacy assists the development of new visions. We emerged from planetary, evolutionary processes, and are members of the biosphere. Significant educational and worldview transformations are needed here.[35] Evolution is not an idea or an ideological

[34] Heather Eaton, 'Global Visions and Common Ground: Biodemocracy, Postmodern Pressures and the Earth Charter', *Zygon: Journal of Religion and Science*, 49 (4), December 2014, 917–937.

[35] Heather Eaton, 'The Revolution of Evolution', *Worldviews: Environment, Culture, Religion*, 11 (1) (2007), 6–31.

claim. It is a scientific reality. We emerged from and are embedded in Earth dynamics. This needs to reform and inform our visions. The project *Journey of the Universe* is an important and multifaceted contribution to these efforts.[36] To integrate evolution and ecological literacy into visions and cultural orientations does not respond directly to all concerns, such as direct violence, oppression and the inter-locking mechanisms of domination. However, it educates, inspires and teaches that we need to live with Earth, on Earth's terms: the Symbiocene.

Wonder

To move towards the Symbiocene, or any of the visions and charters mentioned above, requires fresh insights and energies. Noting the importance and limits of analysis and critique, and that current thinking, politics, economics, education and cultures are mired in inter-locking institutions that are loathe to change, how can any of these new visions be culturally incarnated? This is the difficult question. While a satisfactory answer is elusive, a realm of human experience that is often overlooked could offer something innovative, accessible and potent: that of wonder and awe: 'What we cannot comprehend by analysis, we become aware of in awe'.[37]

The question here is about what inspires, nurtures hope and offers courage, energy and wisdom. Wonder and awe, common human experiences, are rarely included in analysis and critique of current issues. Wonder is assumed to be extraneous to serious deliberations, nonviolent resistance or political pressure. And yet experiences of wonder can be educative, persuasive and transformative: the zeal and zest behind the visions and actions.

Until recently, awe and wonder were the purview of religions and philosophies. Today, they are researched carefully in terms of evolutionary

[36] The *Journey of the Universe* project includes a film, with additional commentaries from leading scientists, a book, an educational series and an informative web site, http://journeyoftheuniverse.com, accessed 1 March 2019.

[37] Samuel Dresner (ed.), *I Asked for Wonder: A Spiritual Anthology—Abraham Joshua Heschel* (New York: Crossroad, 1997), 3.

origins, purpose, existential source, cognitive and/or somatic experiences, interpretations, importance and consequences.[38] The following is a summary of how wonder and awe function, followed by a synopsis of some recent research. Although each is distinct, the meanings here are conflated.

Awe: a direct and initial experience or feeling when faced with something amazing, incomprehensible or sublime.

Wonder: a reflective experience when one is unable to put things into a familiar conceptual framework—leading to open questions rather than conclusions.[39]

Rachel Carson considered wonder to be essential for understanding and protecting Earth life. A key insight of hers was '…the more clearly we can focus our attention on the wonders and realities of the universe about us, the less taste we will have for destruction'.[40] Experiences of wonder and awe educate for, and inspire pro-ecological ideas and practices.

Before delving into the elements of wonder and current research, a word is needed about the capacity of nature films to elicit pro-ecological stances. Questions have been raised about, and critiques levelled at, the magnificent natural world presentations of David Attenborough, for example. For over fifty years, Attenborough has made films bringing the natural world to people's awareness. His motivation has been his love of nature. Renowned social and ecological critic, George Monbiot, wrote that 'David Attenborough has betrayed the living world he loves'.[41] The first issue is that the scenes of a pure, untouched nature are contrived and misleading, as the natural world is often despoiled in regions close to the filming, but never shown. The second is that the outstanding beauty and natural vitality presented lulls the viewer into an ecological optimism, disbelieving there is an ecological crisis or inducing apathy on ecological problems. This appraisal was taken seriously by Attenborough, who is now

[38] Summer Allen, 'The Science of Awe', Greater Good Science Center at UC Berkeley, September 2018, https://ggsc.berkeley.edu/images/uploads/GGSC-JTF_White_Paper-Awe_FINAL.pdf, accessed 1 March 2019.

[39] Shaun Gallagher et al., *A Neurophenomenology of Awe and Wonder: Towards a Non-reductionist Cognitive Science* (London: Palgrave Macmillan, 2015).

[40] Rachel Carson, Speech to Theta Sigma Phi, 21 April 1954.

[41] George Monbiot, 'David Attenborough Has Betrayed the Living World He Loves', *The Guardian*, 7 November 2018, https://www.theguardian.com/commentisfree/2018/nov/07/david-attenborough-world-environment-bbc-films, accessed 8 November 2019.

vocal on ecological issues and calls for significant change. For example, he has presented at the United Nations Climate Summit and the World Economic Forum, and uses his platform to highlight the threats of climate change and more.

The key question is about what creates the conditions for pro-ecological views that lead to action. What is customarily referred to as pro-environmental behaviour is studied from numerous angles and optics. For some, the natural world is the fundamental inspiration, and its destruction is the impetus for action. For others, values of altruism (children, the future), or egoism (lifestyle), are the best motivating strategies. These debates are important and relevant. The perspective taken in this chapter is that wonder can elicit pro-ecological praxis.[42] The claim is not that this will be true for everyone, and it is true that the dynamics of wonder are not well researched. The following provides an overview of some elements of, and research about, wonder that support the claim that it encourages pro-ecological praxis. Examining how wonder functions reveals its potency as an additional resource for a just, peaceful and ecologically sustainable future.

Primarily, wonder is a *capacity*. It is the ability to ponder that every leaf on every tree is forever unique, or to sense the true presence of trees. This capacity allows one to marvel at the songs of birds, at the way that grass is alive with communities of beings, that acorns turn into trees, that elephants mourn, that dolphins play or that whales communicate over vast distances. Wonder increases depth and detail to what one sees and feels.

Wonder *awakens*. Wonder is an invitation: an awakening. To awaken to the natural world is to see beauty, complexity, vitality, and the breadth and depth of nature. Kathleen Deane Moore writes:

> A person with a sense of wonder marvels at a redwood's muscled trunk, at a sword fern's fiddlehead, as if he were seeing it for the first time; hears as if he had never heard before the song of the winter wren or

[42] Huanhuan Zhao et al., 'Relation Between Awe and Environmentalism: The Role of Social Dominance Orientation', *Frontiers in Psychology: Environmental Psychology*, 3 December 2018, https://doi.org/10.3389/fpsyg.2018.02367, accessed 10 November 2019.

> smelled the bracken-fern. Wonder is the opposite of boredom, indifference, or exhaustion—the lapse into unseeing familiarity.[43]

Abraham Joshua Heschel writes that wonder is related to radical amazement:

> Wonder is a state of mind in which … nothing is taken for granted. Each thing is a surprise, *being is unbelieveable*. We are amazed at seeing anything at all; amazed … at the fact that there is being at all.… Amazed beyond words . … To them, all sight is suddenness.[44]

Wonder implies being open, curious and responsive. Deane Moore suggests that 'a sense of wonder is an attitude of openness or receptivity that leads a person from a preoccupation with self into a search for meaning beyond oneself'.[45] It is important to note the shift from amazement to awareness of realms beyond the self: of meanings or horizons outside normal concerns and routines.

Rachel Carson experienced, reflected upon and wrote about wonder.[46] She spent hours studying tidal pools, crabs or communing with the Earth at night.[47] Carson, a scientist, could also experience the natural world as holding mysteries and depth that the human mind and spirit could intuit but not dissect. There are ineffable, compelling and wonderous realms to the Earth that become present if we attend quietly, courteously, and receptively. It is the natural world that elicits this response: 'its sudden beauty, its intricate interrelations, its power, its contingency, the "ingenuity" of its design, the stunning fact of it, and—most important—its ultimate mystery.'[48] In this manner, wonder *reveals*.

[43] Kathleen Deane Moore, 'The Truth of the Barnacles: Rachel Carson and the Moral Significance of Wonder', *Environmental Ethics*, 27, Fall 2005, 265–277, 269.

[44] Abraham Heschel, 'Radical Amazement', in *Man Is Not Alone: A Philosophy of Religion* (New York: Farrar Straus and Young, 1951), 12, 13, 15.

[45] Ibid.

[46] Rachel Carson, *The Sense of Wonder* (New York: HarperCollins, 1965).

[47] Rachel Carson, *The Edge of the Sea* (Boston: Houghton Mifflin, 1955).

[48] Deane Moore, 267.

Wonder allows for liminal experiences: a threshold to insights only gained through experiences of wonder.[49] In this manner, the natural world becomes revelatory, numinous, ineffable and, for some, sacred. Such experiences of wonder are transformative: increasing awareness, expanding interiority, providing insights, reconstructing knowledge, and rousing responsibility. Wonder educates.

Wonder changes our reference points. Discerning realms of existence larger, and other than, humanity stimulates questions about meaning, purpose and the vast dimensions of reality. For some, wonder begets or strengthens realizations that our anthropocentrism is flawed, and that we are not the centre, the main reference point, the key player or the chief concern of the biosphere, or of the grand drama of the universe. Wonder has the capacity to change our attitude and elicit responsibility.

At these difficult crossroads, wonder holds a capacity to transform more effectively than only analysis and critique. Yet, experiences of wonder require recognition and cultivation. Wonder is not superficial or simply a lovely experience. It educates: spiritual sensibilities, mind, ethics and ethos. Wonder and awe lead to appreciation and reverence, and reverence to responsibility. Such values and virtues are desperately needed today. Heschel wrote:

> Humankind will not perish for want of information; but only for want of appreciation. The beginning of our happiness lies in the understanding that life without wonder is not worth living. What we lack is not a will to believe but a will to wonder.[50]

[49] Heather Eaton, 'Ritual Time and Space: A Liminal Age and Religious Consciousness', with Anne Christine Hornborg, in Sigurd Bergmann and Yong-Bock Kim (eds), *Religion, Ecology and Gender: East-West Perspectives* (Studies in Religion and the Environment/Studienzur Religion und Umwelt) (Münster-Hamburg-Berlin-Wien-London-Zürich: LIT 2009), 79–90.

[50] Abraham Joshua Heschel, *God in Search of Man: A Philosophy of Judaism* (Farrar, Straus and Giroux, 1976), 46–47.

Contemporary Research on Wonder and Awe

Numerous researchers from distinct disciplines are studying wonder, such as Dacher Keltner, a psychologist and the founding director of the Greater Good Science Centre in California.[51] He studies the evolutionary significance, the types and structures of such experiences, as well as their effect. These findings confirm that wonder transforms people. Those who experience wonder are more likely to define themselves as a member of a culture, a species, or a moral cause. Awe embeds the individual self in a social identity.

Experiencing wonder and awe make people more inclined to help someone in need, and feel less entitled and self-important.[52] Inquiries are ongoing about an awe-altruism link: that awe creates a more modest, less narcissistic self, and one who demonstrates greater kindness towards others.[53] Studies find that wonder leads people to cooperate, share resources and sacrifice for others. In short, it increases pro-social behaviour.[54]

From another angle, *The Lab of Misfits*, directed by neuroscientist Beau Lotto, also studies wonder.[55] Lotto says that we are 'wired for wonder'. Using EEG headsets, galvanic skin response, psychological testing and heart rate monitoring, Lotto found that after experiences of awe, there are neuro-behavioural changes: the *signature of awe*. Lotto found that after experiences of awe, people are emotionally uplifted, more generous, able

[51] Dacher Keltner and Jonathan Haidt, 'Approaching Awe, a Moral, Spiritual, and Aesthetic Emotion', *Cognition and Emotion*, 17 (2) (2003), 297–314; D. Keltner, 'Why Do We Feel Awe?' *Mindful: Taking Time for What Matters*, 11 May 2016, http://www.mindful.org/why-do-we-feel-awe/, accessed 1 March 2019.

[52] M. Rudd, K. D. Vohs, and J. Aaker, 'Awe Expands People's Perception of Time, Alters Decision Making, and Enhances Well-Being', *Psychological Science*, 23 (10), 2012, 1130–1136.

[53] Paul K. Piff, Pia Dietze, Matthew Feinberg, Daniel M. Stancato, and Dacher Keltner, 'Awe, the Small Self, and Prosocial Behavior', *Journal of Personality and Social Psychology*, 108 (6), 2015.

[54] Other research indicates that experiences of awe strengthen immune response. Some refer to an awe deprivation disorder which can lead to mental and physical health concerns, similar to nature deficit disorder: Jennifer Stellar et al., 'Positive Affect and Markers of Inflammation: Discrete Positive Emotions Predict Lower Levels of Inflammatory Cytokines', *Emotion*, 15 (2), April 2015, 129–133.

[55] *Lab of Misfits*, http://www.labofmisfits.com, accessed 1 March 2019.

to accept differences and feel connected to others. They are more empathetic, courageous, risk-tolerant and able to step into the unknown.[56]

Sean Gallagher, with others, published *A neurophenomenology of awe and wonder: towards a non-reductionist cognitive science*: an important contribution to the significance of awe and wonder. A central study explored the inner experiences of wonder from astronauts who contemplated Earth from space. They found that the overwhelming experience was how the beauty of Earth elicited a need to protect it. They perceived the Earth as alive, fragile, of unlimited beauty, and infinitely precious.[57] Their experience of wonder led them to feel a responsibility towards the planet and the biosphere.

The final example of recent research comes from biologist Andreas Weber who writes about poetic ecology in *The Biology of Wonder: Aliveness, Feeling, and the Metamorphosis of Science*.[58] This beautiful book shows how the biological sciences need to expand their methods to accurately study life. Life must be approached from the mind and the heart. Weber establishes that feelings are the foundation of life—all life—and he does this by demonstrating the 'feelings' in the interior reality to organisms, including in each cell. Wonder is intimately involved at all levels. He writes:

> Biology thus realizes that something identical to our own emotions—something deeply similar to our own longing for continuation, our desire to be—qualifies as the epicenter from which the entire spectrum of nature unfolds. This understanding provides us with a home in the wilderness again, ... that so many people are longing for ... and that they seek to protect.[59]

Emotions, sentient systems, the desire for life, interiority and subjectivity are the organizing principles of life, the biosphere. Wonder is a connective tissue that allows us to see, feel, understand, appreciate and protect.

[56] *Lab of Misfits*. Retrieved at http://www.labofmisfits.com/experiential-experiments.html, accessed 1 March 2019.

[57] Gallagher et al., 2015.

[58] Andreas Weber, *The Biology of Wonder: Aliveness, Feeling, and the Metamorphosis of Science* (Gabriola Island, BC: New Society Press, 2016).

[59] Ibid., 5.

Conclusion

There are choices at the crossroads of this difficult era of Earth life, and of what will be in our crosshairs. This chapter presented three intersecting themes: violence and domination; the roles of analysis, critique and vision; and wonder. All three are important for a just, sustainable, peaceful future. The connections between the three are not linear, but hopefully cogent. The difficulties are monumental, and solutions tentative. It is crucial to grapple more with the dynamics of violence and domination. In tandem, we need to develop the theories and practices of nonviolence. It is equally important to consider how analyses and critiques, while necessary, are insufficient in developing new visions for a viable future. Understanding is only one step in transforming. Furthermore, while we need to resist, protest and grieve, we also need inspiration, energy and insight.

For an ecological future, we need ecological literacy. Awareness of evolution both educates about the Earth community and thwarts human arrogance and ignorance. These must become conventional within institutional fabrics. The Earth Charter, Earth Democracy, Journey of the Universe and other initiatives incorporate ecological literacy and evolution into charting pathways for a sustainable and peaceful future. How such changes occur at the personal and political levels is barely describable or apparent. However, changes are occurring.

What is apparent is that people need visions and hope to guide and sustain the transformations. Hope, insight and energy can be found in the realms of awe and wonder. The dynamics of wonder and awe offer fresh possibilities, and are simple to access: being attentive to the natural world. Wonder expands both awareness and humbleness. Wonder infuses and sustains. Such experiences are an antidote to despair and grief. Wonder soothes and restores, and fuels political action. Personally and collectively, wonder holds power and promise, and may be a life-giving aspect of the journey.

There are choices at these crossroads. There are choices about what we aim at in our crosshairs. Perhaps we will find a way of joining with others trying to find a viable, noble and peace-filled future.

> Those who contemplate the beauty of the earth find reserves of strength that will endure as long as life lasts. *Rachel Carson*

Bibliography

Albrecht, Glenn. 'Exiting the Anthropocene and Entering the Symbiocene'. *Minding Nature*, 9 (2), Spring 2016, 12–16.

Albrecht, Glenn. 'Solastalgia, a New Concept in Human Health and Identity'. *Philosophy Activism Nature*, 3, 2005, 41–55.

Allen, Summer. 'The Science of Awe'. Greater Good Science Center at UC Berkeley, September 2018. https://ggsc.berkeley.edu/images/uploads/GGSC-JTF_White_Paper-Awe_FINAL.pdf.

Berry, Thomas. *The Dream of the Earth*. San Francisco: Sierra Club, 1988.

Berry, Thomas, edited by Mary Evelyn Tucker and John Grim. *The Christian Future and the Fate of Earth*. Maryknoll, NY: Orbis Books, 2009.

Brondizio, E.S., J. Settele, S. Díaz, and H.T. Ngo (eds). *Global Assessment Report on Biodiversity and Ecosystem Services of the Intergovernmental Science-Policy Platform on Biodiversity and Ecosystem Services*. Bonn, Germany: IPBES Secretariat, 2019.

Carson, Rachel. *The Edge of the Sea*. Boston: Houghton Mifflin, 1955.

Carson, Rachel. *The Sense of Wonder*. New York: HarperCollins, 1965.

Cremin, Hilary and Terence Bevington. *Positive Peace in Schools: Tackling Conflict and Creating a Culture of Peace in the Classroom*. Abingdon: Routledge, Taylor & Francis Group, 2017.

Cunsolo, Ashlee and Karen Landman (eds). *Mourning Nature: Hope at the Heart of Ecological Loss and Grief*. Montreal: McGill-Queen's University Press, 2017.

Deane Moore, Kathleen. 'The Truth of the Barnacles: Rachel Carson and the Moral Significance of Wonder'. *Environmental Ethics*, 27, Fall 2005, 265–277.

Dresner, Samuel (ed.). *I Asked for Wonder: A Spiritual Anthology—Abraham Joshua Heschel*. New York: Crossroad, 1997.

Eaton, Heather. 'The Revolution of Evolution'. *Worldviews: Environment, Culture, Religion*, 11 (1), 2007, 6–31.

Eaton, Heather. 'Global Visions and Common Ground: Biodemocracy, Post-modern Pressures and The Earth Charter'. *Zygon: Journal of Religion and Science*, 49 (4), December 2014, 917–937.

Eaton, Heather. 'Gender Injustices'. In *Routledge Handbook of Religion and Ecology*, Willis J. Jenkins and Mary Evelyn Tucker (eds). New York: Routledge, 2016, 348–357.

Eaton, Heather. 'A Spirituality of the Earth'. In *The Nature of Things*, Norman Habel and Graham Buxton (eds). Eugene, Oregon: Wipf and Stock Publishers, 2017, 229–243.

Eaton, Heather with Anne Christine Hornborg. 'Ritual Time and Space: A Liminal Age and Religious Consciousness'. In *Religion, Ecology and Gender: East-West Perspectives*, Sigurd Bergmann and Yong-Bock Kim (eds) (Studies in Religion and the Environment/Studienzur Religion und Umwelt). Münster-Hamburg-Berlin-Wien-London-Zürich: LIT 2009, 79–90.

Eaton, Heather and Lauren Levesque (eds). *Advancing Nonviolence and Social Transformation: New Perspectives on Nonviolent Theories*. London: Equinox, 2016.

Fitz-Gibbon, Andrew. *Positive Peace: Reflections on Peace Education, Nonviolence, and Social Change*. Philosophy of Peace Series, Vol. v. 217. Amsterdam; New York, NY: Rodopi, 2010.

Gallagher, Sean, et al. *A Neurophenomenology of Awe and Wonder: Towards a Non-reductionist Cognitive Science*. London: Palgrave Macmillan, 2015.

Galtung, Johan. 'Violence, Peace, and Peace Research'. *Journal of Peace Research* 6 (3) 1969. TRANSCEND International, a Global Network for Peace, Development and Environment. https://www.transcend.org/galtung/ and the Galtung Institut for Peace Theory and Peace Practice, https://www.galtung-institut.de/en/home/johan-galtung/.

Garside, Juliette and Anthony Watts. 'Environment Reporters Facing Harassment and Murder, Study Finds'. *Global Witness*. https://www.globalwitness.org/en/blog/journalism-battleground-environment-just-tip-iceberg/. Published 17 June 2019.

Heidelberg Institute for International Conflict Research. https://hiik.de/conflict-barometer/current-version/?lang=en.

Heschel, Abraham. *God in Search of Man: A Philosophy of Judaism*. New York: Farrar, Straus and Giroux, 1976.

Heschel, Abraham. *Man Is Not Alone: A Philosophy of Religion*. New York: Farrar Straus and Young, 1951.

Keltner, Dacher. 'Why Do We Feel Awe?' *Mindful: Taking Time for What Matters*, 11 May 2016. http://www.mindful.org/why-do-we-feel-awe/.

Keltner, Dacher, and Jonathan Haidt. 'Approaching Awe, a Moral, Spiritual, and Aesthetic Emotion'. *Cognition and Emotion*, 17 (2), 2003, 297–314.

Latour, Bruno. 'Why Has Critique Run Out of Steam? From Matters of Fact to Matters of Concern'. *Critical Inquiry*, 30, Winter, 2004, 225–248.

Lotto, Beau. *Lab of Misfits*. http://www.labofmisfits.com.

Maguire, Daniel. *The Moral Core of Christianity and Judaism: Reclaiming the Revolution*. Philadelphia: Fortress, 1993.

Merton, Thomas (ed.). *Gandhi on Nonviolence: Selected Texts from Mohandas K. Gandhi on Nonviolence, Peace and War*. New York: New Directions, 1965.

Miliband, Ed. 'Enough of the Climate Nightmare: It's Time to Paint the Dream'. *The Guardian*, International Edition, 4 July 2019. https://www.theguardian.com/commentisfree/2019/jul/04/enough-climate-nightmare-paint-dream-inequality.

Moltmann, Jürgen. *God in Creation*, trans. M. Kohl. New York: Harper and Row, 1985.

Monbiot, George. 'David Attenborough Has Betrayed the Living World He Loves'. *The Guardian*, 7 November 2018. https://www.theguardian.com/commentisfree/2018/nov/07/david-attenborough-world-environment-bbc-films.

Piff, Paul K., et al. 'Awe, the Small Self, and Prosocial Behavior'. *Journal of Personality and Social Psychology*, 108 (6), 2015, 883–899.

Rudd, M., K. D. Vohs, and J. Aaker. 'Awe Expands People's Perception of Time, Alters Decision Making, and Enhances Well-Being'. *Psychological Science*, 23 (10), 2012, 1130–1136.

Sanderson, Frances and Mark Hathaway. 'Golden Rule Interfaith Commentaries', 2019. https://www.scarboromissions.ca/golden-rule/golden-rule-interfaith-commentaries/3.

Stellar, Jennifer, et al. 'Positive Affect and Markers of Inflammation: Discrete Positive Emotions Predict Lower Levels of Inflammatory Cytokines'. *Emotion*, 15 (2), April 2015, 129–133.

Tucker, Mary Evelyn and John Grim. *Journey of the Universe* Project. http://journeyoftheuniverse.com.

Weber, Andreas. *The Biology of Wonder: Aliveness, Feeling, and the Metamorphosis of Science*. Gabriola Island, BC: New Society Press, 2016.

Yates Sexton, Jared. *The Man They Wanted Me to Be: Toxic Masculinity and a Crisis of Our Own Making*. Berkeley: Counterpoint, 2019.

Zhao, Huanhuan, et al. 'Relation Between Awe and Environmentalism: The Role of Social Dominance Orientation'. *Frontiers in Psychology: Environmental Psychology*, 3 December 2018. https://doi.org/10.3389/fpsyg.2018.02367.

Heather Eaton Full Professor, Saint Paul University, Ottawa, Canada, holds an interdisciplinary Ph.D. in ecology, feminism and theology and is engaged in ecological, feminist and liberation theologies, religion and science, and interreligious responses to ecological crisis. Her books include: *Advancing Nonviolence and Social Transformation: New Perspectives on Nonviolent Theories* with Lauren Levesque (2016), *The Intellectual Journey of Thomas Berry: Imagining the Earth Community* (2014), *Ecological Awareness: Exploring Religion, Ethics and Aesthetics*, with Sigurd Bergmann (2011) *Introducing Ecofeminist Theologies* (2005), and *Ecofeminism and Globalization: Exploring Religion, Culture, Context* (2003) with Lois Ann Lorentzen, and countless academic articles. Heather is a board member of *Worldviews: Global Religions, Environment and Culture*, and of the *Religion and Ecology* session of the American Academy of Religion. She is past president of the *Canadian Theological Society*. Her research includes: ecotheology; peace and conflict studies on gender, ecology and religion; animal rights; nonviolence.

CHAPTER 4

'Holding' a Just and Ecological Peace

Ariel Salleh

A just and ecological peace for only part of humanity will not be just. Too often, the analysis of geopolitical mastery overlooks the micro-power relations that support it. This essay examines the consequences of bypassing the politics of everyday life. It begins by addressing the social 'othering' of Women under patriarchalism, of Strangers under colonialism and of Workers under capitalism. In turn, each of these interlocking political systems and its 'othered' identities has an historically given role in the Human domination of Nature. The essay will describe some of the many sociological and environmental costs that result from a dominant global culture based on 'othering'; and it contrasts this conflictual logic with the relational logic of 'holding'. 'Holding' activities are integrative and energise life-affirming exchanges between human and natural processes. They are practised most often at the margins of capitalism, for example in household labours or on peasant farms. Aldo Leopold wrote that 'A thing is right when it tends to preserve the integrity, stability, and beauty of the biotic community. It is wrong when it tends otherwise'.[1] The present

[1] Aldo Leopold, *A Sand County Almanac* (New York: Oxford University Press, 1949), 224.

A. Salleh (✉)
School of Social and Political Sciences,
University of Sydney, Sydney, NSW, Australia

J. Camilleri and D. Guess (eds.), *Towards a Just and Ecologically Sustainable Peace*, https://doi.org/10.1007/978-981-15-5021-8_4

argument accords with Leopold's ethic and with worldwide social movement protests against the death-designed military industrial imaginary of the nation-state.

Increasingly, in the twenty-first century, localism and community are seen as the essential material condition for achieving justice, sustainability and peace. This text introduces a plurality of such alternative proposals, markedly different in origin, yet each endorsing principles of mutuality, cultural autonomy, direct democracy and eco-sufficiency. The chapter expresses a materialist ecological feminist approach: one recognising that 'the personal is political' and that without praxis there is no theory worthy of the name. These two premises are central to the genre of women's scholarship that emerged internationally five decades ago. The grounding of feminist thought in everyday life speaks both to its integrity as sociological critique and to its usefulness as a guide to political action for change. This ecofeminist thought has found a place 'in' the university, but it is not 'of' the university; sincere feminist writing still prioritises the fragile materiality of Life-on-Earth. As a sociology of knowledge, grassroots womanist critique challenges what conventionally passes for theory in academia. In these times, ecofeminism inevitably adopts a meta-position *vis-a-vis* the disciplines it applies its lens to: this includes the literatures of politics and peace studies.

The Act of 'Othering'

Realising a Just and Ecological Peace will not be easy in a globalised world where not only war, but industrialisation, and even representative government rely on a culture of 'othering' whatever is experienced as 'not-self'.[2] The Woman, The Stranger, The Worker and Nature too are objectified as abstractions, mere system components serving to secure:

- Anthropocentric power over animals, plants and inorganic being
- Patriarchal power over women, sexual minorities and children
- Colonial power over peasant and Indigenous communities
- Capitalist power over workers, precarious and enslaved peoples.

[2] Ariel Salleh, 'Body Logic: 1/0 Culture', in *Ecofeminism as Politics: Nature, Marx, and the Postmodern* (London: Zed Books, 2017/1997).

The 'othering' of anthropocentrism prevents humans from grasping how they are themselves material, nature-in-embodied form. It is possible that the primal moment of Western civilisation was just such an act of Human positioning over Nature. Certainly, the treatment of nature as a resource is foundational to the operation of patriarchal, colonial and capitalist power today. The searches for a just peace and for an ecological peace converge at this point.

That said, it is one thing to consider systemic power in the abstract as academics do. To reverse the pervasive effects of everyday 'othering', a person committed to making social change needs to think reflectively about where he or she lives within these universalising structures of domination. Sometimes, only a hands-on experience of activist praxis can teach how societal structure and individual intention are joined. In any event, a peacemaker of good faith will be prepared to examine his or her sociological 'situatedness' by asking:

- Do I identify with nature or see it as the inert ground of my life?
- Do I receive protection from patriarchal institutions or abuse?
- Do I benefit from colonisation of Indigenous lands or do I lose?
- Do I gain from capitalism or am I economically exploited by it?

The negotiation of a just and ecological peace may well threaten an activist's own sex-gendered, racialised or economic privilege. If this intimate personal sacrifice is not accepted, the outcome can be an ambivalent commitment to change, or worse, paralysing political denial. Peacemaking is existential, and as a 'holding' activity, it is materially embodied in our very own lived lives.

Othering 'The Stranger' and 'The Worker'

The systems of 'othering'—anthropocentric, patriarchal, colonial and capitalist—evolved together over time and are interlocking. Capitalism has attracted the most comprehensive theoretical analysis, but in an era of economic globalisation, capitalist and colonial power can be difficult to distinguish from each other. The Global North is increasingly dependent on the Global South for food, minerals, cheap factory labour and a rising middle class of new consumers for its goods and services. The affluent North 'others' its costs by free trade and offshore production and these externalities include displacement of peasants and Indigenous peoples from their traditional land, damaged livelihood resources like forests and water, and broken cultures and families. The United Nations' development agenda does not help when it objectifies workers as 'human capital' and habitat as 'natural capital'. The ecomodernist policy of multilateral agencies is constructed with abstract computable 1/0 indicators that are reductionist. As such, these fail to correspond to the complex materiality of the very ecological cycles they are supposed to assess.[3] Leading up to the 2012 Rio+20 Earth Summit, the corporate sector enlisted governments and the United Nations in an ambitious financial resuscitation plan for global capital known as the Green New Deal. This new ruling class hegemony was generated by the World Bank, the United Nations Environment Programme, neoliberal think-tanks, promotional agencies, cash-starved academic centres and occasionally big NGOs like Greenpeace and World Wildlife Fund (WWF). As anthropologist Jason Hickel and others calculate, the Green economy is a fraud. Moreover, the 2015 UN Sustainable Development Goals endorse environmentally damaging commercial practices based on privatised water management, energy wasteful digital communications and so-called 'smart' genetically engineered seeds.[4] International climate change solutions like the Clean Development Mechanism also put capitalist accumulation before local Indigenous self-sufficiency.[5]

The uneven development of modernity in the twenty-first century destabilises the Workers of obsolete manufacturing regions in the Global North, as much as it strips independence from farmers—The Stranger in the Global South. Wasted communities in the North result in resentment-driven Alt-Right populations, who viciously reject the migrant refugees from wasted communities in the South. A corporate media profits from

[3] Ariel Salleh, 'Body Logic: 1/0 Culture', 61–85.

[4] Ariel Salleh, 'Green Economy or Green Utopia? Rio+20 and the Reproductive Labor Class', *Journal of World Systems Research*, 2012, 18/2: 141–145; Vandana Shiva, *Water Wars: Privatization, Pollution, and Profit* (Cambridge: South End, 2002).

[5] Jason Hickel, 'Ecologise', https://www.ecologise.in/2018/10/12/why-growth-cant-be-green-hickel/, accessed 6 October 2019; Ariel Salleh, 'Climate, Water, and Livelihood Skills: A Post-Development Reading of the SDGs', *Globalizations*, 2016, 13/6: 952–959; Ana Isla, 'Who Pays for Kyoto Protocol? Selling Oxygen and Selling Sex in Costa Rica', in A. Salleh (ed.) *Eco-Sufficiency and Global Justice: Women Write Political Ecology* (London: Pluto Press, 2009). See also *Environmental Justice Atlas*, https://ejatlas.org/, accessed 6 October 2019.

stories of terror and the spread of public fear. In turn, governments use that fear to legitimise 'securitisation' measures like monitoring private citizens. Mainstream and popular social media reportage glamorises violence from spectator sport to high school shootings. The psychology of 'othering' and discrimination is applied against unemployed and landless classes, castes, ethnicities, sexual minorities, women and youth. Other animal species are mere resources for food or recreation. International militaries and local warlords alike, enculturate recruits to an extreme form of masculinity, routinising the prostitution of women 'as bodies'. Girl and boy children are kidnapped into servitude and soldiering, even as suicide bombers.

Contemporary capitalism is managed by technocratic experts, whose goal is growth measured by Gross Domestic Product (GDP). This productivity index is linked to consumer lifestyles but also to the machinery of war—each of which is reliant on mineral extractivism in colonised territories. Italian peace activist Marco Deriu points out that most military missions are designed to enable the control of commercial sites. The banking sector brokers arms deals on behalf of conglomerates like General Electric, Mitsubishi and Hewlett-Packard. At the same time, military hardware is adapted to everyday civilian use in household products from nanotech-based computing and 5G microwave communications, to genetically engineered foods and medical enhancement technologies.[6] Even in peacetime, the US military-industrial-media-complex is the world's greatest single polluter through its spread of carbon emissions, toxic chemical spills, nuclear and electromagnetic radiation. It is not enough simply to judge this system as morally wanting; the prevailing international 'mode of production' is a mode of destruction, driven by people who have grown up accepting the psycho-dynamics of 'othering' as normal. Deriu's analysis bridges psychological and economic materialities, but emphasises Degrowth as the first practical step towards peace.[7]

[6] The literature on health and environmental impacts of new digital technologies is vast. See 'International Appeal Stop 5G on Earth and in Space', https://www.5gspaceappeal.org/the-appeal, accessed 6 October 2019. A further analysis is provided by Sue Pocket, 'Public Health and the Radio Frequency Radiation Emitted by Cellphone Technology, Smart Meters and Wifi', *NZMJ*, 2018, 131/1487: 96–106.

[7] Marco Deriu, 'Pacifism', in A. Kothari et al. (eds), *Pluriverse: A Post-Development Dictionary* (New York: Columbia University Press and New Delhi, 2019); Aude Fleurant et al., *SIPRI Fact Sheet: Top 100 Arms-Producing and Military Services Companies*, x, https://www.sipri.org/sites/default/files/The-SIPRI-Top-100-2015.pdf, accessed

Othering 'The Woman'

The critique of capitalism and colonialism usually fails to deal with the role that patriarchalism plays in maintaining each of these systems. The micro-power relations of sex-gender penetrate all institutions. Wherever women are mere objects, the property of men, they are valued only as sexual and reproductive labour. The fact that both capitalist and colonial power rely on the hegemony of masculine superordination and entitlement is affirmed by the continuing criminalisation of abortion. The systems of social domination—anthropocentric, patriarchal, colonialist, capitalist—can function together or separately. For instance, the overlap is very clear in the lucrative exploitation of women garment workers in the Asian fast-fashion industry. Again, the hiring of Asian women as womb surrogates for childless or gay couples in the Global North exemplifies objectification by sex-gender, race and class.

United Nations' statistics tell that women do 65 per cent of the world's work for only 10 per cent of world wages paid. After two generations of feminist protest in the Global North and South, the majority of women are exploited internationally as 'not quite labour', as 'a condition of production' or as 'a natural resource'. The #MeToo phenomenon in the United States is drawing attention to women's sexualised plight in the workforce, but one suspects that the success of this campaign is due to the fact that it faults only the individual 'bad apple', while keeping the economic system intact. Liberal values notwithstanding, when capitalism is replaced by communism, or even by hippie communes, the masculine norm still flourishes.

The cultural and material 'othering' of women would seem to be all but universal, even if its expression varies across societies. But even the now dominant global ideology of liberalism was influenced by the great monotheistic religions, and indeed, by 'the humanism' of classical Greece. The patriarchal mindset has justified its exercise of political power by projecting an imaginary in which God, then Kings, then men generally, occupied a hierarchy of dominion over Indigenes, women, children—all deemed 'closer to nature'—down the line (Fig. 4.1).

The Great Chain of Being ideology is still tacitly accepted as 'the natural order of things'. It shapes people's thinking in everyday life,

8 October 2019; Marty Branagan, *Global Warming, Militarism, and Non-Violence* (Sydney: UNSW Press, 2014).

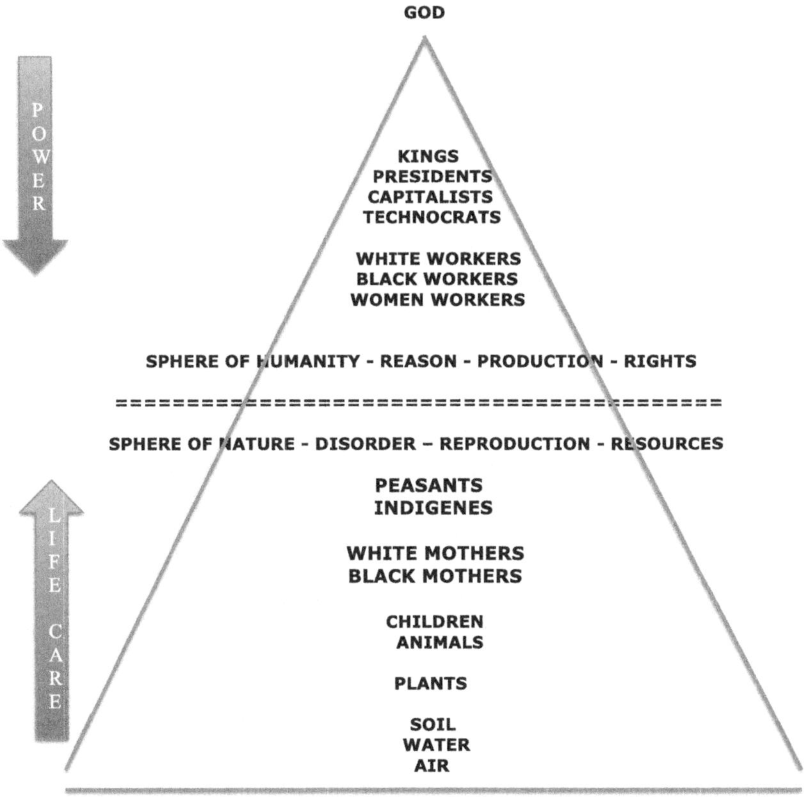

Fig. 4.1 The Great Chain of othering

controlling the status quo through paired dualisms such as Humanity over Nature, White over Black, Masculine over Feminine, where the first term is always valued over the second (1/0). These divisions function unconsciously as metaphors, rationalising the manipulation of non-valued 'others' through domestic labour, colonial slavery or natural resource extraction. Further, 'the natural order' has penetrated theoretic constructs in economics, law, science and philosophy. The recent emergence of social movement politics is easy to read from the Great Chain schema. Top down, it shows: the eventual usurpation of kings by a capitalist merchant class; working-class labour resistance; decolonial struggles; the rise of feminism; youth rebellion; animal rights; and ecology. Power and privilege continue to be enjoyed by those at the top of the status hierarchy; those below carry material responsibilities but enjoy few rights. The effects of 'othering' are seen in worker precarity, refugee exclusion, domestic violence, child abuse, species extinction, peak water and global warming. That said, patriarchalism long predates capitalism and it is found in many non-Western cultures as well.

Historical analyses of systemic power relations have been offered by North American theologians Rosemary Ruether and Elizabeth Dodson Gray. Their biblical exegesis suggests that early monotheistic leadership drew a clear line between Humanity and Nature, and a clear line between men as humanity-proper versus fertile-embodied-women as 'natural resources'.[8] This institutional power is said to enact, indeed to sublimate, the 'self versus other' dualism, as each generation of boy children breaks away from the mother's body. In onto-epistemological terms: a psychology of 'holding' is replaced by 'othering'. This subjective move is repeated in the existential struggle of the infant from embodied dependency to individual self-awareness. Societies in which misogynist traditions already prevail, reinforce the individual separation with violence. Conversely, if identity formation is unresolved, an unconscious war of 'othering the feminine' can persist within the masculine psyche. In ancient times, war provided an armoury of self-images to bolster men's distinctive difference from Women and Nature. To an extent, this is assisted today through the celebration of technology.

In political and academic circles, much attention is paid to the class war of workers versus capitalists and to colonial wars fought by Indigenous peoples; but an unnamed war plays out closer to home. As the World Council of Churches noted in 2003: 'Violence, intimidation, abuse, and exploitation thrive in the shadows of division and inequality. Domestic violence is a hidden tragedy in societies everywhere …'.[9] In July 2018, an Internet Discussion List circulating among feminist scholar-activists ran the following message:

[8] Rosemary Ruether, *New Woman: New Earth* (New York: Seabury, 1975); Elizabeth Dodson Gray, *Green Paradise Lost* (Cambridge, MA: Round Table Press, 1979).

[9] World Council of Churches, *Statement on the Way of Just Peace*, 10th Assembly Report of the Public Issues Committee, 8 November 2013.

> In the 10 days since we began this conversation, Australian local news has reported:
>
> 3 women's bodies dumped in bush near highway
> Woman burned alive at home by husband
> Torture and mother-rape by 17 year old youth
> Pregnant woman with multiple stabs after home invasion
> Indigenous community grieving after rape of infant ...[10]

In Australia, an affluent country of only 25 million people, one and often two women a week are murdered by their life-partner. Sex-gender statistics for India, Argentina, New Guinea, Poland and South Africa, are comparable.[11]

In Saudi Arabia, women as a sexual and reproductive 'resource' are infantilised by the customary legal guardianship of fathers and brothers. Those who seek 'human' rights meet with arbitrary imprisonment or disappearance. Women's International League for Peace and Freedom reports interviews in Israel with 700 Jewish and Palestinian women who say they are afraid of personal attack more than they are afraid of bombs. Women in post-War Iraq are trafficked. Women in Afghanistan must beg to join their country's peace negotiations. Women of the Syrian opposition are systematically raped by government troops in a strategy to destroy family honour. Once disgraced, their husbands throw them on to the streets. The Middle East has its unique turmoil, but no region of the world is free of the war on women.[12]

English peace activist Cynthia Cockburn is another who recognises war as affirming the culture of masculinity. War nourishes the very 'othering' that sustains systems of social power. The Worker is perceived as inferior and exploitable. The Stranger is made vulnerable by the stigma of race or ethnicity. The Woman is simply sub-human or 'closer to nature'.[13] Terror

[10] Ariel Salleh, Personal Communication on 'Time to Focus' International Email Discussion Group on the Global Epidemic of Violence on Women, 17 July 2018.

[11] For international statistical surveys, see the database WomanStats, http://www.womanstats.org, accessed 6 October 2019.

[12] 'Women Peace and Security', *KOFF Newsletter*, 2015, No. 141; BBC World Service News, 28 November 2018, https://www.bbc.com/news/world-middle-east, accessed 28 November 2018.

[13] Cynthia Cockburn, *From Where We Stand: War, Women's Activism and Feminist Analysis* (London: Zed Books, 2007).

in both private and public spheres is how men, demeaned by other men, re-make diminished ego-identity.

The war on women must be part of the conversation about a just and ecological peace. In the twenty-first century, a combination of female foeticide, domestic abuse and militarised collateral damage on civilian populations has resulted in a falling demographic ratio of women to men globally. In Asia alone, one and a half million women have lost their lives in the last decade due to masculinist attitudes. Are governments and international agencies keeping statistics on sex-gender violence? Why did the very effective Australian Royal Commission into Child Sexual Abuse not name its work 'an investigation into patriarchal power'? In England, an official record of domestic homicides is maintained by the Office for National Statistics, and it even calls for international standardisation of measures. But no government has developed a strategy to end the killing. As British socialist Judith Watson observes:

> Liberal and reformist approaches are not enough or may even be positively harmful. They have led some feminists to concentrate efforts on working with the police and judicial system to argue for increased prison sentences ... [Yet there are] cases in Britain where judges argued that little girls were temptresses and let rapists off with a slap on the wrist ... [On the other hand,] a revolutionary attitude that says we just have to wait until the overthrow of capitalism and/or patriarchy and/or the petro-economy, is no use either.[14]

To reiterate: the four systems of oppression discussed in this essay are historically interconnected and compound each other in everyday situations. But the activist citizen is not helped by media reportage framed in terms of the single issue 'othering' of liberal pluralism. The public focus on competing brands of 'identity politics' suffocates debate in a series of double binds. Black communities respond by refusing to work with white feminists; or socialists condemn environmentalists for being middle class. Ecofeminist activist Samantha Hargreaves has experienced 'othering' in

[14] Judith Watson, Personal Communication with Author on 'Time to Focus' International Email Discussion Group on the Global Epidemic of Violence on Women, 30 July 2018; UK Statistics, 2017, https://www.ons.gov.uk/peoplepopulationandcommunity/crimeandjustice/bulletins/domesticabuseinenglandandwales/yearendingmarch2017#domestic-homicides, accessed 6 October 2019. For a blog on Femicide, https://kareningalasmith.com/, accessed 6 October 2019.

the guise of decolonial 'political correctness', deflecting attention from sex-gender violence by default.

> There have been public revelations recently here in South Africa and well beyond, concerning problems of sexism and sexual harassment in our very organisational 'homes'. When the problem steps out of the shadows where it has been for many decades and into the domain of the 'public', then we people who identify as Left are challenged to take a stand. Instead, some comrades turn aside and, if they don't deliberately obstruct, say nothing and in so doing protect the perpetrators.[15]

Patriarchal attitudes are intrinsic to capitalism and colonisation, power structures that work together by what sociologists call 'an elective affinity'. For this reason, masculinist entitlement is over-determined, multilayered, readily misinterpreted and too often perversely theorised. Abuse of children and cruelty to animals are further aspects of the widespread prerogative traditionally held by men of the dominant global culture over what they see as 'lesser' life forms. Such activities are a kind of extractivism; a gratification of self through energies drawn off 'othered' bodies. Sometimes liberal academics argue that it is time to stop talking about the traditional sex-gender dualism, but it is plain from this discussion that 'peace on earth' will never be achieved as long as 'othering' is the foundational logic of global institutions and ideologies.

Othering 'The Earth'

Moving beyond the anthropocentric framework of liberal, socialist and postmodern feminisms, the Indian ecological feminist Vandana Shiva has written: 'The construction of women as "the second sex" is linked to the same inability to cope with difference as is the development paradigm that leads to the displacement and extinction of diversity in the biological world'.[16] As this quote illustrates, a materialist ecofeminism is transnational and concerned with a multiplicity of oppressions. Activists like Shiva step up wherever the reproduction of Life-on-Earth is threatened by

[15] Samantha Hargreaves, Personal Communication with Author on 'Time to Focus' International Email Discussion Group on the Global Epidemic of Violence on Women, 19 July 2018.

[16] Maria Mies and Vandana Shiva, *Ecofeminism* (London: Zed Books, 2014/1993), 64.

war, toxic waste, deforestation, or dispossession of livelihoods. In seeing society and ecology as integrally joined, ecofeminists deepen simplistic liberal notions of feminism as being about 'women having the right to do the same as men'. That emancipation could only worsen prospects for a Just and Ecological Peace. Ecofeminists observe that masculinist power is derived increasingly from technological transformation of the given Earth into a 'man-made' one—regardless of consequences. Historian Carolyn Merchant's research in *The Death of Nature* traces this process back to the seventeenth century English 'witch-hunt', a time when the murder of millions of knowledgeable women herbalists and midwives was ruled necessary to clear the way for Europe's scientific revolution.[17] The consequence was a re-definition of the Earth and the human body as functional machines—so giving rise to the nature-alienated instrumentalism of 'Enlightened industrial modernity'.

Women theologians from Ivonne Gebara in Brazil to Heather Eaton and Anne Elvey writing in the Anglo world, are others who help show how the political struggle for nature is sex-gendered.[18] Another key ecofeminist thinker is the late Rosalie Bertell, Grey Nun, biometrician, epidemiologist and recipient of several honorary doctorates. Bertell was concerned with contradictory assumptions in the 'security discourse' promoted by governments, multilateral agencies and the business sector. Food security, water security and climate security are international buzzwords, but the security mission is self-defeating once war becomes part of it. Under the securitisation discourse, billion dollar defence programs perpetuate the very human deprivation that fosters mundane social conflict. The tension between embodied wellbeing versus the use of Agent Orange, nuclear or biological weaponry is plain. Then there is the military research into electromagnetic devices, able to produce man-made storms, earthquakes, or tidal waves. The United States maintains an installation in Alaska known as the High-frequency Active Auroral Research Programme (HAARP), which experiments with natural planetary processes for defence purposes. But disturbed magnetic fields

[17] Carolyn Merchant, *The Death of Nature: Women, Ecology, and the Scientific Revolution* (San Francisco: Harper and Row, 1990/1980).

[18] Ivonne Gebara, *Longing for Running Water: Ecofeminism and Liberation* (Minneapolis: Fortress, 1999); Heather Eaton, *Introducing Ecofeminist Theologies* (London: Bloomsbury, 2005); Anne Elvey, 'The Material Given: Bodies, Pregnant Bodies and Earth', *Australian Feminist Studies*, 2003, 18/4: 199–209.

interfere with life-support systems like the navigation of fish and bees. In humans, electromagnetic radiation is associated with cataracts, leukemia, altered brain and body chemistry, blood pressure and heart rate.[19] Canadian geographer Emily Gilbert believes that governments interested in the militarisation of climate include England, France, Germany, possibly India and China. Citing the *Quadrennial Defense Review* she notes that by 1978, a number of nations had signed an Environmental Modification Convention (ENMOD), but by 1991 the Pentagon had waived the requirement for impact assessment.[20]

US marines in Afghanistan have been reported using 800,000 gallons of fuel per day, with considerable CO_2 emissions. Expending massive amounts of oil in wars to secure oil is plainly irrational. According to feminist researcher Betsy Hartmann, the US defines the securitisation of oil and gas reserves as 'an aid-military complex' for civilian development, and it claims to be fighting in the Middle East for women's rights.[21] The Department of Defense has committed to 'green and sustainable' technologies, with multiple collaborators, assisted by the World Trade Organization (WTO), Global Environment Facility (GEF) and United Nations Framework Convention on Climate Change (UNFCCC). Defence contractor Lockheed Martin, manufacturer of stealth bombers, ballistic missiles, unmanned drones, depleted uranium weapons and logistical support systems, now focuses on solar thermal energy, synthetic fuel cells, ocean energy technology and smart grids. A biofuel-driven Great Green Fleet of aircraft carriers was scheduled for 2016.[22] Department of Defense energy generators—some 100,000 around the world dependent on oil—are to be replaced by solar technology and the 'green military'

[19] Rosalie Bertell, *Planet Earth as a Weapon of War* (London: Women's Press, 2000).

[20] Emily Gilbert, 'The Militarization of Climate Change', *ACME: An International E-Journal for Critical Geographies*, 2012, 11/1: 1–14; US Department of Defense, 2010. *Quadrennial Defense Review Report*, http://www.defense.gov/qdr/, accessed 8 October 2019.

[21] Betsy Hartmann, 'Rethinking Climate Refugees and Climate Conflict: Rhetoric, Reality and the Politics of Policy Discourse', *Journal of International Development*, 2010, 22/2: 233–246.

[22] *The Guardian*, UK, 2009, http://www.darpa.mil/Our_Work/STO/Programs/Biofuels.aspx; http://www.guardian.co.uk/environment/2009/aug/13/us-marines-afghanistan-fuel-efficiency, accessed 6 October 2019.

will use portable micro-grids.[23] The 'smart' trend now brings households under the ambit of Electronic Grids and Information Systems, while images of 'the other' will be manipulated by advanced NetWar techniques.

Just and Ecological Paths

Is there a life-affirming alternative to this misbegotten civilisation of the fathers? Certainly, a diversity of voices around the globe is speaking out with proposals for change. One of these, Australian environmentalist Ted Trainer, outlines a model for justice and sustainability called The Simpler Way. As his prescient and pioneering 1985 book *Abandon Affluence!* puts it: 'The conventional definition of "development" in terms of striving endlessly for growth and affluence … accelerates ecological destruction, inequality and poverty, social breakdown and armed conflict over resources and markets'.[24] Trainer adopts a basic 'limits to growth' case, celebrating homemade simplicity, creative frugality and eco-sufficient provisioning. He judges an ideal *per capita* resource consumption to be 10 per cent of present rich world levels, a future of zero growth with a fraction of current production and trade—so as to allow time for conviviality. Trainer challenges eco-socialists who assume that the state can raise living standards by ironing out what Marx saw as internal contradictions of capitalism. Likewise, he would question a centrally administered Basic Income Scheme. His eco-anarchist vision is premised on small-scale local economies. Free of top-down bureaucracy, these are governed by citizen initiatives like town meetings, voluntary committees, working bees and time for spontaneous discussion.

The Simpler Way converges with recent international experiments in Transition Towns, Degrowth, Permaculture, Eco-Villages and the practice of such politics is well underway in Catalonia and parts of Greece. Meanwhile, in the Global South, social movement networks like Systemic Alternatives, South America and Radical Ecological Democracy, India,

[23] Gary Leech, 'Greening Warfare? Lockheed Martin's Sustainability Agenda', *Beyond Capital*, Blog, 10 September 2010, http://beyondcapital.org/2012/09/10/greening-warfare-lockheed-martins-sustainability-agenda/, accessed 10 September 2012.

[24] Ted Trainer, *Abandon Affluence!* (London: Zed Books, 1985); Compare also, https://simplicityinstitute.org/, accessed 6 October 2019; http://transitionnetwork.org, accessed 6 October 2019; https://systemicalternatives.org, accessed 6 October 2019; https://radicalecologicaldemocracy.org, accessed 6 October 2019.

are encouraging peasant and tribal societies to resist the colonial narrative of affluence. They promote an ethic of Commoning that honours traditional economies, provisioning methods that already prefigure the self-governance and ecological integrity that activists in the Global North are striving for.[25] Materialist ecofeminists also refuse the band-aid solutions of a nature-killing productivist economy with its ecological modernist ideology, false promises of Earth System Governance, corporate social responsibility and state reform. The logic of 'holding' is to rebuild communitarian societies from the ground up, revitalising the lost ethic of care.

Another vision of reciprocity with nature is the *ubuntu* value system that permeates many Indigenous oral traditions from South Africa. In contrast to the West's divisive dualist logic of 'othering', Lesley Le Grange from Stellenbosch University explains that Zulu, Xhosa and Ndebele peoples use an expression: *Umuntu ngumuntu ngabanye Bantu* to account for human unfolding in the biophysical web. As a morality and philosophy, *ubuntu* expresses a relational logic that contrasts markedly with the so-called 'civilised' ideal of an individual with hard 'othering' ego boundaries. Africans will say: 'We are, therefore I am; thus, by harming others, I harm myself'. Le Grange writes that: 'The Zulu word for human being is *umuntu*; a materiality constituted by: *umzimba* (body, form, flesh); *umoya* (breath, air, life); *umphefumela* (shadow, spirit, soul); *amandla* (vitality, strength, energy); *inhliziyo* (heart, centre of emotions); *umqondo* (head, brain, intellect) and so on.[26] In the *ubuntu* world, harm to nature rebounds socially, so care must be taken in common to protect the conditions of Life-on-Earth. This reasoning converges with the youth-driven movement for Commons in Europe, and with an emerging interest in the feminist ethic of care among Degrowth scholars, in Germany, Scandinavia and Portugal. It honours Indigenous Andean concepts of the good life such as *sumak kawsay* and its Spanish reiteration *buen vivir*, wherein wholesome livelihood is prized before money.

[25] For an exhaustive compilation of alter-globalisation movement politics see: Ashish Kothari, Ariel Salleh, Arturo Escobar, Federico Demaria, and Alberto Acosta (eds), 2019, *Pluriverse: A Post-Development Dictionary* (New York: Columbia University Press, 2019 and New Delhi: Tulika/AuthorsUpFront).

[26] Lesley Le Grange, 'Ubuntu, Ukama, Environment and Moral Education', *Journal of Moral Education*, 2012, 41/3: 329–340; See also Inge Konik, 'Ubuntu and Ecofeminism', *Environmental Values*, 2018, 27/3: 269–288.

Earth-based philosophies are enabling once-colonised peoples to resist the false promise of consumption as progress—including the Green Economy engineered by multilateral agencies and their business sector sponsors.

The cause of a just and ecological peace is powerfully reinforced by yet a further social movement convergence between ecofeminism, The Simpler Way, Commoning, *ubuntu*, *buen vivir* and eco-theology. The Irish Columban missionary Seán McDonagh has done much to advance this counter-colonial turn. As Catholic Social teaching came face-to-face with the industrial revolution, the encyclical *Populorum Progressio* set out to humanise the meaning of 'development'. Liberation Theology got underway in 1970s Brazil, but the Church tended to dismiss ecology as a middle-class concern, much as Marxists did in that era. Father Thomas Berry argued for grounding life within the natural world as antidote to the 'technological trance'. McDonagh enlarged on this, inspired by his time among the T'boli highlanders of Mindanao. By the late 1990s, Leonardo Boff's *Cry of the Earth* spoke to climate change, species loss, deforestation and water poisoning, while proposing Indigenous templates for sustainable economic provisioning.[27] The Christian analyses of capitalism, colonialism and ecology culminated in 2015 with Pope Francis' encyclical *Laudato Si': On Care for Our Common Home*. In the words of the Holy Father: 'deterioration of the environment and of society affects the most vulnerable people on the planet' (No. 48), 'evasiveness serves as a license to carry on with our present lifestyles and models of production and consumption' (No. 59).

[27] Thomas Berry, *The Dream of the Earth* (San Francisco: Sierra Club Books, 1988); Sean McDonagh, *To Care for the Earth: A Call to a New Theology* (London: Chapman, 1986); Leonardo Boff, *Cry of the Earth, Cry of the Poor* (New York: Orbis Books, 1997).

The Act of 'Holding'

What kind of life experience enables a person to turn away from indiscriminate notions of 'progress'? In *Laudato Si'* (No. 128), Pope Francis insists that 'work' is central to the meaning of life, but that fulfilment is undermined by technological change.[28] People in consumer societies have less reciprocal give and take with their material environment than non-industrials do. Moreover, the popular conflation of 'high tech' with social advancement, alienates any sense that humanity is itself nature-in-embodied form. Capitalist 'othering' undermines all facets of its resource base, but it is particularly hard on women. As Delphine Yeyet from Gabon describes it:

> In a subsistence economy, men are obliged to earn their livelihood in cooperation with women without exploiting them. In a monetary economy, however, the thirst for comfort and profit pushes men to exploit women and chase them from the domains of political and social action.[29]

In another telling post-development insight, the late Chilean economist Manfred Max Neef noted how the engineered satisfiers of 'advanced societies' cost great effort and end by sabotaging the very convenience they were designed for; whereas self-managed provisioning meets multiple needs at once—subsistence in partnership with habitat, conviviality, learning, identity and belonging.[30]

Following research on West Africa and South Asia, the German sociologist Maria Mies also spelled out the benefits of a community 'subsistence perspective'. Her ecofeminist analysis would anticipate the current European advocacy of Degrowth and Solidarity Economies, and it resonates with the political movement ideals of Andean *buen vivir*;

[28] Pope Francis, *Laudato Si': On Care for Our Common Home*, 2015, http://w2.vatican.va/content/francesco/en/encyclicals/documents/papa-francesco_20150524_enciclica-laudato-si.html, accessed 6 October 2019.

[29] Quoted in Trinh Minh-ha, *Woman, Native, Other* (Bloomington: Indiana University Press, 1989), 108. See also: World March of Women, www.marchemondiale.org, accessed 6 October 2019.

[30] Ariel Salleh, 'A Barefoot Epistemology', in *Ecofeminism as* Politics; Manfred Max-Neef et al., *Human Scale Development* (New York: Apex, 1991).

African *ubuntu*; and Indian *swaraj*.[31] Mies complements the Marxist analysis of production by acknowledging the contribution of 'reproductive workers'—mothers, peasants, Indigenous gatherers. In fact, we might well argue that these labour groupings form a distinct 'meta-industrial class', whose regenerative work is indispensable for 'holding' the world's humanity-nature metabolism in relative stability.[32]

In a paradigmatic statement of such 'agency in complexity', Vandana Shiva has described how:

> It is in managing the integrity of ecological cycles in forestry and agriculture that women's productivity has been most developed and evolved. Women transfer fertility from the forests to the field and to animals. They transfer animal waste as fertilizer for crops and crop by-products to animals as fodder. They work with the forest to bring water to their fields and families. This partnership between women's work and nature's work ensures the sustainability of sustenance ...[33]

Shiva, as a mother and a quantum physicist, is acutely aware of the sophisticated rationality that these care-giving activities instil. The First Peoples of Australia also use a form of 'holding labour' to achieve sustainability. They may cultivate the Earth but often as not they will move nomadically through country in the knowledge that it will replenish and provide again when they return. Indigenous Australians talk of their gathering economies as having 'good fit with country'. Meanwhile, the Indonesian peasant union Via Campesina asserts that their own organic provisioning model 'cools down the Earth'.[34]

Meta-industrial labour nurtures energetic flows between humanity and nature in various ways. Ulla Terlinden links the logic of house-keeping and eco-sufficient farming skills:

[31] Maria Mies, *Patriarchy and Accumulation on a World Scale* (London: Zed Books, 1986); Vandana Shiva, *Staying Alive: Women, Ecology, and Development* (London: Zed Books, 1989); Giacomo D'Alisa, Federico Demaria, and Giorgios Kallis (eds), *Degrowth: A Vocabulary for a New Era* (London: Routledge, 2014).

[32] Ariel Salleh, 'Global Alternatives and the Meta-Industrial Class', in R. Albritton et al. (eds), *New Socialisms: Futures Beyond Globalization* (London: Routledge, 2004).

[33] Shiva, 1989, 45.

[34] Via Campesina, https://viacampesina.org/en/, accessed 6 October 2019.

> [Each] requires ... a broad range of knowledge and ability. The nature of the work itself determines its organization. The work at hand must be dealt with *in its entirety* ... The worker must possess a high degree of personal synthesis, initiative, intuition and flexibility.[35]

Another rich account of vernacular labour immersed in the biophysical world is developed in philosopher Sara Ruddick's book *Maternal Thinking*. She writes that maintaining a family requires harmonising a complex of sub-systems, as well as considerable decision-making and diplomatic skills:

> The value of objects and accomplishments turns on their usefulness in satisfying needs ... [Such labour is] continuously involved with connection, separation, development, change, and the limits of control.[36]

Ruddick's concept of 'holding' is especially relevant to a just and ecological peace:

> To hold means to minimize risk and to reconcile differences rather than to sharply accentuate them. Holding is a way of seeing with an eye toward maintaining the minimal harmony, material resources, and skills necessary for sustaining a child in safety. It is the attitude elicited by 'world protection, world-preservation, world repair'.[37]

In ecofeminist peace movements like Madre or the Code Pink network, this precautionary attitude is applied beyond home and neighbourhood to the global politics of Life-on-Earth. Theologian Gerald Schlabach's principles of a just peace resonate with 'holding' practice when he says:

[35] Ulla Terlinden, 'Women in the Ecology Movement, Ecology in the Women's Movement', in E. Altbach et al. (eds), *German Feminism* (Albany: SUNY Press, 1984), 320. Italics added.

[36] Sara Ruddick, *Maternal Thinking: Toward a Politics of Peace* (Boston: Beacon, 1989), 131.

[37] Ruddick, 1989, 79.

> Peace is not the absence of conflict. It is the presence of creative processes by which human communities become ever more skilful and habituated at working through their conflicts without recourse to violence and without demeaning one another's dignity.[38]

To appraise the insights of women thinkers in this context is neither to advocate an 'identity politics' that argues from victimhood, nor fall back on essentialist myths of some 'innate feminine morality'. Rather, it is to advance an onto-epistemological thesis about cognitive capacities that women—or men—derive from active engagement in nurturing the metabolism of humanity and nature. Like the word 'othering', the word 'holding' has multiple dimensions—psychological, epistemological, biophysical and political.

Today, as noted above, a pluriverse of new political movements is speaking out. To add yet further exemplars: the Yasuni 'leave the oil in the soil' initiative; the communitarian zeal of Rojava women; the Extinction Rebellion of London youth. Again, the rejection of anthropocentrism, patriarchalism, colonialism and capitalism by these voices does not mean 'walking backwards in history' as Western liberal technocrats like to claim. It means a heightened awareness of how centuries of 'othering' has led to a destructive global regime of Humanity against Nature and endemic wars among ethnic, sex-gendered and class identities. As a confluence of initiatives from every continent, these visions of local sufficiency, direct democracy and cultural autonomy, are the logical corollaries of a just and ecological peace. For in everyday 'holding labours', the self encounters its true self in 'the other'.

[38] Gerald Schlabach, 'What Is Just Peace? A Collation and Synthesis Based on Catholic and Ecumenical Christian Sources', *Catholic Nonviolence Initiative* (University of St. Thomas, MN, 2018).

Bibliography

BBC World Service News, 28 November 2018. https://www.bbc.com/news/world-middle-east.

Berry, Thomas. *The Dream of the Earth.* San Francisco: Sierra Club Books, 2018.

Bertell, Rosalie. *Planet Earth as a Weapon of War*. London: Women's Press, 2000.

Boff, Leonardo. *Cry of the Earth, Cry of the Poor.* New York: Orbis Books, 1997.

Branagan, Marty. *Global Warming, Militarism, and Non-Violence.* Sydney: UNSW Press, 2014.

Cockburn, Cynthia. *From Where We Stand: War, Women's Activism and Feminist Analysis.* London: Zed Books, 2007.

Defense Advanced Research Projects Agency (DARPA). http://www.darpa.mil/Our_Work/STO/Programs/Biofuels.aspx.

D'Alisa, Giacomo, Federico Demaria, and Giorgios Kallis (eds). *Degrowth: A Vocabulary for a New Era.* London: Routledge, 2014.

Deriu, Marco. 'Pacifism'. In A. Kothari et al. (eds), *Pluriverse: A Post-Development Dictionary.* New York: Columbia University Press, 2019, and New Delhi: Tulika/AuthorsUpFront.

Dodson Gray, Elizabeth. *Green Paradise Lost.* Cambridge, MA: Round Table Press, 1979.

Eaton, Heather. *Introducing Ecofeminist Theologies.* London: Bloomsbury, 2005.

Elvey, Anne. 'The Material Given: Bodies, Pregnant Bodies and Earth'. *Australian Feminist Studies*, 2003, 18/4: 199–209.

Environmental Justice Atlas. https://ejatlas.org/.

Femicide Blog. https://kareningalasmith.com/.

Gebara, Ivonne. *Longing for Running Water: Ecofeminism and Liberation.* Minneapolis: Fortress, 1999.

Gilbert, Emily. 'The Militarization of Climate Change'. *ACME: An International E-Journal for Critical Geographies*, 2012, 11/1: 1–14.

Guardian (The), UK, 2009. http://www.darpa.mil/Our_Work/STO/Programs/Biofuels.aspx; http://www.guardian.co.uk/environment/2009/aug/13/us-marines-afghanistan-fuel-efficiency.

Hargreaves, Samantha. Personal Communication with Author on 'Time to Focus' International Email Discussion Group on the Global Epidemic of Violence on Women, 19 July 2018.

Hartmann, Betsy. 'Rethinking Climate Refugees and Climate Conflict: Rhetoric, Reality and the Politics of Policy Discourse'. *Journal of International Development*, 2010, 22/2: 233–246.

Hickel, Jason. 'Ecologise'. https://www.ecologise.in/2018/10/12/why-growth-cant-be-green-hickel/.

International Appeal to Stop 5G on Earth and in Space. https://www.5gspaceappeal.org/the-appeal.

Isla, Ana. 'Who Pays for Kyoto Protocol? Selling Oxygen and Selling Sex in Costa Rica'. In A. Salleh (ed.). *Eco-Sufficiency and Global Justice: Women Write Political Ecology*. London: Pluto Press, 2009.

Kothari, Ashish, Ariel Salleh, Arturo Escobar, Federico Demaria, and Alberto Acosta (eds). *Pluriverse: A Post-Development Dictionary*. New York: Columbia University Press and New Delhi: Tulika/AuthorsUpFront, 2019.

Konik, Inge. 'Ubuntu and Ecofeminism'. *Environmental Values*, 2018, 27/3: 269–288.

Leech, Gary. 'Greening Warfare? Lockheed Martin's Sustainability Agenda'. *Beyond Capital*, Blog, 10 September 2012. http://beyondcapital.org/2012/09/10/greening-warfare-lockheed-martins-sustainability-agenda/.

Le Grange, Lesley. 'Ubuntu'. In A. Kothari et al. (eds). *Pluriverse: A Post-Development Dictionary*. New York: Columbia University Press and New Delhi: Tulika/AuthorsUpFront, 2019.

Le Grange, Lesley. 'Ubuntu, Ukama, Environment and Moral Education'. *Journal of Moral Education*, 2012, 41/3: 329–340.

Leopold, Aldo. *A Sand County Almanac*. New York: Oxford University Press, 1949.

Merchant, Carolyn. *The Death of Nature: Women, Ecology, and the Scientific Revolution*. San Francisco: Harper and Row, 1990/1980.

Max-Neef, Manfred, et al. *Human Scale Development*. New York: Apex, 1991.

McDonagh, Sean. *To Care for the Earth: A Call to a New Theology*. London: Chapman, 1986.

Mies, Maria. *Patriarchy and Accumulation on a World Scale*. London: Zed Books, 1986.

Mies, Maria, and Vandana Shiva. *Ecofeminism*. London: Zed Books, 2014/1993.

Pocket, Sue. 'Public Health and the Radio Frequency Radiation Emitted by Cellphone Technology, Smart Meters and Wifi'. *New Zealand Medical Journal*, 2018, 131/1487: 96–106.

Pope Francis. *Laudato Si': On Care for Our Common Home*, 2015. http://w2.vatican.va/content/francesco/en/encyclicals/documents/papa-francesco_20150524_enciclica-laudato-si.html.

Radical Ecological Democracy. https://www.radicalecologicaldemocracy.org/.

Ruddick, Sara. *Maternal Thinking: Toward a Politics of Peace*. Boston: Beacon, 1989.

Ruether, Rosemary. *New Woman, New Earth*. New York: Seabury, 1975.

Salleh, Ariel. 'Body Logic': 1/0 Culture in *Ecofeminism as Politics: Nature, Marx, and the Postmodern*. London: Zed Books, 2017/1997.

Salleh, Ariel. 'Global Alternatives and the Meta-Industrial Class'. In Robert Albritton, John Bell, Shannon Bell, and Richard Westra (eds). *New Socialisms: Futures Beyond Globalization*. London: Routledge, 2004.

Salleh, Ariel. 'Green Economy or Green Utopia? Rio+20 and the Reproductive Labor Class'. *Journal of World Systems Research*, 2012, 18/2: 141–145.
Salleh, Ariel. 'Climate, Water, and Livelihood Skills: A Post-Development Reading of the SDGs'. *Globalizations*, 2016, 13/6: 952–959.
Salleh, Ariel. Personal Communication on 'Time to Focus' International Email Discussion Group on the Global Epidemic of Violence on Women, 17 July 2018.
Schlabach, Gerald. 'What Is Just Peace? A Collation and Synthesis Based on Catholic and Ecumenical Christian Sources'. *Catholic Nonviolence Initiative*. University of St. Thomas, MN, 2018.
Shiva, Vandana. *Staying Alive: Women, Ecology, and Development*. London: Zed Books, 1989.
Shiva, Vandana. *Water Wars: Privatization, Pollution, and Profit*. Cambridge: South End, 2002.
Simplicity Institute. https://simplicityinstitute.org/.
SIPRI. *Top 100 Arms-Producing and Military Services Companies*, 2015. https://www.sipri.org/sites/default/files/The-SIPRI-Top-100-2015.pdf.
Systemic Alternatives. https://systemicalternatives.org.
Terlinden, Ulla. 'Women in the Ecology Movement'. In E. Altbach, J. Clausen, D. Schultz, and N. Stephan (eds), *German Feminism*. Albany: State University of New York, 1984.
Trainer, Ted. *Abandon Affluence!* London: Zed Books, 1985.
Trainer, Ted. 'Eco-Anarchism'. In A. Kothari et al. (eds). *Pluriverse: A Post-Development Dictionary*. New York: Columbia University Press and New Delhi: Tulika/AuthorsUpFront, 2019.
Transition Network. https://transitionnetwork.org.
Minh-ha, Trinh. *Woman, Native, Other*. Bloomington: Indiana University Press, 1989.
UK Statistics. 2017. https://www.ons.gov.uk/peoplepopulationandcommunity/crimeandjustice/bulletins/domesticabuseinenglandandwales/yearendingmarch2017#domestic-homicides.
US Department of Defense. *Quadrennial Defense Review Report*, 2010. http://www.defense.gov/qdr/.
Via Campesina. https://viacampesina.org/en/.
Watson, Judith. Personal Communication with Author on 'Time to Focus' International Email Discussion Group on the Global Epidemic of Violence on Women, 30 July 2018.
WomanStats. http://www.womanstats.org.
Women, Peace, and Security. *KOFF Newsletter*, 2015, No. 141.
World Council of Churches. *Statement on the Way of Just Peace*. 10th Assembly Report of the Public Issues Committee, 8 November 2013.
World March of Women. www.marchemondiale.org.

Ariel Salleh is a sociologist of knowledge with activist experience in water and mining politics, biodiversity conservation and mothering. Her books include *Ecofeminism as Politics: Nature, Marx, and the Postmodern* (2017/1997); *Eco-Sufficiency & Global Justice* (2009); and *Pluriverse: A Post-Development Dictionary* (co-edited 2019). She is a founding editor of the journal *Capitalism Nature Socialism* and has held teaching and research appointments at New York University, Friedrich Schiller University Jena, University of Sydney and Nelson Mandela University: www.arielsalleh.info.

CHAPTER 5

'Walking the Land': An Alternative to Discourse as a Path to Ecological Consciousness and Peace

Freya Mathews

Let us start with some of the received verities of environmental thought, as viewed through the lens of environmental philosophy over the last several decades. Modern civilization is urgently in need of a change of worldview. The anthropocentric and dualist view of nature that we in the West have inherited from the scientific revolution of the seventeenth century, and that represents nature as brute object—a realm of mere matter—relative to the sovereign human subject—the realm of mind—is

Note This chapter is an adaptation and extension of an essay that was posted on the ABC Religion and Ethics website on 17 June 2019 under the title, **Why has environmental ethics failed to achieve a moral reorientation of the West?** https://www.abc.net.au/religion/why-has-environmental-ethics-failed-to-achieve-a-moral-reorient/11216540.

F. Mathews (✉)
La Trobe University, Melbourne, VIC, Australia
e-mail: F.Mathews@latrobe.edu.au

J. Camilleri and D. Guess (eds.), *Towards a Just and Ecologically Sustainable Peace*, https://doi.org/10.1007/978-981-15-5021-8_5

the very cornerstone of modern civilization. With its implied instrumentalism towards nature, this view laid the ideological ground for the environmental crisis, now shaping up as a total catastrophe for human civilization as well as for the rest of life on Earth.[1]

One of the persistent themes of environmental philosophy since its early days in the 1980s has been that the domination of nature serves as an ideological template for political domination across many axes: a default modality of brutal exploitation and callous indifference towards the Earth community to which we as humans belong will inevitably set the tone for our social and political relations with one another. For this reason, it has long been argued, particularly by ecofeminists and environmental philosophers in the critical theory tradition, that peace and justice can never be achieved in society until we learn to live in ecological harmony with our Earth community.[2] This is a theme that also echoes through Pope Francis' Encyclical on the Environment, *Laudato Si'*.[3]

Of course, the actual causes of human conflict, as well as of oppression, lie not merely in ideology but in underlying economic inequalities and other material circumstances. Ultimately, these must be addressed if the conditions for conflict and oppression are to be ameliorated. Nevertheless, ideology is a powerful legitimator of aggression, oppression, exploitation and injustice. So the view of nature as moral nullity must, for both environmental and political reasons, be replaced by a new, Earth-centred or ecocentric outlook. Instead of construing nature as devoid of the kinds of qualities—mind, intelligence, sentience, meaning and purpose—that confer moral significance, we need to re-construe it as fully imbued, like ourselves, with those very qualities. We need to recognize in the natural world a larger field of intelligence, meaning and purpose, kinship and communicativity—a field in which our own human consciousness is embedded, as a sub-set, and to which it is inalienably referenced.

For more than forty years, environmental philosophers have been advocating such a moral rehabilitation of nature. They have rigorously laid

[1] David Wallace-Wells, 'The uninhabitable Earth', *New York Magazine*, 10 July 2017, http://nymag.com/intelligencer/2017/07/climate-change-earth-too-hot-for-humans.html, accessed 30 June 2019.

[2] Val Plumwood, *Feminism and the mastery of nature* (New York: Routledge, 1993); Arne Vetlesen, *The denial of nature* (London: Routledge, 2015).

[3] Pope Francis, *Encyclical letter Laudato si' of the Holy Father Francis: On care for our common home* (Vatican City: Vatican Press, 2015).

out arguments for Earth-centredness, ecocentrism, biocentrism or bio-inclusiveness in ethics. The case for such positions has been detailed in a variety of ways—from arguments concerning the intrinsic value of non-human life to Kantian defences of living things as ends in themselves (as opposed to mere means to the ends of others) to critiques, like those of ecofeminism, of the ideologically loaded human/nature, mind/matter and culture/nature dualisms that have framed most thinking in the Western tradition.[4]

In addition to formulating new ethical categories such as biocentrism and the intrinsic value of non-human life, environmental philosophers in the 1980s and 1990s asked an array of foundational questions. Who or what is ultimately to qualify as morally significant? Living things? But if so, what counts as a living thing? Do individual organisms alone count as living things, or do larger living systems also count, in which case how are we to decide who has moral priority when the interests of individual organisms conflict with those of systems—as when feral animals threaten the integrity of ecosystems? Should an environmental ethic cover all living things? Should plants and fungi count as morally considerable in their own right? If so, how considerable? As considerable as animals? Should a distinction be drawn, morally speaking, between higher and lower animals? But which animals are higher and which lower? And what about microbes? Single cells? Viruses? Species? And natural features of the landscape that are not alive, such as rocks and rivers? Should an environmental ethic also cover these? Philosophers teased out such questions, without of course reaching final agreement on them.[5]

Meanwhile, the categories and arguments that had been developed in environmental philosophy were taken up across a range of academic disciplines. In the humanities, discourses such as ecocriticism, eco-cultural studies, animal studies, multispecies studies, biosemiotics, cultural geography and the new materialisms emerged, reconfiguring their disciplines-of-origin through the lenses of the new eco-ethical categories. Social theorists had also, from the start, been working with these categories to develop new detailed blueprints for ecological societies. From Murray Bookchin in the 1970s to bioregionalists and ecosocialists through the

[4] Andrew Brennan and Yeuk-Sze Lo, 'Environmental ethics', in *The Stanford encyclopedia of philosophy*, ed Edward N. Zalta, 2016. https://plato.stanford.edu/archives/win2016/entries/ethics-environmental, accessed 1 November, 2019.

[5] Ibid.

1980s and 1990s to present-day theorists from the Biomimicry Institute and Simplicity Institute, thinkers have been offering detailed scenarios for societies organized around paradigm-shifting bio-inclusive values. In the sciences, conservation biology has also (until recently) organized its research around this new value axis.

Did any of this visionary and revisionary thinking, this evocation of an alternative ecological worldview, avail? No. The state of the environment, as every scientific indicator confirms, only continued to worsen.[6] And although many forms of environmental activism and innovation budded forth in the 1980s in response to calls for an Earth-centred paradigm shift, offering any number of radical, organic, countercultural alternatives, these alternatives rarely attained mainstream status. They either lingered on, as minority movements, or faded only to be reinvented by later generations, still as minority movements. By the turn of the century, 'radicalism' itself seemed to have gone out of fashion as a mode of progressivism, acquiring sinister new undertones associated with violent right-wing extremism. Younger people staked their claims to cutting-edginess and the generational high ground not so much on politics, let alone on the 'social movements' of the earlier counter-culture, as on the great digital revolution that was taking place in *their* time, as *their* turf.

Progressive individuals, including younger ones, certainly acknowledged the need for environmental sustainability as part of the bottom line of capitalist economies, while sincerely bemoaning the 'biodiversity crisis', but environmental issues were just one part of a much larger portfolio of social issues that, it was hoped, could be addressed within the moral, political and economic parameters of our liberal democracies. The call for a specifically environmental ethic, for a new Copernican Revolution in ethics that would overturn anthropocentrism—the veritable cornerstone of Western civilization—and reconfigure human identity as essentially ecological, requiring epochal reform of economies and polities, seemed by the early twenty-first century to have pretty much fizzled out.

That, at any rate, is how the state of play appeared to myself, as a keen long-time observer and commentator, until just two years ago. At that moment, there was something of an awakening in parts of Europe and the Anglophone world. A genre of essays heralding near-term climate

[6] IPBES Report, https://www.ipbes.net/news/Media-Release-Global-Assessment, accessed 30 June 2019.

chaos and consequent civilizational collapse suddenly erupted, sending shock waves far and wide. These essays were authored not by the usual doomsday malcontents and conspiracy theorists but by highly reputable journalists and scholars, such as David Wallace-Wells,[7] Jem Bendell[8] and Rupert Read.[9] Their appalling forecasts are based on behind-the-scenes conversations with climate scientists apparently too constrained by professional conservatism to publish their own real inferences. The direness of these essays has been compounded by a spate of dumbfounding reports, such as the WWF Living Planet Index[10] that shows a 60% decline in wildlife populations from 1970 to 2014 and the Intergovernmental Platform on Biodiversity and Ecosystem Services Global Assessment Report announcing that a million species are currently facing extinction.[11]

Arguably, it is this new literature that has thrown up, just within the last year, vibrant new movements such as Extinction Rebellion and the School Climate Strikes that have broken the mould of older, now outworn forms of environmental activism. Something, it seems, is finally starting to shift.

However, momentary upsurges in climate change consciousness have occurred in the past—in 1989, when the IPCC was first established and ambitious emissions targets were widely set, including in Australia, and in 2006, when Al Gore's *Inconvenient Truth* galvanized world opinion. But in both those cases the upsurge was short-lived. Public opinion proved not sufficiently greened to mandate strong environmental and climate change policy initiatives from governments. Green parties remained, for the most part, minority parties, and in liberal democracies electorates chose governments that continued to put short-term economic interests very far ahead of environmental reforms. Emissions accordingly continued relentlessly to rise, global temperatures to soar, biodiversity to unravel.

[7] David Wallace-Wells, 'The uninhabitable Earth', *New York Magazine*, 10 July 2017, http://nymag.com/intelligencer/2017/07/climate-change-earth-too-hot-for-humans.html, accessed 30 June 2019.

[8] Jem Bendell, 'Deep adaptation', IFLAS Occasional Paper 2, 27 July 2017, http://lifeworth.com/deepadaptation.pdf, accessed 1 November 2019.

[9] Rupert Read, 'This civilization is finished: So what is to be done?', IFLAS Occasional Paper 3, 10 December 2018, http://lifeworth.com/IFLAS_OP_3_rr_whatistobedone.pdf, accessed 30 June 2019.

[10] WWF Living Planet Report, http://www.livingplanetindex.org/projects?main_page_project=LivingPlanetReport&home_flag=1, accessed 30 June 2019.

[11] IPBES Report, https://www.ipbes.net/news/Media-Release-Global-Assessment, accessed 30 June 2019.

Admittedly, it is different this time. Climate change is now undeniably upon us, as fires bear down on our homes, droughts wither our crops and floodwaters lap at our doorsteps. So the current upsurge is unlikely wholly to subside. But will the way societies address the climate emergency, when they do finally address it, be inspired and led by an environmental ethic as opposed to mere human expedience? Since environmental ethics has had so little traction to date, it is perhaps hardly to be hoped that now, when humanity itself is under serious threat, the interests of other-than-human species will figure prominently on government agendas. On the contrary, although amelioration of climate change will indeed incidentally benefit many species, economic resources currently invested in conservation may be redeployed, in the new state of climate emergency, to human protection, and the interests of mere 'nature' relegated once more, indeed more than ever, to the moral background.

So despite the ever-deepening climate catastrophe and the ever-more-visible extinction crisis, it seems we are really no closer to the kind of ecological consciousness and culture that could release us from dominator patterns of thought and afford us genuine foundations for habits of peace with Earth and one another. The current upsurge of activism seems primarily a last ditch response to the threat that climate change is now palpably posing to our own security. Mixed in with this anxiety is grief and anger at the catastrophic decline and disappearance of wildlife and biodiversity. But saving (some) species from outright extinction seems to be the sum of the agenda in this connection. There seems little awareness in the rhetoric of current activism that civilization itself, in its current form, is so deeply rooted in anthropocentrism that its trajectory in relation to nature cannot be altered by a few ad hoc changes of policy. Even in the heartland of conservation itself—as evidenced in the pages of conservation biology journals—a push to prioritize human interests over the interests of threatened species has recently found favour, in the name of 'eco-modernism'.[12]

One is inevitably led then to ask *why*. Why has the groundbreaking thinking spearheaded a generation or two ago by environmental philosophy failed? Why has impeccably reasoned argument done so little to shift the wider society off its anthropocentric moorings? Why have the

[12] Breakthrough Institute, *Eco-modernist manifesto*, April 2015, www.ecomodernism.org, accessed 30 June 2019.

ecological counter-cultures of past decades remained just that—counter-cultures, despite their demonstrably sound rational credentials? Thirty years ago, I personally thought that if we could only demonstrate that the anthropocentric bias of the Western tradition was rationally indefensible, a value reorientation would ensue that would begin comprehensively to re-draw the moral map. Environmental philosophers did, in my opinion, demonstrate this, but moral change on a significant scale has not ensued.

One way of making sense of this failure of uptake of environmental ethics and its many discursive corollaries is to revisit the Marxist theory of historical materialism. One by no means has to be a Marxist to find historical materialism a compelling explanation for why societies adopt the particular value-sets they do.

According to historical materialism, the values that characterize a given society—which is to say, the consciousness, culture and identities which prevail therein—arise from its underlying 'modes of production', its basic economic modalities, in particular the practices or *praxes* whereby its members act upon nature in their efforts to wrest a livelihood from it.[13] So, for example, hunter-gatherers dwelling in a rainforest might be expected to hold very different views of self, society and world from, say, industrial workers in a nineteenth-century factory town. All that counts as culture and consciousness in a particular society is basically, from an historical materialist perspective, ideological, in the sense that it reflects and legitimizes more basic economic conditions. These ideological structures cannot be changed by argument (philosophy, science, discourse) or exhortation (moral persuasion). They can only change when the underlying praxes of the society in question change. So, one would not expect to persuade a hunter gatherer to become anthropocentric in outlook just by engaging them in philosophical argument any more than one would expect by the same method to persuade workers in, say, a factory farm or on an assembly line in a twentieth-century factory to embrace an Earth-centred or ecocentric perspective.

While historical materialism is not a cast-iron or fully comprehensive explanation of values in society, it definitely helps to explain why occasional upsurges of environmental protest historically tend to peter out and why even the best-intentioned individuals fail to sustain other-than-token environmental reforms to their life styles. No amount of

[13] For an account of historical materialism in the sense in which I am reading it here, see Alison M. Jaggar, *Feminist politics and human nature* (Brighton: Harvester, 1983).

wilderness workshops or classroom discussion or even public debate will genuinely induce ecological consciousness or identity in us, as members of modern societies, if we have to return to the twenty-first-century landscape of capitalism—the shopping malls and mills of industry, commerce and corporatism—after leaving our green rallies and conferences and eco-retreats. And return to these most of us do, since alternative opportunities for making a living are exceedingly thin on the ground.

To allow that historical materialism largely explains why value-sets vary from society to society and from one historical moment to another need not imply that all such value-sets are *merely* relativist. I think certain value-sets can be demonstrated by reason to be sound and others not so if they are considered against a background of agreed further ends. But what historical materialism does teach us is not to expect a society to *adopt* new value-sets, no matter how rationally preferable to the old ones, if the new values are inconsistent with the basic praxis of that society.

While historical materialism might then go a long way towards explaining why value-sets vary from society to society, it does not in itself solve the problem of how actually to bring about new value regimes when required. For to replace a set of underlying praxes such as our own with praxes that would induce ecological consciousness would require massive economic and political investment in new eco-compatible modes of production. Such investment could presumably not occur unless those very ecological values were already in place.

It is this vexing circularity—which I call the Hard Problem of environmental reform—that makes such reform in contemporary societies so unachievable.[14]

[14]From a Marxist point of view, changes of consciousness occur when an economic system encounters 'contradictions' between the requirements of labour and the logic and conditions of production. These contradictions lead to failure of the institutions on which the system is based, where such failure will necessitate a re-set of the economy. As part of the process of re-setting, new praxes, modes of production and institutional arrangements will emerge. In due course these will give rise to new forms of consciousness. The ecological crisis is undoubtedly a contradiction of capitalism, but in this instance we surely cannot merely wait for the contradiction to play itself out. Since the 'failure' to which this crisis is likely to lead might be wholesale biosphere collapse, we must try to avert that collapse by instituting in advance new values that will guide the economy towards eco-compatible goals.

In the light of such circularity, and the moral inertia it entails, what is to be done? Are there other motivating forces as powerful as materialist or economic ones that might be mobilized to shift entrenched outlooks? Marx would presumably have said no. Materialist or economic forces are determinative, for him, precisely because they ultimately govern our survival, and nothing is more fundamental, in terms of impact on consciousness, than the survival imperative. But perhaps this presumed priority of material forces is debatable. Perhaps in the human context there are other forces, further to materialist ones, that also impact on our survival. The imperative to affiliate, for instance. The need to belong to a community or group or troupe is perhaps as core to our survival, in evolutionary terms, as our need for food and shelter. This seems to be borne out not only in our own present-day experience of life in society, with its many imperatives to conform, but also when we look to our evolutionary reflection in primate societies. In chimp troupes, affiliation is important not only for the purpose of sharing resources but also for social legitimacy—misfits and stragglers are policed and killed. In evolutionary terms, affiliation may be as powerful a determinant of identity and consciousness as praxis is.

Perhaps then identity, in the sense conferred by affiliation, is a potential site of value shift, one that could serve as prelude and impetus to the longer term, ecological 'transvaluation of values' that may indeed, as historical materialists insist, require an overhaul of economic praxis.

How might this strategy work? What loci of affiliation might serve to embed Earth-friendly values in society? One major possibility is religion. For Marx, of course, religion was a prime instance of ideology, different religions serving merely to prop up and legitimate different economic regimes. In industrial societies, characterized by intensely instrumentalist relations with the entire biosphere, only religions that reinforce an anthropocentric orientation might be expected to achieve social traction.

But in a twenty-first-century context in which the star of secularism is rising, at least in some parts of the West, and that of traditional religions appears to be waning, the bedrock formations of belonging and hence of identity that religions afforded may be eroding.[15] Perhaps for

[15] If Australia is at all representative of Western democracies (with the exception perhaps of USA), it is demonstrable that secularism is on the rise, though the terrain of religion in society is shifting and complex. Although in 2016 the majority of Australians were still affiliated with a religion or spiritual tradition, according to the census for that year, 'about one third of all Australians (30 per cent, or 7 million people) indicated

many people in this new context the basic need for affiliation is ceasing to be satisfactorily met. The forms of identity and consciousness that emanate from capitalist-industrial modes of praxis are possessive, individualist and self-centric as well as instrumentalist: they untether the self from any larger—social or environmental—meanings or responsibilities.[16] Possibilities of affiliation do of course exist, but, arising as they do from contingent interests or causes rather than from the moral or metaphysical core of people's existence, they arguably leave members of modern societies morally and metaphysically marooned and accordingly at existential risk.

A new formation, introduced to take the place of traditional religions but serving Earth-friendly values rather than the anthropocentric ones served, to varying degrees, by major present-day religions, might in this context prove as powerful a determinant of consciousness as are materialist or economic forces. Of course, a preliminary step in this connection would surely be to attempt to green existing religions. Such efforts may not suffice to turn the moral tide in the modern West, however, because (a) religion in its currently prevailing forms might already be too discredited—as inimical to science; as authoritarian rather than democratic; and latterly as riven with sexual and other scandals. And (b) those world religions with the greatest currency and influence in the West, viz the Abrahamic faiths, may resist being greened to any significant extent in any case, having arisen as expressions of an agrarianism that set humanity outside and above nature, as domesticator, ruler and engineer of hitherto sovereign (i.e. self-ruling) lands. Strenuous efforts are admittedly at present being made within the Abrahamic faiths to overcome the dualist

either "No Religion" or a secular belief such as Atheism, Humanism or Agnosticism. The number of people indicating they had "No Religion" has increased by almost 50 per cent from 2011 to 2016'. Australian Bureau of Statistics, Media Release, 18 January 2018, https://www.abs.gov.au/AUSSTATS/abs@.nsf/mediareleasesbyReleaseDate/8497F7A8E7DB5BEFCA25821800203DA4?OpenDocument, accessed 1 November 2019. In 1966, almost 90% of Australians identified as Christian, while fifty years later in 2016 a mere 52% did so; 'other religions' accounted for 1% in 1966 and 8% in 2016; 'no religion' accounted for close to 0% in 1966 and 30% in 2016. Bernard Salt, 'Faith no more as a nation goes godless', *The Australian*, 1 July 2017. https://www.theaustralian.com.au/nation/inquirer/faith-no-more-churches-in-trouble-as-the-nation-goes-godless/news-story/4e5c0d079888964c1ad871c6a7c61623, accessed 1 November 2019.

[16] C. B. Macpherson, *The political theory of possessive individualism (Hobbes to Locke)* (Oxford: Oxford University Press, 1962).

tendencies of their traditions and reinterpret those traditions in ecological terms.[17] In Christianity, for example, a rich discourse of eco-theology has been evolving over decades—with offerings ranging from the 'cosmic Christ' of Matthew Fox[18] to the Earth Bible of Norman Habel[19] to the 'Christology' and 'deep incarnation' of Niels Gregersen[20] to Pope Francis' own beautiful recent paean to an inspirited cosmos, *Laudato Si*.[21] This discourse has tremendous value for committed Christians seeking to reconcile their pre-existing faith with a growing sense that environmental catastrophe is charged with spiritual as well as practical significance. But to people not already strongly committed to Christianity, the appeal of an ecological message wrapped in the anthropomorphisms typical of theisms, Christian or otherwise, is likely to be weak.

In any case, whatever progress may be made in this matter of greening existing religions, it may nevertheless be strategic to introduce in addition an entirely new formation, a formation which creates a new narrative of identity and belonging, a narrative devoid from the start of anthropomorphic undertones. We might perhaps choose not to call this Earth-friendly formation a 'religion' at all, even though it might be socially organized into communities of interest as religions are. For it would differ from theistic traditions inasmuch as it would not feature notions of godhead, popularly construed in highly anthropomorphic forms, but would instead feature science as integral to its notion of the universe. Mind, however, in some larger sense, might be re-construed as immanent in matter,[22] and the universe itself might thus be perceived as intrinsically alive, inherently communicative, and accordingly as the ultimate wellspring of

[17] Joseph A. Camilleri, 'Ecology and peace: Responding to the ethos of exclusion', in *Ecological aspects of war: Religious and theological perspectives from Australia*, eds Anne Elvey, Deborah Guess, and Keith Dyer (Adelaide: ATF Press, 2016).

[18] Matthew Fox, *The coming of the Cosmic Christ* (San Francisco: Harper and Row, 1998).

[19] Norman Habel, 'Introducing The Earth Bible', in *The Earth Bible, Volume one: Readings from the perspective of Earth* (Sheffield: Sheffield Academic Press, 2000), 25–37.

[20] Deborah Guess, 'Oil beyond war and peace: Rethinking the meaning of matter', in *Ecological aspects of war: Religious and theological perspectives from Australia*, eds Anne Elvey, Deborah Guess, and Keith Dyer (Adelaide: ATF Press, 2016).

[21] Pope Francis, *Encyclical letter Laudato si' of the Holy Father Francis: On care for our common home* (Vatican City: Vatican Press, 2015).

[22] Thomas Berry and Brian Swimme, *The universe story* (San Francisco: Harper San Francisco, 1992).

meaning.[23] There would be no need, from the point of view of this new formation, for texts or scriptures, nor hence for spiritual interpreters or authorities. Earth, as microcosm of the living universe, would provide the 'scripture'; transactions with Earth-mind or mind-in-nature would be a personal affair, a personal locus of revelation.[24] The orientation of this Earth-friendly formation might be described as eco-spiritual, but unlike spirituality generally, which is often taken to connote value-sets held outside of formal institutions, the Earth-friendly values of the new formation could be held collectively, since the purpose of this formation would be to constitute powerful new loci of affiliation. Members would be allied not only to Earth itself and the larger community of all life, as implied by eco-spirituality, but also to organized local 'congregations' comprised of people whose allegiance was likewise to Earth.

Actually, I think it would be important not to call this value-set, and the new narrative of identity attending it, a religion (or faith or credo) because the term, *religion*, has long been used to disparage environmentalists—to imply they act from irrational motives. But nor would it count merely as philosophy, since it would betoken much more than philosophy does—a whole-hearted commitment to care for Earth-life and identification with the human community sharing that commitment. Perhaps the term, *cosmology*, might serve best—cosmologies can be exclusively scientific or exclusively mythopoetic or a combination of both. The very term, *cosmos*, after all is directly derived from the Greek, *kosmos*, meaning order, and is in this sense inherently normative, implying that the physical universe as we encounter it does not merely hang together contingently but is self-conforming to some kind of inner principle of integrity or goodness. Such a cosmos is immanently lawful in its configuration not merely in a causal but in a normative sense. An ecological cosmology would thus have much more in common with the Earth-based cosmologies of Aboriginal Australia than with major religions such as the

[23] Freya Mathews, 'Panpsychism: Position statement', in *Inter-religious philosophical dialogues*, eds Graham Oppy and Nicholas Trakakis (London: Routledge, 2017): 45–71; 136–159; 221–238; Freya Mathews, 'Living cosmos panpsychism', in *Routledge handbook on panpsychism*, ed William Seager (New York: Routledge 2019).

[24] Freya Mathews, 'Panpsychism: Position statement', in *Inter-religious philosophical dialogues*, eds Graham Oppy and Nicholas Trakakis (London: Routledge, 2017), 45–71; 136–159; 221–238.

Abrahamic faiths, since it, like Aboriginal cosmologies,[25] would be organized around an immanent, normative axis of ecological Law rather than around worship of gods.

In the absence as yet of widespread economic praxes conducive to an ecological orientation, day-to-day practices that could anchor the Earth-friendly values of the new cosmology in actual experiences of reality could include the practice of conservation. Through in situ activities such as revegetation, restoration and rewilding, people could gradually begin to decode, and become implicated in, actual ecologies, gaining in ecological literacy and becoming initiated into the intricacies—the myriad minds and mysteries—of actual life communities.

Indeed, I would suggest that hands-on practices of conservation, undertaken not in a purely utilitarian spirit but as devotional service—as the defining telos of one's community and as the perceived end-point of human agency—afford new ways for us to re-enter reality and find our normative direction therein. To practice conservation effectively requires the closest attention to the particularities of a given place, to the lie of its land and the patterns of its weather, to the minutiae of the manifold identities and relationships that are forever forming and reforming there.

The practice of conservation also involves push and pull: we make interventions, such as plantings, thinnings, weedings and perhaps, in some circumstances, baitings and sprayings. We must pay attention to the consequences of those interventions for ecosystems, including all the vertebrate and invertebrate actors therein, rapidly adjusting our actions in light of often unintended outcomes. Our activities may expose us to risk, as we immerse ourselves in life-worlds outside the blind bubble of modern civilization. In these normally overlooked life-worlds, venoms and wild antagonists, hidden perils of many kinds lie in wait for us. Such threats, as much as our ministrations, force us to cultivate attentiveness and responsiveness, and little by little this attentiveness, together with the respect that grows from our engagement with a multitude of inscrutable agencies, opens our eyes. It opens our eyes to worlds within worlds within worlds of astonishing embodiments of life, all cohering and conforming to one another—insofar as they are not derailed by the industrial juggernaut of modernity—in accordance with the manifest principles of creation and

[25] Christine F. Black, *The land is the source of the law* (London: Routledge, 2011); Irene Watson, *Aboriginal peoples, colonialism and international law: Raw law* (London: Routledge, 2014).

regeneration that Indigenous peoples signal when they say, 'the Law is in the land'.[26] In this way, right under our very noses, the land may begin to open to us, to come alive, and a whole new horizon of relationship, presence, communicativity, enthralment, mystery and indeed revelation may come into view.

To approach this Law in something like the way Aboriginal people approach it, which is to say affectively and not merely abstractly or theoretically—and hence to inhabit this Law normatively, as one's very consciousness—requires precisely such a shift in epistemology as I have been intimating here. This is a shift from an epistemology based on theoretical reason, as in Western discourses of philosophy and science, to one based on feelingful awareness. Traditional teachers of high repute, such as Senior Law Men, Bill Neidjie of Kakadu[27] and David Mowaljarlai[28] and Paddy Roe,[29] both of the Kimberley, emphasize repeatedly that Aboriginal ways of knowing cannot be extricated from feeling. One arrives at this kind of knowledge not by adopting a stance of detached observation and inference, as Western scientists and philosophers do, but by, as Mowaljarlai puts it, 'walking the land'.[30] By this I take him to mean that we should walk the land not merely in a literal sense but in a paradigm-shifting epistemological sense as well. Rather than stepping back from the land, as the observer and theoretician do, we have actively to enter it, address it and engage it as a collaborator that can and will join forces with us in some vital venture.

While such collaboration will undoubtedly require a high degree of empirical attentiveness on our part, it will also challenge our agency. We will feel the land's resistance to efforts which go against the 'grain', so to speak, of its own tendencies or conativity, and a kind of smoothing-of-the-way for efforts which are conatively aligned with it. One may find oneself leaning into these latter 'openings' and promptly correcting one's behaviour in face of resistances simply because doing so feels

[26] Christine F. Black, *The land is the source of the law* (London: Routledge, 2011).

[27] Bill Neijie, *Story about feeling* (Broome: Magabala, 1989).

[28] David Mowaljarlai and Jutta Malnic, *Yorro yorro: Everything standing up alive* (Broome: Magabala, 1993).

[29] Jim Sinatra and Phin Murphy, *Listen to the people, listen to the land* (Carlton: Melbourne University Press, 1999).

[30] David Mowaljarlai and Jutta Malnic, *Yorro yorro: Everything standing up alive* (Broome: Magabala, 1993).

supremely 'right'—it feels supremely right to find oneself slipping along in a kind of groove of assent while it feels downright wrong to find oneself pushing against some invisible grain. In other words, in walking the land, not merely as an observer but as a collaborator in ventures of mutual concern to the land and oneself, one starts to experience the land's responsiveness to one's presence, where that cannot leave one other than profoundly moved. One has perhaps begun indeed to experience the kind of feeling—a kind of vital, participative awareness—that Senior Law Men such as Neidjie and Mowaljarlai have been trying to explain to non-Indigenous countrymen for many years. Once we have discovered this way of knowing, there will be no question of remaining ethically indifferent to the living world, marooned inside a plastic bubble of anthropocentrism. To care for it—and seek solace in its presence—will be as natural as doing so is for Aboriginal people, because caring for it will be what makes us feel attuned and alive ourselves. To sense that one is noticed and intimately acknowledged by country is to experience a metaphysical affirmation that anchors one's existence to a level of reality that is outside human time and beyond the reach of scepticism.

To walk the land in this new way, adapted to the praxical possibilities of our own time and also to the ecological imperatives of a wounded planet, is not merely the prerogative of privileged landholders. Place-specific conservation activities are surely in principle available to everyone. Those with disposable capital might join with friends to purchase an ecologically strategic property, then safeguard its future with a conservation covenant and prepare to embark on what might become, in its quiet way, a depth-initiation that few anticipate. Those without such financial means can still commit to an ecologically strategic place, by volunteering for caring-for-country type programmes on public or private, urban or rural estates, or by creating such programmes themselves. Undertaken collectively, via congregations of commitment to an ecological cosmology, such activities not only implicate us, through our own sweat and care, in a cherished place, but bond us to the colleagues-in-care who likewise find themselves drawn into its larger significance. Before too long we may start to feel like keepers of the place in question—the particular woodland or mountain, river or creek, rocky outcrop or arid shrubland—its interests gradually overtaking our own, our allegiance to it outgrowing our narrower, more personal perspectives.

Since commitment to sites pulls keepers into affinity not only with land but with one another, the bonds it forges may emanate not only in care of land but also in regimes of social solidarity and material mutual aid, just as associations based on religious conviction do. This dimension of mutual aid, at a time of escalating insecurity and ever-diminishing levels of state, community and family support, would reinforce the affiliative value of the new congregations, fitting them even more closely to the needs of our era.

Such bonds would moreover transcend other markers of difference, such as religion, race and class, since one's colleagues-in-care become, again in something a little akin to the Aboriginal sense, one's countrymen, knitted together by shared loyalty to, and identification with, the ecological interests of a cherished place. This is moreover an identification that, unlike identifications based on ideology, does not divide one's own group from others: the fact that my group cares for country in a particular locality in no way sets us against groups who care for other localities. On the contrary, it is understood, via our shared eco-cosmology, that all places need to be cared for, and hence that all groups are engaged in a common work. The loyalty of each congregation to its own living country ensures that none covet the country of another and none wish to leave the country in which they have become, through their practice, so deeply invested. In this way, commitment to country, brought about by practices of 'walking the land', may help to settle societies down.

While the hands-on conservation activity of these land-based congregations may not in itself be strictly praxical, in the historical materialist sense, as it does not constitute a mode of production, it is in-service to the praxical, inasmuch as it helps to repair the biospheric fabric that is a condition for any and all ongoing economic activity.

The efficacy of such activity also perhaps brings out a deeper truth of historical materialism itself, which is that people undergo changes of normative consciousness not as a result merely of discourse but of personal—and particularly communicative—immersion in the realities which discourses merely represent. From this point of view, no amount of exposure to environmental philosophy or science—or for that matter, art or literature or theatre—will ever really change the consciousness and normative habits of the public, for the very reason that such discourse is addressed to people as audiences, which is to say as spectators (viewers, readers), as opposed to participants in environmental realities. The model of learning inherent in the common-sense assumption that public consciousness can be changed via discourse is in fact rooted in dualism, inasmuch as it assumes that understanding is ultimately a matter of grasping ideas or theories at an abstract level, rather than immersing oneself in relevant real-world situations and cultivating the skills of attentiveness, acuity and accommodation that would enable one to negotiate those situations sensitively and responsively.[31] Environmentalists may in this sense have been working all along with a faulty model of consciousness change: one does not shift public consciousness merely by telling people environmental truths nor even by representing those truths via arts or literature. The most that people can learn, deep down, from being part of an audience is how to be part of an audience, where this is a matter of bracketing their own agency and remaining, as spectators, distanced from the action and hence from their responsibility for it. In order for people genuinely to come to inhabit ecological or Earth-centred consciousness, attuned interactivity with environmental realities may be required.

By creating new congregations to 'walk the land' then, not so much in pursuit of livelihood, as was the case with hunter gatherers, but rather in a spirit of conservation, we might start to break the circularity that has foiled our attempts, as environmentalists, to shift society towards Earth-centredness. Such a new social formation would represent a response to our current discursive disquiet regarding the environment, evidenced in the current upsurge of activism, but its deeper, motivating appeal—the hook that would draw people, and bind them, into the congregations—would not be merely discursive but rather the promise of affiliation. Within the framework of this new social formation, the practice of conservation, understood as a cosmological practice emanating from an epistemology of walking the land, would ensure that knowledge of the land was inevitably charged with feeling, where this would in turn lay down a deep and abiding foundation for genuine ecological fealty.

The environmental anxiety and ferment that is currently sweeping parts of the Western world might provide an unprecedented historical opportunity for such a movement towards deeper change to emerge. Within this historical milieu, an environmental vanguard informed with genuinely lived ecological consciousness might indeed be capable of configuring itself. The existence of such a vanguard might in turn suffice eventually to motivate greater scrutiny of existing techno-economic arrangements in society, where this could lead to investment in new economic arrangements. Such a change of direction at an economic level would then truly

[31] Freya Mathews, 'The dilemma of dualism', in *Routledge international handbook on gender and environment*, ed Sherilyn MacGregor (New York: Routledge, 2017).

begin to dismantle dualism and launch us towards a civilization based not on domination but on adaptive co-existence with all life, human and other-than-human alike.

Bibliography

Australian Bureau of Statistics. 'Media Release'. 18 January 2018. https://www.abs.gov.au/AUSSTATS/abs@.nsf/mediareleasesbyReleaseDate/8497F7A8E7DB5BEFCA25821800203DA4?OpenDocument, accessed 1 November 2019.

Bendell, Jem. 'Deep adaptation'. IFLAS Occasional Paper 2. 27 July 2017. http://lifeworth.com/deepadaptation.pdf, accessed 1 November 2019.

Berry, Thomas, and Swimme, Brian. *The universe story*. San Francisco: Harper San Francisco, 1992.

Black, Christine F. *The land is the source of the law*. London: Routledge, 2011.

Breakthrough Institute. 'Eco-modernist manifesto'. April 2015. www.ecomodernism.org, accessed 30 June 2019.

Brennan, Andrew, and Lo, Yeuk-Sze. 'Environmental ethics'. In *The Stanford encyclopedia of philosophy*, ed Edward N. Zalta. 2016. https://plato.stanford.edu/archives/win2016/entries/ethics-environmental, accessed 1 November 2019.

Camilleri, Joseph A. 'Ecology and peace: Responding to the ethos of exclusion'. In *Ecological aspects of war: Religious and theological perspectives from Australia*, eds Anne Elvey, Deborah Guess, and Keith Dyer. Adelaide: ATF Press, 2016.

Fox, Matthew. *The coming of the Cosmic Christ*. San Francisco: Harper and Row, 1998.

Guess, Deborah. 'Oil beyond war and peace: Rethinking the meaning of matter'. In *Ecological aspects of war: Religious and theological perspectives from Australia*, eds Anne Elvey, Deborah Guess, and Keith Dyer. Adelaide: ATF Press, 2016.

Habel, Norman. 'Introducing the Earth Bible'. In *The Earth Bible, Volume one: Readings from the perspective of Earth*. Sheffield: Sheffield Academic Press, 2000, 25–37.

IFLAS Occasional Paper 3. 10 December 2018. http://lifeworth.com/IFLAS_OP_3_rr_whatistobedone.pdf, accessed 30 June 2019.

IPBES Report. https://www.ipbes.net/news/Media-Release-Global-Assessment, accessed 30 June 2019.

Jaggar, Alison M. *Feminist politics and human nature*. Brighton: Harvester, 1983.

Macpherson, C. B. *The political theory of possessive individualism (Hobbes to Locke)*. Oxford: Oxford University Press, 1962.

Mathews, Freya. 'Living cosmos panpsychism'. In *Routledge handbook on panpsychism*, ed William Seager. New York: Routledge, 2019.
Mathews, Freya. 'Panpsychism: Position statement'. In *Inter-religious philosophical dialogues*, eds Graham Oppy and Nicholas Trakakis. London: Routledge, 2017.
Mathews, Freya. 'The dilemma of dualism'. In *Routledge international handbook on gender and environment*, ed Sherilyn MacGregor. New York: Routledge, 2017.
Mowaljarlai, David, and Malnic, Jutta. *Yorro yorro: Everything standing up alive*. Broome: Magabala, 1993.
Neijie, Bill. *Story about feeling*. Broome: Magabala, 1989.
Plumwood, Val. *Feminism and the mastery of nature*. New York: Routledge, 1993.
Pope Francis. *Encyclical letter Laudato si' of the Holy Father Francis: On care for our common home*. Vatican City: Vatican Press, 2015.
Read, Rupert, and Alexander, Samuel. *This civilization is finished: So what is to be done?* Melbourne: Simplicity Institute, 2019.
Salt, Bernard. 'Faith no more as a nation goes godless'. *The Australian*. 1 July 2017. https://www.theaustralian.com.au/nation/inquirer/faith-no-more-churches-in-trouble-as-the-nation-goes-godless/news-story/4e5c0d079888964c1ad871c6a7c61623, accessed 1 November 2019.
Sinatra, Jim, and Murphy, Phin. *Listen to the people, listen to the land*. Carlton: Melbourne University Press, 1999.
Vetlesen, Arne. *The denial of nature*. London: Routledge, 2015.
Wallace-Wells, David. 'The uninhabitable Earth'. *New York Magazine*. 10 July 2017. http://nymag.com/intelligencer/2017/07/climate-change-earth-too-hot-for-humans.html, accessed 30 June 2019.
Watson, Irene. *Aboriginal peoples, colonialism and international law: Raw law*. London: Routledge, 2014.
WWF Living Planet Report. http://www.livingplanetindex.org/projects?main_page_project=LivingPlanetReport&home_flag=1, accessed 30 June 2019.

Freya Mathews is Adjunct Professor of Environmental Philosophy at Latrobe University, Australia. Her books include *The Ecological Self* (1991), *Ecology and Democracy* (editor) (1996), *For Love of Matter: A Contemporary Panpsychism* (2003), *Journey to the Source of the Merri* (2003), *Reinhabiting Reality: Towards a Recovery of Culture* (2005), *Without Animals Life Is Not Worth Living* (2016) and *Ardea: A Philosophical Novella* (2016). She is the author of over a hundred articles in the area of ecological philosophy. Her current special interests are in ecological civilization; indigenous (Australian and Chinese) perspectives on (so-called) sustainability and how these perspectives may be adapted to the context of contemporary global society; panpsychism and critique of the metaphysics of modernity; ecology and religion; and conservation ethics and rewilding in the context of the Anthropocene. In addition to her research activities, she co-manages a private conservation estate in northern Victoria. She is a fellow of the Australian Academy of the Humanities.

PART II

Cosmological and Religious Perspectives

CHAPTER 6

An Islamic Approach to Environmental Protection and Ecologically Sustainable Peace in the Age of the Anthropocene

Zuleyha Keskin and Mehmet Ozalp

Introduction

The need for positive change in the world is becoming more topical each day. While various global issues are often dealt with independently, it is time to stop compartmentalising the various issues and start seeing the strong common currents that run through them all. Not only will this lead to synergy as we seek to address growing global problems, but it will also facilitate the establishment of a worldview which benefits every inhabitant that calls this earth its home.

The number of wars and conflicts in the world in the last century is of grave concern, with 108 million people being killed due to war in

Z. Keskin · M. Ozalp (✉)
Charles Sturt University, Melbourne, VIC, Australia
e-mail: mozalp@csu.edu.au

Z. Keskin
e-mail: zkeskin@csu.edu.au

J. Camilleri and D. Guess (eds.), *Towards a Just and Ecologically Sustainable Peace*, https://doi.org/10.1007/978-981-15-5021-8_6

the twentieth century.[1] With the advancement in technology, the ease of killing a mass number of people in a short period of time becomes easier than ever before. Conflicts are emerging frequently in almost all parts of the world, leaving negative effects for years and even decades after they emerged.

However, peace in itself is not sufficient. Justice is also needed whereby every individual enjoys the rights they deserve. While human rights are often spoken about at the local and global level, many atrocities take place around the world. Out of many statistics, two will suffice to highlight the extensive problems and inequalities that exist globally: 66% of the world's population lives in poverty and 12 million women, men and children are enslaved around the world with 600,000–800,000 being trafficked each year.[2]

The need to protect the environment is also pressing. The main stimulus for a call for action in this area is the findings of climate science on the alarming harm human activity is causing on the planet to such an extent that geologists have named the present era the Anthropocene, an epoch where human activity has reached the scale of affecting the very geology of the planet.[3] The level of environmental awareness is at its peak: growing attention on environmental and climate science and its findings is widely covered in the media, and included in the educational curriculum.

When global issues are collectively analysed through an Islamic theological lens, the need to change one's worldview becomes apparent. It is not sufficient to simply attempt to 'fix' problems that have been created over the years, but to change the way one views every part of creation that exists on this earth. This would lead to justice and an ecologically sustainable peace. In the light of this, this chapter will discuss the notion of a God-centric worldview which aims to connect all of creation to the Creator. In this way, the potential, the value and the purpose of creation become elevated since everything is seen in the name of the Creator.

[1] Chris Hedges, *What Every Person Should Know About War* (New York: Free Press, 2003), 1.

[2] *Social Justice Resource Centre*, https://socialjusticeresourcecenter.org/facts-and-figures/ (accessed 1 September 2019).

[3] Jan Zalasiewicz et al., 'The Anthropocene: A New Epoch of Geological Time?' *Philosophical Transactions of the Royal Society A* 369 (2011): 835–1111.

Potential

The potential embedded in all of creation is an important element of Islamic theology and spirituality. Everything is perceived through what it can be, not only through what it is. This perspective not only applies to human beings, it also encompasses all living things including the natural environment.

According to the Qur'an, the potential for human development is far reaching, 'Verily We have created humankind in the fairest form (*ahsan al-taqwim*), then sent him down to the lowest of the low, except for those who believe and do good deeds.'[4] The verse underscores the variable nature of human beings and the possibility of attaining the 'the fairest form' (*ahsan al-taqwim*) suggesting that humans have the potential to grow and attain perfection. The natural human disposition (*fitra*) has been created in such a way that it can move its way up from the 'lowest of the low' to express 'the fairest form.'

The description of the fairest form has been understood in different ways by exegetical scholars. According to Muhammad ibn Jarir al-Ṭabari (d. 923), humankind has been created as having the best character and a beautiful form.[5] *Fakhr al-Din al-Razi* (d. 925), has a similar understanding, describing the fairest form to be inner and outer beauty.[6] Both these explanations stress the beauty found within human beings. Ismail Haqqi Bursawi (d. 1725) focusses more on the value of human beings rather than their beauty when commenting on this verse. He upholds the value of human beings to such a level that he asserts human beings have the highest intrinsic worth of all creation on earth.[7]

The fact that Muslim scholars have used the tree analogy to discuss the human potential has multiple implications. This analogous comparison is significant as it is a connecting point between human beings and

[4] Qur'an, 95:4–6.

[5] Muhammad ibn Jarir Tabari, 'Tafsir al-Tabari' [Tafsir of Tabari], http://www.altafsir.com/Tafasir.asp?tMadhNo=1&tTafsirNo=1&tSoraNo=95&tAyahNo=4&tDisplay=yes&UserProfile=0&LanguageId=1 (accessed 15 July 2019).

[6] Abu ʿAbd Allah Muḥammad Razi, 'Mafatih Al-Ghayb' [The Keys to the Unseen], http://www.altafsir.com/Tafasir.asp?tMadhNo=1&tTafsirNo=4&tSoraNo=95&tAyahNo=4&tDisplay=yes&UserProfile=0&LanguageId=1 (accessed 15 July 2019).

[7] Bursawi, 'Ruhu'l-Bayan,' https://www.altafsir.com/Tafasir.asp?tMadhNo=3&tTafsirNo=36&tSoraNo=95&tAyahNo=4&tDisplay=yes&UserProfile=0&LanguageId=1 (accessed 15 July 2019).

the natural environment. The tree analogy is used by the theologian and mystic Ibn Arabi (d. 1240), where the tree denotes the universe.[8] This perspective resonates with Qurtubi's (d. 1273) exegesis of the Qur'anic verse which describes human beings as created in the fairest form. Qurtubi describes human beings as like a small universe in order to highlight the potential inherent within them.[9] Twentieth-century Muslim scholar Said Nursi (d. 1960) uses the tree analogy to describe the potential found within human beings: 'Yes, however many degrees there are from a seed to a huge tree, the abilities lodged in human nature are more numerous.'[10] The tree analogy is useful to express the tremendous potential in human nature as the reader can appreciate the vast range of development involved from a seed to a fully grown majestic tree. The analogy, at the same time, highlights the tremendous development within a natural object like a tree and by extension in all created forms.

This inherent potential is understood to apply to all human beings. That is, every single individual on this earth is viewed as a seed which has the potential to become a 'tree.' This approach to human beings creates a platform where every person should have the opportunities it needs to grow and realise their potential. This is only possible in a peaceful environment where their human rights are also protected. Just as a seed cannot grow unless it has the right environment, no human being can grow unless they live in a peaceful and just environment.

The shared potential between humans and other living beings also creates a genuine connection between human beings and the environment since one identifies with a tree or other plant life in a way that fosters a sense of empathy. Every person would be devastated if their freedom to unleash their potential were to be prevented or, worse, destroyed. Similarly, seeing a tree, which has a similar potential, not realise its potential and flourish would be equally painful, as though one were seeing their own potential being impeded. Such powerful analogies, where the environment is seen as overlapping with the needs and desires of human beings, are a crucial part of the conversation needed in acting to protect

[8] Said Mentak, 'The Tree,' in *Islamic Images and Ideas: Essays on Sacred Symbolism*, ed. John A. Morrow (USA: McFarland & Inc. Company Publishers, 2014), 128.

[9] Qurtubi, 'al-Jami' li-Ahkam al-Qur'an,' http://www.altafsir.com/Tafasir.asp?tMadhNo=1&tTafsirNo=5&tSoraNo=95&tAyahNo=4&tDisplay=yes&UserProfile=0&LanguageId=1 (accessed 15 July 2019).

[10] Said Nursi, *The Flashes*, trans. Şükran Vahide (Istanbul: Sözler, 1995), 104.

the environment. Highlighting the great potential of human beings may appear to have limited ecological benefits, but when that potential is so intimately connected to the environment, it changes the way the environment is perceived.

An appreciation for the environment is not just effected through these powerful analogies. Just like human beings, all living things are understood to have a potential yearning to be fulfilled. As Nursi points out, all of creation goes through the process of 'expanding from the potential to the actual, through great effort and exertion.'[11] This principle of realising potential is visible throughout the entire universe and is referred to as the 'divine practice' (*sunnatullah*) where everything is given an opportunity to experience eagerness and pleasure in fulfilling its natural duty.[12] The pleasure experienced by all beings, animate and inanimate, is due to the ability to perform their duty through the potential that has been instilled within them. It resembles a 'wage' given to all beings that makes them eager and motivated to fulfil their duties.[13]

Nursi explains that even seeds have a natural urge to realise their duty by sprouting and germinating: 'Like someone imprisoned in a constricted place longs to go out into a garden or open space, such a longing, such a joyful state, is also apparent in seeds, in their duty of sprouting.'[14] The personification of a seed is exceedingly powerful as it instils feelings not only towards human beings, but also towards all living beings. This is an approach that does not normally appear in works of theology, but when it is done, it makes a compelling case. It makes it much more difficult to violate the rights of any person or thing as they are all seen as entities yearning to realise their potential.

Discussion of the potential that exists in all life forms naturally raises the notion of the value inherent in all beings.

[11] Nursi, *The Flashes*, 171.

[12] Ibid.

[13] Cüneyt Şimsek, 'The Problem of Animal Pain: An Introduction to Nursi's Approach,' in *Justice and Theodicy in Modern Islamic Thought*, ed. Ibrahim M. Abu-Rabi' (Farnham: Ashgate, 2010), 121.

[14] Nursi, *The Flashes*, 171.

Value

In the Islamic tradition, attaining knowledge of God is seen as one of the most important aspects of faith. This is indicated by the Qur'anic verse: 'I have not created the unseen beings and humankind but to (know and) worship Me (exclusively).'[15] While worship is given as the prime purpose, the exegete Ali Unal (b. 1955) explains that knowledge of God and love of God is entailed in the bounds of worship.[16] That is, quality worship of God is not truly possible if knowledge and love of God is not present preceding the act of worship. For this reason, there has been extensive literature written throughout Islamic history about how one can attain knowledge of God. In this goal, creation, particularly the natural environment, plays a critical role as it mirrors God's names and attributes. All parts of the natural world have an instrumental value because they facilitate knowledge of God. This approach to creation, by which it is connected to the Creator, strongly influences, and perhaps completely changes, the way the environment is viewed.

In Islamic theology, God is not of the substance of this universe. He is unembodied and has an essence unlike any of His creation. In Qur'anic stipulation, 'none is like Him.'[17] A famous statement with regard to God's nature has dominated Islamic theology for centuries: 'Whatever comes to your mind about His nature, God is different to that.'[18] A prominent Islamic theologian, Imam al-Tahawi (d. 935), states: 'Imagination cannot attain Him, comprehensions cannot perceive Him, and creatures do not bear any similarity to Him.'[19] The only aspect of God we are introduced to in the Qur'an is the names of God, 'God – there is no deity save Him; His are the All-Beautiful Names (*asma al-husna*).'[20] Thus, while it is God's essence (*dhat*) that is unknowable, knowledge of God is possible through appreciating God's names and attributes. 'God's

[15] Qur'an, 51:56.

[16] Ali Unal, *The Qur'an with Annotated Interpretation in Modern English* (Somerset, NJ: The Light, 2006), 1062.

[17] Qur'an, 112:4.

[18] Oliver Leaman, *The Qur'an: An Encyclopedia* (London: Routledge, 2006), 36.

[19] Ahmad ibn Muhammad al-Tahawi, *The Creed of Imam al-Tahawi*, trans. Hamza Yusuf (USA: Zaytuna Institute, 2007), 50.

[20] Qur'an, 20:8.

relationship with the creation is mediated by "His beautiful names"...the pillars upon which the phenomenal world rests.'[21]

The names of God are an important means to conceive, conceptualise and understand an otherwise transcendent God, to garner an opportunity to relate to Him. Humankind is believed to have a particularly important role, as humankind was created in 'God's own image' or, translated more literally, 'upon His form.'[22] Furthermore, the Qur'anic verse 'Then He proportioned him, and He blew into him of His spirit'[23] has been interpreted by exegetes to mean that the human being has the greatest potential to mirror the names of God. This concept of relating to God through divine names is also prevalent in the work of Ghazzali (d. 1111). Ghazzali states the path for conceptualising God is through understanding His names by their manifestation on humankind: 'it is conceivable for man to be characterised (by these names) to the extent that they may be spoken of him.'[24] Humans are given this great ability to perceive and understand the names of God by witnessing them but also by being the most comprehensive entity which can mirror the names of God.

This approach to humankind naturally leads to the idea that all human beings have utmost value since every single person was created 'upon His form.' Holding this worldview leads to a desire to fulfil the rights of all individuals and ensure social justice is implemented at its best.

Furthermore, the Qur'an conjoins the Beautiful Names of God as glorifications expressed in the cosmos and the earth: 'He is God, the Creator, the All-Holy Maker, the All-Fashioning. To Him belong the All-Beautiful Names. Whatever is in the heavens and on the earth glorifies Him. He is the All-Glorious with irresistible might, the All-Wise.'[25]

The manifestations of God's names occur in the form of detectable signs in the environment. Interestingly, the Qur'an uses the word *ayat*

[21] Colin Turner, *The Qur'an Revealed: A Critical Analysis of Said Nursi's Epistles of Light* (Berlin: Gerlach, 2013), 22.

[22] William Chittick, *Sufism A Beginner's Guide* (Oxford: Oneworld Publications, 2008), 94.

[23] Qur'an, 32:9.

[24] Abu Hamid Ghazzali, *Ninety-Nine Names of God in Islam*, trans. Robert Charles Stade (Ibadan: Daystar Press, 1970), 39.

[25] Qur'an, 59:24.

(signs) to refer to the actual verses of the Qur'an as well as the signs God has placed in the natural world for the reflecting human mind:

> ...And it is He who spread out the earth and set thereon mountains standing firm and (flowing) rivers: and fruit of every kind He made in pairs: He draws the night as a veil over the Day. Behold, verily in these things there are signs (*ayāt*) for those who think and reflect.[26]

The environment contains signs of God and the universe (particularly the natural environment), acts as the third element linking humanity and God. As stated by Ghazzali, God gave humankind an 'abridged form that brings together every sort of thing found in the universe'[27] so that the universe acts as a mirror, with all objects within it reflecting and manifesting God's names and attributes.[28] On this point, Nursi adds that a single living thing manifests or mirrors as many as twenty names of God.[29] It is the natural world where the greatest creativity of God is displayed and witnessed by humans. This makes the earth, together with its inhabitants and ecological environment, the greatest mirror to God's names and therefore is the most important source of attaining knowledge of God. This renders earth and its lifeforms extremely valuable within the divine plan of the universe and human capacity to relate to God.

In this context, life including the environment is an arena where God displays valuable works of art. By virtue of art being valuable, this theological perspective supports the protection of life as a show of respect towards the Artful Maker, God. In the words of Yunus Emre (d. 1321), a famous mystic and poet, 'we love the created, for the Creator.'[30] There is a natural human affinity towards all of creation as it has an intrinsic value as a result of being the creation of the Creator. This enhances the relationship between the Qur'an, the universe and humankind, making

[26] Qur'an, 13:3.

[27] *Mishkat al-Anwar*, edited and translated by David Buchman as *The Niche of Lights* (Provo, UT: Brigham Young University Press, 1998), 31.

[28] Said Nursi, *The Words*, trans. Şükran Vahide (Istanbul: Sözler, 1993), 221.

[29] Ibid., 655.

[30] Yunus Emre in Zekeriya Baskel, *Yunus Emre: The Sufi Poet in Love* (Lanham: Blue Dome Press, 2013), 56.

everything sacred because it is fulfilling the duty of mirroring[31] or manifesting God's names. When creation is viewed with this lens, what is witnessed is no longer an easily destroyable worthless thing so that peace becomes the natural state that one seeks to be in. A tree is no longer a 'wooden skeleton but an artwork made by God,'[32] and a flower is no longer a natural entity which can be destroyed but a beautiful creation reminding the observer of the Creator's Beauty.

The value of all creation is further supported by the Qur'anic verse which states, 'The seven heavens and the earth, and all beings, therein, declare His glory: there is not a thing but celebrates His praise….'[33] In other words, everyone and everything is praising God within the bounds of its own natural disposition, making them all valuable creations of God. The Hadith 'God is beautiful and He loves beauty' further reinforces this point. God describes Himself as loving beauty and therefore loving the cosmos. 'Hence, there is nothing more beautiful than the cosmos.'[34] Knowing that God loves the universe leads to a natural inclination for humankind to also love the universe and therefore value it. Since the creation (humankind, animals and plants) displays the creativity of God and is a mirror to reflect God's names and attributes, it needs to be protected. When people are killed, animals become extinct and the environment is harmed, we deprive future generations of the opportunity to get to know God at a deeper level.

The concepts of potential and value are intrinsically linked to the idea that all of creation has an interdependent purpose. This supports a call to action that moves from protecting the environment to ensuring a just and ecologically sustainable peace.

[31] According to Jami, everything is a coloured window by which everything manifests itself depending on its colour, but the source of light is God *Mirsad al-Ibad*, ed. Muhammad A. Riyahi (Tehran: Bungah-i Tarjama wa Nashr-i Kitab, 1973).

[32] Salih Yucel, 'Said Nursi's Approach to the Environment: A Spiritual View on the Book of Universe,' *The Islamic Quarterly* 55, no. 1 (2011): 3.

[33] Qur'an, 17:44.

[34] Chittick, *Sufism A Beginner's Guide*, 79–80.

Interdependent Purpose

In the Qur'anic cosmology purpose plays a crucial part and is linked to the potential and value of creation. The Qur'an declares, 'And We did not create the heavens and earth and that between them in play and fun. We only created them for a purpose....'[35] Everything has a purpose which it yearns to fulfil, and serving that purpose renders it valuable. In Islamic theology, God's plan for creation elicits an interdependent design of life by establishing ecosystems of flora and fauna, much in the way that humans develop interdependent communities.

In the wake of Hossein Nasr's (b. 1933) earlier work, one of the most important steps taken in Islamic environmental thinking has indeed been in the area of its teaching and attitudes with regard to animals.[36] The Qur'an clearly talks about living beings existing in ecological systems: 'No living creature is there moving on the earth, no bird flying on its two wings, but they are communities like you.'[37] The comparison of animal species with human communities is significant. Since human societies are complex systems made up of numerous interdependent individuals, this comparison points to the modern concept of ecosystems. The phrase 'communities like you' positions ecosystems in the same league as human societies. The existence of plural 'communities' leads to the conclusion that there are many concurrently existing and independent ecosystems. Responsible treatment of ecosystems and exerting an effort to prevent their damage or destruction can be seen as part of the general Qur'anic prohibition against causing corruption on earth.

While Islam treats the life of all creatures as valuable and recognises ecosystems as communities worthy of protection, it allocates status of a higher degree to human life. Human beings have been 'honoured with goodness'[38] in that men and women are created with the innate capability to recognise goodness and to respect virtue. Human beings are created with a sound 'natural disposition (*fitrah*) of God upon which He

[35] Qur'an, 44:38–39.

[36] Richard Folz, *Animals in Islamic Traditions and Muslim Cultures* (London: Oneworld Publications: 2014); Al-Hafiz Basheer Ahmad Masri, *Animal Welfare in Islam* (Markfield: The Islamic Foundation, 2016); Sarra Tlili, *Animals in the Qur'an* (Cambridge: Cambridge University Press, 2012).

[37] Qur'an, 6:38.

[38] Qur'an, 17:70.

has modelled the humans.'[39] Ultimately, human beings are created as a 'vicegerent (*khalifah*) on earth'[40] with the power and privilege of exercising command over earth's life forms and utilising its resources. Not only are they charged with the responsibility of protecting the rights of others through peace and social justice but they also have the responsibility of protecting the natural world so that corruption is not caused on earth[41] by destroying either its order or its beauty. Whenever the Qur'an puts responsibility onto humans, it comes with an obligation to follow through with the responsibility and the resultant accountability before God. Hence, humans should expect to be judged on how they treat other living creatures and the environment.

Linked with the discussion on potential and value, there is an acknowledgement that all of creation has an interdependent purpose which needs to be maintained in order for there to be ongoing harmony on earth. This is only possible with peace. What unites all three concepts is the Islamic notion of harmony and balance which transcends any one creation or entity. Balance is important in the Islamic worldview and is a key to providing a just and ecologically sustainable peace.

Towards an Islamic Just and Ecologically Sustainable Peace

A key part of understanding a just and ecologically sustainable peace is to understand justice in the Islamic context which can be defined as 'putting a thing in its proper place' [42] or as 'balance, equilibrium and equilibration.'[43] A Qur'anic verse which emphasises the importance of balance and maintaining balance is as follows:

[39] Qur'an, 30:30.

[40] Qur'an, 2:30.

[41] Qur'an, 2:27, 5:32.

[42] Bilal Kuşpınar, 'Justice and Balance in Creation,' in *Justice and Theodicy in Modern Islamic Thought*, ed. Ibrahim M. Abu-Rabi' (Farnham: Ashgate, 2010), 228.

[43] Nursi, *The Flashes*, 400.

> And the heaven – He has made it high (above the earth), and He has set up the balance. So that you may not go beyond (the limits with respect to) the balance. And establish weight in justice and do not make deficient the balance.[44]

This verse brings together four elements of justice and balance and underscores human responsibility in ensuring that balance and justice is achieved and maintained in all domains of life. Bursawi quotes Prophet Muhammad when explaining this verse: 'The Prophet said justice is a pillar of the earth and heaven'[45] thereby highlighting the strong link between balance and justice when dealing with earth and its inhabitants. Qushayri interprets this Qur'anic verse to mean that justice must be implemented in all acts of life. He also understands it to mean that one must be sincere, truthful, in that one must have equality outwardly and inwardly.[46] This is particularly important in social justice where the rights of all are fulfilled whether it be for individuals within one's community or outside one's community. Therefore, it is necessary to have a sincere desire to maintain the balance that exists within the universe, striving towards a peaceful world which offers justice for all.

Justice is a very strong theme in the Qur'an, so much so that it became customary to repeat the verse: 'Behold, God enjoins justice, and devotion to doing good, and generosity towards relatives, and He forbids you indecency, wickedness and vile conduct. He exhorts you (repeatedly) so that you may reflect and be mindful!',[47] an exhortation which is said at the end of every Friday sermon in all mosques in the Muslim world. It would suffice to note within the scope of this chapter that the way humankind and animals are fairly dealt with has been the topic of Muslim scholars for centuries. A large part of Islamic law and jurisprudence deals

[44] Qur'an, 55:6–9.

[45] Bursawi, 'Ruhu'l-Bayan,' http://altafsir.com/Tafasir.asp?tMadhNo=3&tTafsirNo=36&tSoraNo=55&tAyahNo=7&tDisplay=yes&UserProfile=0&LanguageId=1 (accessed 15 July 2019).

[46] 'Abd al-Karim Qushayri, 'Laṭa'if al-Isharat' [Subtleties of the Illusions], http://altafsir.com/Tafasir.asp?tMadhNo=3&tTafsirNo=31&tSoraNo=55&tAyahNo=7&tDisplay=yes&UserProfile=0&LanguageId=1 (accessed 15 July 2019).

[47] Qur'an, 16:90.

with human transactions known as *muamalat* (dealings).[48] The legal and ethical discussion of human rights and animal rights is based on *hadith* (narrations of Prophet Muhammad). One such example of protecting animals is when the Prophet's companions took baby birds from a nest, to which Prophet Muhammad said: 'Who has hurt the feelings of this bird by taking its young? Return them to her.'[49]

Such examples emphasise the great importance of justice, equilibrium and balance that exist on earth through ecosystems and the need to maintain them. Nursi notes that justice 'is the principle by which the whole universe and all beings act' and therefore if humankind acts against this justice, they become the 'object of anger and disgust'[50] of the universe. This equilibrium and balance can be found on earth at the micro- and macro-level where everything is 'ordered and weighed with so sensitive a balance, so fine a measure, that the human mind can nowhere see any waste or futility.'[51] Such balance and order is seen as a manifestation of God's name All-Just.[52] Therefore, destroying peace and justice in the human domain and destroying the ecosystem could be seen as distorting the manifestation of God's name, All-Just which would be a profound violation towards God.

Furthermore, humans, as the most comprehensive mirrors of God, need to be able to manifest God's name All-Just by being just in their treatment of everything that surrounds them so that the equilibrium set out on earth is not irreversibly tampered with.

When creation is viewed with its intrinsic value and potential, the way it is treated is positively affected. No longer can a single human life be discarded so easily since it has such great value in the eye of the Creator, as well as creation. The Qur'an verse 'whoever kills a soul…it is as if he had slain humankind entirely. And whoever saves one – it is as if he has saved humankind entirely'[53] reinforces this notion. In a way, this Qur'anic

[48] Refer to Mohammad Hashim Kamali's 'Chapter 13 Maslahah Mursalah (Considerations of Public Interest),' in *Principles of Islamic Jurisprudence* (Cambridge: The Islamic Texts Society, 2011) for details.

[49] Abu Dawud, Book 18, Hadith 1610.

[50] Nursi, *Flashes*, 402.

[51] Ibid., 401.

[52] Al-Adl (The Just), 'Questions on Islam,' https://questionsonislam.com/article/al-adl-just (accessed 10 July 2019).

[53] Qur'an, 5:32.

verse is suggesting that a single soul is equal to all of humankind, without any mention of the faith or ethnicity of the individual. This makes justice an essential part of one's worldview where all human life is sacred.

Human life is not the only end. All of creation is seen to have a purpose, a value and a potential to be realised. All life is a way of expressing this value and potential. Such a theological understanding will provide the foundation needed to establish a just and ecologically sustainable peace for humans and all other living beings on earth.

Conclusion

An Islamic theological assessment examining three concepts of potential, value and interdependent purpose illustrates that the earth, with all its inhabitants and its ecosystems, is an intrinsic part of God's plan for humans in realising their potential, garnering their value and achieving their purpose. While seeing the potential in all humankind leads to a desire to establish peace and justice, seeing the potential in the environment generates affinity and empathy towards the environment, so that seeking its protection becomes a natural response. Creation displays the infinite creativity of God; everything in the natural world worships God in a unique way. Furthermore, not only humankind but also animal species along with their ecosystems form communities and have a right to live peacefully within their communities.

Humans are endowed with intelligence and ingenuity to exert power over the rest of the creation. With this power comes accountability in the treatment of all living creatures and the environment. Hence, all forms of life on earth must be preserved as extremely valuable and humans are charged with that responsibility. Ultimately, the Islamic concept of accountability in achieving justice and balance on earth charges human beings to establish a just and ecologically sustainable peace.

Bibliography

Al-Adl (The Just). 'Questions on Islam.' https://questionsonislam.com/article/al-adl-just (accessed July 10, 2019).

Baskel, Zekeriya. *Yunus Emre: The Sufi Poet in Love*. Lanham: Blue Dome Press, 2013.

Bursawi, Ismail Haqqi. 'Ruhu'l-Bayan' [The Spirit of the Commentary]. http://www.altafsir.com/index.asp (accessed September 1, 2019).

Chittick, William. *Sufism: A Beginner's Guide*. Oxford: Oneworld Publications, 2008.

Folz, Richard. *Animals in Islamic Traditions and Muslim Cultures*. London: Oneworld Publications: 2014.

Ghazzali, Abu Hamid. *Ninety-Nine Names of God in Islam*. Translated by Robert Charles Stade. Ibadan: Daystar Press, 1970.

Hedges, Chris. *What Every Person Should Know About War*. New York: Free Press, 2003.

Kamali, Mohammad Hashim. *Principles of Islamic Jurisprudence*. Cambridge: The Islamic Texts Society, 2011.

Kuşpınar, Bilal. 'Justice and Balance in Creation.' In *Justice and Theodicy in Modern Islamic Thought*, edited by Ibrahim M. Abu-Rabi', 227–240. Farnham: Ashgate, 2010.

Leaman, Oliver. *The Qur'an: An Encyclopedia*. London: Routledge, 2006.

Masri, Al-Hafiz Basheer Ahmad. *Animal Welfare in Islam*. Markfield: The Islamic Foundation, 2016.

Mentak, Said. 'The Tree.' In *Islamic Images and Ideas: Essays on Sacred Symbolism*, edited by John A. Morrow, 124–132. USA: McFarland & Inc. Company Publishers, 2014.

Mishkat al-Anwar, edited and translated by David Buchman as *The Niche of Lights*. Provo, UT: Brigham Young University Press, 1998.

Nursi, Said. *The Flashes Collection*. Translated by Şükran Vahide. Istanbul: Sözler, 1995.

Nursi, Said. *The Words*. Translated by Şükran Vahide. Istanbul: Sözler, 1993.

Qurtubi, Abu ᶜAbd Allah. '*Tafsir Li Jami Ahkami'l Qur'an*' [The Collection of Qur'ānic Injunctions]. http://www.altafsir.com/index.asp (accessed September 1, 2019).

Qushayri, 'Abd al-Karīm. 'Lata'if al-Isharat' [Subtleties of the Illusions]. http://www.altafsir.com/index.asp (accessed September 1, 2019).

Razi, Abu ᶜAbd Allah Muhammad. 'Mafatihh al-Ghayb' [The Keys to the Unknown]. http://www.altafsir.com/index.asp (accessed July 15, 2019).

Şimsek, Cüneyt. 'The Problem of Animal Pain: An Introduction to Nursi's Approach.' In *Justice and Theodicy in Modern Islamic Thought*, edited by Ibrahim M. Abu-Rabi', 111–134. Farnham: Ashgate, 2010.

'Social Justice Resource Centre.' https://socialjusticeresourcecenter.org/facts-and-figures/ (accessed September 1, 2019).

Tabari, Muhammad ibn Jarir. 'Tafsir al-Tabari' [Tafsir of Tabari]. http://www.altafsir.com/index.asp (accessed September 1, 2019).

Al-Tahawi, Ahmad ibn Muhammad. *The Creed of Imam al-Tahawi*. Translated by Hamza Yusuf. USA: Zaytuna Institute, 2007.

Tlili, Sarra. *Animals in the Qur'an*. Cambridge: Cambridge University Press, 2012.

Turner, Colin. *The Qur'an Revealed: A Critical Analysis of Said Nursi's Epistles of Light*. Berlin: Gerlach, 2013.

Unal, Ali. *The Qur'an with Annotated Interpretation in Modern English*. Somerset, NJ: The Light, 2006.

Yucel, Salih. 'Said Nursi's Approach to the Environment: A Spiritual View on the Book of Universe.' *The Islamic Quarterly* 55, no. 1 (2011): 1–17.

Zalasiewicz, Jan, Mark Williams, Alan Haywood, and Michael Ellis. 'The Anthropocene: A New Epoch of Geological Time?' *Philosophical Transactions of the Royal Society A* 369 (2011): 835–1111.

Zuleyha Keskin is Course Director at the Centre for Islamic Studies and Civilisation, Charles Sturt University. She is a senior lecturer, lecturing on Islamic spirituality and contemporary Islamic studies. Zuleyha is also the managing editor of the *Australian Journal of Islamic Studies* and the Vice President of the Australian Association of Islamic and Muslim Studies (AAIMS). Her research interests include inner peace, Islamic spirituality and the intersection of Islamic theology with contemporary issues. Zuleyha has authored various books, chapters and articles, including *Contesting the Theological Foundations of Islamism and Violent Extremism* (2019), *Causes of Radicalisation: Theological Arguments as the Ultimate Trigger* (2019), *Theologisation of Volunteerism: An Islamic Perspective* (2019) and *Said Nursi's Tawḥid-Centric Worldview and Inner Peace* (2019).

Mehmet Ozalp is an Associate Professor of Islamic studies and the Director of the Centre for Islamic Studies and Civilisation (CISAC) which he founded at Charles Sturt University. He is the founder and Executive Director of ISRA (Islamic Sciences and Research Academy of Australia). Under his leadership, CISAC and ISRA pioneered Islamic Studies courses at the undergraduate and postgraduate level in 2011. Mehmet is the editor-in-chief of the *Australian Journal of Islamic Studies*. He serves on the executive committee of the strategic research centre Public and Contextual Theology (PaCT) linked to CSU. He serves on the Human Research Ethics Committee at the University of Sydney.

Mehmet was awarded the Australian Bluestar Award in Education (2011) and Charles Sturt University Professional Excellence Award (2012), Australian Muslim Role Model of the Year award (2012) and Australian Muslim Lifetime Achievement Award (2018). He is the author of twenty-eight publications including the books: *101 Questions You Asked About Islam*, *Islam in the Modern World* and *Islam between Tradition and Modernity: An Australian Perspective.*

CHAPTER 7

Islamic Ethics and Truth Commissions in the Muslim World: Towards a Just and Ecologically Sustainable Peace?

Salim Farrar

Introduction

The continuation of wars in Afghanistan, Syria and Yemen and dangerous instability in the Gulf, other parts of the Middle East and indeed across the Muslim World, paint a very bleak future. They raise the question whether we (in the Muslim World) can ever move towards a just peace and, if so, how. Tribal wars, factional interests, sectarian differences and superpower meddling (arguably the most potent factor) foment conflict situations and present impasses in resolving conflicts and achieving peaceful solutions. Aggravating the situation still further, much of the Muslim World is experiencing first-hand the effects of climate change. From the oil-rich Gulf and the starved deserts of Yemen, to the rapidly sinking Maldives and even Indonesia, environmental degradation

S. Farrar (✉)
Sydney Law School, University of Sydney, Camperdown, NSW, Australia
e-mail: salim.farrar@sydney.edu.au

J. Camilleri and D. Guess (eds.), *Towards a Just and Ecologically Sustainable Peace*, https://doi.org/10.1007/978-981-15-5021-8_7

is posing a huge problem in terms of food security, water scarcity, sustainable development and equitable access to resources, providing reasons for conflicts and complicating their resolution.[1]

There are no quick-fire solutions to such complex problems. In reality, they have been with us for centuries though their current scale seems unprecedented. Part of the answer, however, must surely lie in values—towards both justice and peace—and in our relationship with the Environment. Although religious values appear to be in decline even in the Arab world,[2] religious justifications continue to influence Muslim populations globally and religion remains a very important factor in their day-to-day lives.[3] I will argue that Islam's recognition and essential ethic that we, as human beings, are accountable to God for our actions in relation to our fellow human beings and the environments which they inhabit, and that we are *entrusted* with the care of God's creation, is a starting point towards achieving a just and ecologically sustainable peace. But values alone are not enough. They must be given concrete form and practical substance through laws, rules, procedures and flexible mechanisms that respond to local needs for present and future generations.

Outside of the Muslim World, notwithstanding decades (in some instances, centuries) of interminable fighting, injustice and discrimination, parties to conflicts have experimented with methods of transitional justice to navigate the horrors of the past and continuing tragedies of the present as a means of attaining a just peace. As a procedural mechanism linking local values to aspirations, Truth and Reconciliation Commissions (TRCs) have provided one method to assist historically divided communities to lay

[1] Mostafa Tolba and Najib Saab (eds), 'Arab Environment, Future Challenges', *Report of the Arab Forum for Environment and Development* (AFED) 2008, http://www.fao.org/fileadmin/user_upload/rome2007/docs/Arab_Environment_Future_Challenges.pdf, accessed 2 September 2019. See, in particular, Hasan Partow, 'Environmental Impact of Wars and Conflicts', 159–173.

[2] According to the BBC News Arabic and Arab Barometer (a network based at Princeton University), there has been a rise in people describing themselves as 'not religious' and a 'plummeting of trust' in religious leaders (see Kate Hodal, 'Arab World Turns Its Back on Religion—And Its Ire on the US', *The Guardian*, 24 June 2019).

[3] According to the Pew Research Center (Religion and Public Life), the most religious countries are in Africa, the Middle East, South Asia and Latin America. In the Middle East and North Africa (MENA), at least 70% of people say that religion is very important to them. In the Muslim-majority countries of Pakistan, Indonesia and Afghanistan, this figure rises to 90%. https://www.pewforum.org/, accessed 21 July 2019.

the building bricks of peace. More recently, TRCs have also recognised that methods of reconciliation also need to respect the land.[4]

The experiments have also started in the Muslim World, albeit more slowly and hesitantly. To date, there have only been two TRCs in the Arab-Muslim World: Morocco's Equity and Reconciliation Commission (ERC) 2004–2005 and Tunisia's Truth and Dignity Commission (TDC) 2013–2019. In addition to these two examples, we now must add Indonesia (the largest Muslim country in the world) and its ongoing TRC experiment in Aceh (since 2013), a development of the earlier initiative in Timor-Leste (2008 Indonesia-Timor-Leste Commission of Truth and Friendship).[5] Given their unique characteristics, my focus and case studies will be on Morocco's ERC and the Aceh TRC.

The chapter is divided into six sections. Parts one and two define and explain the Islamic concepts of ethics and stewardship and their relationship towards justice, peace and the protection of the environment. Part three examines the concept of transitional justice and its relationship to Islamic ethics. Given the importance and motivation of the state in establishing TRCs and all forms of transitional justice, part four explores the role of the state and Islamic governance. Part five defines and takes a detailed look at TRCs in general and part six introduces the Moroccan ERC and Aceh TRC as case studies from the Muslim World.

Islamic Ethics, Stewardship ('*Ri'ayat*'/'*Khilafah*') and the Environment

The core Islamic ethic is right conduct informed by the belief in *Al-Tawhid* or 'Oneness'; that God has no partner in His Divinity and is the One, True Owner of the Universe and everything that is in it. God is

[4] Anishina Bek News, 2015. 'Call to Action Report: Truth and Reconciliation Commission Recommends Environmental Stewardship and Protection of Our Natural Resources', http://anishinabeknews.ca/2015/12/20/truth-and-reconciliation-commission-recommends-environmental-stewardship-and-protection-of-our-natural-resources/, accessed 2 September 2019.

[5] Although this TRC applied to a predominantly Catholic territory, it had the full support and endorsement of Muslim scholar and then Indonesian President, Susilo Bambang Yudhoyono. See further: Megan Hirst, 'An Unfinished Truth: An Analysis of the Commission of Truth and Friendship's Final Report on the 1999 Atrocities in East Timor', 2009, https://www.ictj.org/sites/default/files/ICTJ-TimorLeste-Unfinished-Truth-2009-English.pdf, accessed 3 November 2019.

'*Rabb al-'Alameen*'[6]—the Lord of the Universe. Human ownership is contingent, partial and temporary and can only eventuate through Divine Creating and Grace. What we have is but a trust from God: from looking after our own bodies, families and communities to caring for the environment. According to Joseph Kaminski, this is a fundamental Islamic precept. He writes: 'one will find a common thread that transcends sectarian differences—an inalienable duty to protect the environment'.[7] The Qur'an (Islam's Holy Book) and the Hadith (the Prophetic Sayings), the primary Islamic sources, refer to human beings, normatively, as both trustees ('*khala'if*') and stewards ('*ra'oon*'), implying this is not our world to do as we like, but to follow the limits which God has set. The Qur'an mentions: 'Believe in Allah and His Messenger and spend of that whereof He hath made you *trustees* (*mustakhlafeen*); and such of you as believe and spend (aright), theirs will be a great reward'.[8] This applies both to interpersonal human relations and to dealings with the environment. Humanity has a responsibility to preserve the equilibrium or right measure ('*al-mizan*') in God's creation and to apply His limits. As the Qur'an explains: 'And the sky He hath uplifted; and He hath set the measure, that ye exceed not the measure, but observe the measure strictly, nor fall short thereof'.[9] This responsibility is shared by *all* of humanity, for Prophet Muhammad stated: 'Every one of you is a steward and is responsible for his charges: the Ruler of the people is a steward and is responsible for his subjects; a man is the steward of his household and is responsible for it; a woman is the steward of her husband's home and is responsible for it; and the servant of a man is a steward of his master's property and is responsible for it. All of you is a steward and is responsible for his charges'.[10] As trustees and stewards, we are each responsible in our particular ways for bringing about the conditions for maintaining and preserving the human and physical environment. From

[6] Qur'an, Surah *al-Fatihah*, verse 1.

[7] Joseph Kaminski, 'The OIC and Paris 2015 Climate Change Agreement: Islam and the Environment', in *Global Governance and Muslim Organisations*, Leslie A. Pal and M. Evren Tok (eds), 171–195 (173) (Cham: Palgrave Macmillan, 2018).

[8] Qur'an, Surah *al-Hadid*, verse 7.

[9] Qur'an, Surah *al-Rahman*, verses 7–9.

[10] 'Abdullah ibn Sa'ad, ibn Abi Jamrat al-Azdi, *Mukhtasar Sahih al-Bukhari* (Beirut: Dar Ibn al-Hazm, 1993), Kitab al-Jumu'at, Hadith No. 51, 50.

an Islamic perspective, bearing and taking charge of that responsibility should be an act of worship (*'ibadah*).

The terms of these trusts and stewardship are embedded in the Islamic Legal System, the Shari'a. According to Zoli, Bassiouni and Khan: 'The Islamic legal system is both normatively comprehensive and, yet, flexible in adapting to different contexts and challenges, as shari'a principles orient the legal system in changing times and radically different contexts'.[11] Not everything is clear-cut—not all Qur'anic verses and Hadiths are clear—they frequently require juristic interpretation to elicit the normative rule (*al-ijtihad*) and in which individual scholars may differ significantly (*al-ikhtilaf*). When performed by competent, humble scholars of the appropriate level, their interpretations (*al-fiqh*) provide a guide in different times and places to the practical operations of life in even the most minute of details and difficult of situations. This includes managing our societies in such ways that are productive of peace and which respect our physical environments. It is instructive that the Islamic humanitarian law (*al-siyar*) envisaged how a breakdown in law and order could also result in environmental degradation and imposed strict guidelines. It is reported in the early precedent of the most pious of Prophet Muhammad's Companions, Abu Bakr, the first Caliph, that he warned Yazid upon going to war: 'Do not uproot or burn palms or cut down fruitful trees'.[12] By outlawing a 'scorched earth policy', the early Muslim rulers not only enabled the economies and communities of societies previously at war to pick themselves up and sustain their livelihoods, they also complied with the essential ethic to protect the land of God as it was not theirs to destroy.

Justice and the Pursuit of Peace

It is not Islamic for rulers to exploit the Earth for their profit or pleasure. Nor is it Islamic for rulers and leaders to exploit the people for their own ends, even where ostensibly in the guise of peace. While the Qur'an recognises war as sometimes necessary, it also proclaims the need

[11] Corri Zoli, M. Cherif Bassiouni, and Hamid Khan, 'Justice in Post-Conflict Settings: Islamic Law and Muslim Communities as Stakeholders in Transition', *Utrecht Journal of International and European Law*, 2017, 33(85): 38–61 (40).

[12] Kaminski, 'The OIC and Paris 2015 Climate Change Agreement: Islam and the Environment', 173.

to end hostilities: 'if they incline to peace, you also incline to it, and (put your) trust in Allah'.[13] The pursuit of peace, however, must be just. As mentioned elsewhere in this volume, without both peace and justice, there can be no meaningful peace. The Qur'an proclaims: 'Verily, Allah prescribes justice (*al-'adl*)'.[14] The Prophet's Sunnah also informs that Allah has prohibited injustice for Himself and forbids injustice between His creation.[15]

This begs the question as to the definition of justice. Linguistically, justice or '*al-'adl*' is defined as the opposite of '*al-jawr*'; namely tyranny, cruelty and going beyond one's powers ('*al-dzulm*').[16] Clearly, as God is all-powerful, injustice is a purely human affair and would include violating any of God's prohibitions and failing to perform obligations towards Him, whether they relate to failures to protect His Rights (*huquq Allah*), which would include protection of the Earth and its resources, or to failures to protect the rights of persons (*huquq al-nas*). Scholars of Qur'anic exegesis further defined justice as: 'the *settling* of rights between you, *omitting* oppression and *delivering* each right to the person to whom it belongs'.[17] Note the administrative context of this definition and the gerund form of the verbs (*al-masdar*), emphasising implementation and active realisation of 'justice'. This aspect of justice is the job of rulers.

But justice is also the responsibility of the ordinary person, an aspect of communal living and common social responsibility, as the Prophet informed: 'The one who sees something unlawful, let him change it with his hand, and if he cannot, then with his tongue. If he cannot do that, the least he must do is to hate it with his heart'.[18] While elsewhere in the Sunnah, tolerance is praised, this is not the case where it would result in violating God's holy injunctions. This includes the responsibility of coming forward to tell the truth, even where to do so might practically work against you. The Prophet urged us to 'speak the truth even when

[13] Surah *al-Anfal*, verse 61.

[14] Surah *al-Nahl*, verse 90.

[15] There are numerous authentic hadith which call out injustice and demand action to address it.

[16] Muhammad ibn Abi Bakr ibn 'Abd al-Qadir, Al-Razi, *Mukhtar al-Sihah* (Beirut: Dar al-Fikr, 2001), 177.

[17] 'Abdullah ibn Ahmad, Al-Nasafi, *Madarik at-Tanzil*, vol 2 (Beirut: Dar al-Nafaes, 2014), 429.

[18] Hadith reported in *Sahih Muslim*, No. 49.

bitter'[19] and the Qur'an warns, 'do not mix up the truth with falsehood, nor hide the truth while you know (it)'.[20] For the victims of injustice, the question is intrinsically ethical and moral. While the victim is given the right to respond proportionately and can insist on a similar injury to be inflicted or even death (in murder cases), the Qur'an and Sunnah (the Prophetic template) urge restraint, mercy and forgiveness.[21] Books on Islamic ethics also recall what the Prophet confirmed from Allah as He said in the Torah: 'When an injustice has been done to you, be patient because My Victory is better for you than your victory'.[22]

Islamic Ethics and Transitional Justice

According to Stan and Nedelsky, 'transitional justice' refers to 'a variety of judicial and nonjudicial means through which states and societal groups seek to come to terms with past human rights violations by providing truth, justice, redress and reconciliation'.[23] According to the late Cherif Bassiouni, the notion of justice being 'transitional' does not translate well into Arabic, the language of the Qur'an and the Prophet's Sunnah. A matter is either 'just' (*'adl'*)—that is, a matter has been *delivered* to where it belongs (see above) or it is not. In effect, with the introduction of *transitional* justice, we are saying justice is 'on the way'. While Bassiouni noted transitional justice is a new concept to the Arab world, he added: 'in some respects it is congenial to Arab culture and more particularly is in tune with Islamic law'.[24]

[19] This is a portion of a longer hadith reported by Ibn Hibban in his Saheeh (vol 1, 228); Al-Tabarani in *Mu'jam al-Kabeer* (vol 2, at 158); Ibn Majah and others.

[20] *Surah al-Baqarah*, verse 42.

[21] See Susan C. Hascall, 'Restorative Justice in Islam: Should Qisas Be Considered a Form of Restorative Justice?', *Berkeley Journal of Middle Eastern & Islamic Law* 4:35, 48. In a hadith, the Prophet mentioned: 'The one who is not merciful, Allah may not be merciful to him; and the one who is not forgiving, Allah may not be forgiving to him' (reported by Al-Tabarani, in his *Mu'jam al-Kabeer*, vol 2, 351).

[22] Abu al-Qasim, Al-Tabarani, *Makarim alAkhlaq* (Beirut: Dar al-Kutub, 1989) 'Ilmiyyah (DKI), Hadith no. 39.

[23] L. Stan and N. Nedelsky, *Encyclopedia of Transnational Justice* (Cambridge: CUP, 2013), at xli.

[24] Editorial, *International Journal of Transitional Justice*, 2014.

As we have already explained, truth and justice lie at the heart of Islamic law and ethics, but so too do redress and reconciliation (*Islah bi-l 'adl*).[25] How that reconciliation takes place and the form it takes, necessarily will be contextual, but will be framed on the lines laid out above. In the context of warring factions, the Qur'an provides: 'And if two groups or parties among the believers fall to fighting, then make peace between them both. But if one of them outrages against the other, then fight you (all) against the one that which outrages till it complies with the Command of Allah. Then if it complies, then make reconciliation with them justly, and be equitable. Verily, Allah rewards those who are equitable'.[26] In the search for a 'just' reconciliation, the arbitrator/peace-maker will seek both to avoid harm and the reciprocation of harm.[27] This will entail a recognition of the harm that has been done, of the rights of the parties (as defined in the Islamic texts) and the strength and ease of proof against those who have allegedly wronged. It may also permit a minor harm to avoid a greater one.[28]

Although victims may seek redress in an Islamic court where, unlike in Common Law or even Civil Law jurisdictions, they would have the rights to participate in the trial, including a right to confront their abuser and to enforce punishment against them or receive compensation (*diya*) in its place, those rights depend on having the requisite Islamic proofs. Proof in the Islamic law of evidence is typically stringent because of the consequences of a miscarriage of justice against an accused.[29] Victims, therefore, may not get the justice they seek in a traditional Islamic court, so alternative methods of resolving the dispute and addressing the grievances (no matter how serious and deeply felt) are often required (and have been historically).[30]

[25] Muyhideen Abu Zakariyya, Al-Nawawi, *Al-Arba'oona Al-Nawawiyyah fi al-Hadith al-Nabawiyyah* (Beirut: Dar al-Mashari'a, 2002), Hadith 26, 24–25.

[26] Surah *al-Hujrat*, verse 9.

[27] This comes from a famous hadith of the Prophet and is also a well-known maxim in Islamic jurisprudence.

[28] This is also one of the axioms of Islamic jurisprudence. See further: Abdul Salam, al-'Izz ibn, *Qawa'id al-Ahkam* (Damascus: Dar al-Qalam, 2000), vol 1, 8.

[29] See further: S.A. Farrar, 'Islamic Jurisprudence and the Role of the Accused', *Legal Studies*, 2003, 23(4): 587–604.

[30] See Muhammad Khalid Masud, 'The Doctrine of Siyasa in Islamic Law', *Recht van de Islam*, 2001, 18: 1–29.

Those alternative methods would not necessarily exclude methods of transitional justice and could come under what has been traditionally termed, *Siyasa al-Shari'a*, or the politics of the Shari'a. Those responsible for law and order in the relevant territory would have to decide what method and approach would best fit the 'public interest' (*al-maslahah*), as the Shari'a would define it, taking into account the overriding purposes of the Divine Law (*Maqasid al-Shari'a*): the protection of religion, life, property, intellect, lineage and honour. Although they might formally avoid 'legislating', leaders could draft 'regulations' (*qawanoon/nidzamat*) to facilitate administration and enforcement so long as the methods they chose did not permit a prohibition or prohibit an obligation, or otherwise contradict a clear text of the religion.

A common method of transitional justice, from the middle of the twentieth century onwards, has been the use of amnesty (e.g. in South Africa, see below), often in the form of granting immunity from prosecution. In an Islamic context, an amnesty can come about in one of two ways. First, it may result from a decision of the victims of the crime. In cases involving the rights of a person (*huquq al-nas*), the victim can volunteer to waive their rights and remit punishment (with or without a compensatory payment). This is what the Islamic texts refer to as a '*sulh*' or a reconciliation. Not only must the offenders be identified in this context, but they also need to admit to their wrongdoing. Second, it may arise from an exercise of discretion by the ruler (or the state representatives) in crimes where there is no fixed penalty in the religious sources (*ta'zirat*). Again, this presupposes identification of the perpetrators.

Non-disclosure or concealment of a known offender (or harbouring) is a prohibited act in Islam, where it is done to prevent the exercise of another's rights. Non-disclosure can only be justified where deemed necessary to avoid the commission of more serious crimes (see above), or where proof of the crime (and therefore identity of the perpetrator) is still lacking. Accusing a Muslim of a crime without proof (i.e. making a statement in public) is a crime and sinful in itself. The Prophet warned: 'If the people were given in accordance with their claims, then men would claim the properties and blood of people. Rather, the proof is upon the claimant and the oath on the one who denies'.[31] For the ruler or government

[31] Hadith reported by al-Bayhaqi, *al-Sunan*, vol 10, 252.

to reveal the names of alleged perpetrators would also tarnish reputations and their sense of honour (one of the fundamentals underlying the Shari'a).[32] Non-disclosure in this context, therefore, would be deemed a praiseworthy act. The Prophet stated: 'The one who covers (the faults) of a Muslim, Allah will cover his faults in this world and in the Hereafter'.[33]

From this brief discussion, it can be seen that an administration running on Islamic lines would have a lot of freedom of manoeuvre to meet the requirements of the situation. Where, however, the perpetrators admit their actions or if there were witnesses to the acts, an amnesty would not be ethical *unless the victims themselves*, as opposed to the authorities, agreed (voluntarily and not through compulsion). Nor would it be ethical to bury the truth, whether in the name of some nebulous conception of public interest or not. Rather, it would be a further act of injustice and contrary to pronouncements from both the Qur'an and the Sunnah.

Islamic Governance, the 'State' and Globalisation

Originally, these discretionary powers set out above and given to a Muslim ruler via the *Siyasa al-Shari'a* were rights *in personam* and received in trust. They derived from his official capacity as the Imam/Caliph (*Khalifah*), or Chief Steward, to implement Islamic law and teachings across Muslim lands, and to formulate policy for the good of the people. This position came with its own pre-requisites (as inferred by scholars rather than by explicit Hadith).[34] In theory, there was only meant to be one Caliph or Imam to act as Chief Steward for all Muslims but given the geographical breadth and diversity of the Islamic empire (not to mention conflicts), in practice Islamic rule was given effect by local governors and

[32] Disclosure of the names of alleged assailants without necessity would also contravene the confirmed hadith: 'All of the Muslim in respect of his fellow Muslim is sacred: his blood, his property and his honour' (*Sahih Muslim*, No. 2564).

[33] Hadith reported in *Sahih Muslim*, No. 2699.

[34] The 'ideal' Islamic governance would be one that imitated the framework of the Prophet, but the complexity and scale of Muslim societies requires more complex governmental arrangements than persisted in the time of the Prophet. There are several tracts on Islamic governance, but for one of the most referred to Sunni versions, see al-Mawardi's '*al-Ahkam al-Sultaniyyah*'.

sultans with little Caliphal oversight. They would act, formally, as his 'delegate' and exercise his powers in those regions. The extent to which their rule accurately reflected Islamic law and its values depended substantially on the local governor and the operating culture which he encouraged and the local culture of which he was a part. Justice mechanisms were often hybrid, frequently incorporating local customs (*'urf* or *'adat*) to facilitate implementation which may or may not have complied with the 'letter' of the Islamic law as applied in Mecca or Medina.[35]

This system of governance obtained for more than one thousand years in the Muslim World but was first impacted by imperialism and colonialism in the eighteenth and nineteenth centuries, and then dismantled after the First World War, with the emergence of Muslim 'nation states'. Notwithstanding the so-called and now defunct 'Islamic State' and its claims to have resurrected this political office, for our practical purposes, these powers are now exercised by Muslim nation *states* (be they monarchies or 'Islamic republics') and have been for close to a hundred years.[36]

We have assumed to this point that the *state* is regarded as a legitimate source of law and order and of authority. But in many states across the Muslim World, the 'state' itself is one of the sources of the problem and may even be the main perpetrator of injustice. The state, in this context, is not a disinterested or neutral party, but has a vested interest in preserving its reputation and achieving a certain outcome. In the contemporary world, where corruption is rife, including in the Muslim countries, it is difficult for common people to accept that arbitrators or peacemakers appointed by the state will be what they claim (notwithstanding we are meant to presume Muslims to be virtuous unless there is confirmed evidence to the contrary). In the contemporary context, the reality of many different sects even within one state and with each regarding the other with inherent suspicion complicates the situation even further (the

[35] A good example of this is found in the *Undang-Undang Melaka* ('the Laws of Malacca') from the late fifteenth century in which the Malay Sultans mixed local customary rules with the Islamic rules taken from the Shafi'i school of Sunni jurisprudence.

[36] See Masud, 'The Doctrine of Siyasa in Islamic Law'. For an introduction to the topic, see also: Asifa Quraishi-Landes, 'Who Says Shari'a Demands Stoning of Women? A Description of Islamic Law and Constitutionalism', *Berkeley Journal of Middle Eastern & Islamic Law*, 2008, 1: 163–177.

Salafist and Shi'a influences in Saudi Arabia and Iran, respectively, are the most obvious examples).

There exists, therefore, a very real crisis of authority in the Muslim World[37] in relation to who has the legitimacy to act for Muslims, nationally or on a global stage, a matter that relates not just to peace-keeping but also to protection of the environment. Since 1924, the only global body to officially attempt to speak in the name of 57 predominantly Muslim states has been the Organisation for Islamic Cooperation (formerly the Organisation of the Islamic Conference), originally established in the wake of the arson attack on the al-Aqsa Mosque in 1969. Notwithstanding the occasional success, the general consensus is that this organisation has been a failure: a talking shop for deeply flawed Muslim democracies or dictatorships, with no enforcement mechanisms to carry through its agreements or credibility to speak with a uniform Muslim 'voice'.[38] Nevertheless, these known weaknesses have prompted critical reflection and since 2010 sparked a process of reform to promote its relevance and garner more support.

For our purposes, in 2013 the OIC set up a specialised 'Peace, Security and Mediation Unit' (PSMU) together with a 'Wise Persons Council' (WPC) that would operate as a 'mechanism for conflicts resolution and peace building, strengthening the role of the Organization in the field of mediation and preventive diplomacy'.[39] It is still in the process of development, but the OIC itself has prioritised the development of more effective dispute resolution processes as one of its strategic goals (OIC 2025 Programme of Action).[40] As to action on climate change, in 2015, the OIC passed the 'Islamic Declaration on Global Climate Change', a

[37] On the causes of this crisis and fragmentation of authority in the Muslim World, see further: Francis Robinson, 'Crisis of Authority, Crisis of Islam?', *JRAS*, 2009 Series 3, 19(3): 339–354.

[38] See Zhongmin Liu and Peng Fan, 'Islamic Factors in Inter-State Cooperation of the OIC Members', *Asian Journal of Middle Eastern and Islamic Studies*, 2018, 12:1, 1–15, https://doi.org/10.1080/25765949.2018.1439616; see also: Salim Farrar, 'The Organisation of Islamic Cooperation: Forever on the Periphery of Public International Law?' *Chinese Journal of International Law*, 2014, 13(4): 787–817.

[39] See Abdullah al-Ahsan and Stephen Young (eds), *Qur'anic Guidance for Good Governance* (Kuala Lumpur: Palgrave Macmillan, 2017), 194.

[40] OIC, SESRIC, 'Achieving Peace and Security in a World of Turmoil: An Arduous Challenge for the OIC', 2019, www.sesric.org/files/article/659.pdf, accessed 8 April 2019.

responsible effort to frame environmental stewardship in both scientific and religious terms, and to limit carbon emissions and temperature rises to between 1.5 and 2 degrees centigrade. To date, however, not all of the OIC member states have signed the declaration (including Iran, one of the biggest polluters), and of those who have signed, few are living up to their agreements.[41]

In the absence of a viable and credible global Islamic authority, the impetus for change and action remains at the state and local level. The issue, as our focus on TRCs will also illustrate, is that modern Muslim states are guardians of their own legacy and private interests, as well as conduits for international and corporate interests, and tend to manipulate public processes to serve private ends.

Truth and Reconciliation Commissions (TRCs)

TRCs have their origins in Truth Commissions: 'temporary state-sanctioned bodies investigating large-scale human rights violations committed over a specific period'.[42] They comprise one of a panoply of options (including amnesty, see above) for new governments to choose from, where they are transitioning a period of dictatorship, oppression and instability (or war). They provide an alternative to criminal prosecution and mechanisms of legal accountability. Generally speaking (though the form they take is dictated by context, see below), they are temporary bodies established by the state, which make recommendations retrospectively over a particular period. They evaluate patterns of past human rights abuses rather than individual cases and publish their opinions in a final report. All of their recommendations are unenforceable and require further state implementation.[43] The only difference with TRCs is that the latter are set up with the explicit aim of bringing torn communities together and not merely the investigation of historical human rights violations.[44]

[41] Kaminski, 'The OIC and Paris 2015 Climate Change Agreement: Islam and the Environment', 191.

[42] Nokukhanya Mnewabe, 'Truth Commissions', in L. Stan, and N. Nedelsky, *Encyclopedia of Transnational Justice*, 99.

[43] Ibid.

[44] Ibid., 101.

As a procedural mechanism primarily for mediation, TRCs have no necessary substantive content, philosophy or a priori guiding law. While their establishment is often on the back of a shift towards democracy and away from dictatorship, it is the parties to the conflict themselves, with backing from the state, who devise the procedure and terms of reference, reflecting their image, interests, values and objectives. This can include addressing environmental questions and climate change,[45] not merely addressing human rights violations. According to Yasmin Sooka,[46] notwithstanding local particularities and peculiarities, viewed from the perspective of *Transitional Justice*, there are still certain prerequisites for successful and legitimate TRCs. These include: the rights of all parties, including civil society to participate; holding perpetrators of violent abuse accountable to civil society through statutory and administrative measures with clearly defined goals; non-discrimination—that all parties be treated justly irrespective of the side of the conflict they belong to; and empowerment of local actors and civil society, with linkages to democratic initiatives and institutions. She notes, importantly, the transitional justice measures must not be considered an end in themselves and that all levels of the affected society need to 'buy in'. According to Saloman Lerner, what is important is popular acceptance and legitimacy in the eyes of all of the society impacted by the troubles.[47]

The first TRC was set up in 1994 in South Africa as part of the healing process for the dismantling of apartheid (promoted with the motto, 'Revealing is Healing'). Controversially, it also included amnesties for perpetrators of human rights abuses in exchange for their testimony before the TRC where the motivation for their crimes was political. While

[45] See further: Werner Scholtz, and Gerrit Ferreira, 'Climate Change Negotiations and Transitional Justice: The Advent of a Carbon Truth and Reconciliation Commission?', *Comparative and International Law Journal of South Africa*, 2015, 48(1): 42–58.

[46] Y. Sooka, 'Dealing with the Past and Transitional Justice: Building Peace Through Accountability', *International Review of the Red Cross*, 2006, 88(862): 311–325 (313).

[47] S. Lerner, 'Interview with Saloman Lerner', *International Review of the Red Cross*, 2006, 88(862): 225–233 (226).

arguably contrary to International Law,[48] it was (and still is) suggested the simultaneous offers of amnesties were a necessary price to pay to achieve reconciliation, justice and peace for a country still in transition.[49] Thereafter, TRCs have been used in South America (including Chile, Guatemala and Peru), Africa (including Kenya, Nigeria and Sierra Leone, and most recently proposed for the Gambia and Zimbabwe), South East Asia (Aceh and East Timor) and other parts of the world, including even the USA where local communities, blighted by histories of racism and segregation, have found it has offered measures of accountability and closure to victims, as well as an alternative to vengeance.[50] TRCs have not been an unequivocal success—Nigeria's *Oputa Panel* was an explicit failure[51]—but their global presence and continuing popularity point to their ultimate flexibility and the importance of tailoring their form and substance to the particular context.

TRCs in the Muslim World

Elements of a TRC in the Muslim World were first experimented in the aftermath of the Bosnian War and the break-up of the former state of Yugoslavia. As part of the Dayton Peace Accord in 1995 and with pressure from the international community, the Human Rights Chamber for Bosnia and Herzegovina was established. Although this focused primarily on legal and individual accountability and reflected a new constitutional 'rights' framework in which the European Convention on Human Rights

[48] This depends on the nature of the crimes granted amnesty. The UN Declaration on the Protection of All Persons from Enforced Disappearance (1992) prohibits granting amnesties to perpetrators of enforced disappearances. The UN Committee on Human Rights Resolution on Impunity (2005) has also stated that amnesties should not be given to those who commit violations of human rights and humanitarian law. See Carlos Cosa, 'Amnesty', in L. Stan and N. Nedelsky, 2013, *Encyclopedia of Transnational Justice*, 12.

[49] John Dugard, 'Dealing with Crimes of a Past Regime: Is Amnesty Still an Option?' *Leiden Journal of International Law*, 1999, 12(4): 1001–1015; N. Barney Pityana, 'The Truth and Reconciliation Commission in South Africa: Perspectives and Prospects', *Journal of Global Ethics*, 2018, 14(2): 194–207 (195).

[50] See Joshua Inwood, 'Dealing with Hate: Can America's Truth and Reconciliation Commissions Help?', *The Conversation*, 1 March 2017.

[51] See further: Yusuf O. Hakeem, *Transitional Justice, Accountability and the Rule of Law* (Abingdon and New York: Routledge, Glasshouse Books, 2010).

was central, the Chamber had a broad mandate that included truth-seeking.[52] Of particular note were the '*Srebrenica cases*', which examined the right of families to know the truth about the fate and whereabouts of some 7500 missing men and boys. Being an international construct, and located in the heart of Europe, there was neither a formal nor implicit recognition of Islamic law or Islamic ethics in the procedural mechanism (although genocide of Muslims had taken place).

The first TRC in the Arab-Muslim World was the *Equity and Reconciliation Commission* (ERC) of Morocco. It was formally established on 7 January 2004 after King Mohammed VI, the recently inaugurated ruler, supported Morocco's national Human Rights Council's recommendations to set up a truth commission that would investigate the human rights violations perpetrated during the rule of his father, King Hassan II. It ran until 2006. Although not stated explicitly in its title, the new king proclaimed the ERC would be a mechanism to 'uncover the truth' and act as a 'truth and fairness body'. The holding of the ERC would 'deepen the democratic transition' and 'turn the page' following decades of brutal suppression against opposition elements, especially during the 'years of lead' in the 1960s and 1970s.[53] According to the ERC itself, thousands were tortured, held in secret detention centres, and many simply 'disappeared'. Hundreds were killed, allegedly the result of extra-judicial executions.[54]

The ERC did not suddenly 'appear' but was a culmination of efforts from local human rights NGOs, the national human rights agency and even the previous administration to right the wrongs of the past.[55] It was an improved version of the Independent Arbitration Panel (IAP) set up in 1999, shortly after King Hassan's death, that had begun to receive

[52] See further: J. David Yeager, 'The Human Rights Chamber for Bosnia and Herzegovina—A Case Study in Transitional Justice', *International Legal Perspectives*, 2004, 14(1): 44.

[53] Mark Freeman, 'Transitional Justice in Morocco: A Progress Report', *ICTJ*, 2005, https://www.ictj.org/sites/default/files/ICTJ-Morocco-Progress-Report-2005-English.pdf, accessed 27 June 2019.

[54] Pierre Hazan, 'The Nature of Sanctions: The Case of Morocco's Equity and Reconciliation Commission', *International Review of the Red Cross*, 2008, 90(870): 399–407 (403). See also: Freeman, ibid.

[55] King Hassan had approved a request from the Human Rights Council in 1999 two weeks before his death to set up a commission to investigate historical human rights abuses: Freeman, 'Transitional Justice in Morocco: A Progress Report', 10.

and investigate claims for compensation. Operating between 1999 and 2003, the IAP received 5127 applications before its self-imposed deadline, heard testimony from 8000 witnesses at 196 general hearings and included more than 400 individual hearings. More than 5000 judgments were handed down, with claimants successful in 3681 cases, and the Panel awarding a combined total of $100 million (USD) in compensation. As Mark Freeman notes, this was a big advance for the Muslim-Arab World, a significant acknowledgement of state responsibility and substantial compensation.[56] Nevertheless, the IAP was heavily criticised for the method of evaluation of compensation, reflecting lost income rather than the hurt and harm actually caused to victims. The focus was also on money rather than the broader concept of 'reparations' that would have entailed production of death certificates, return of body remains and offers of pastoral support for the mental and physical well-being of victims. The IAP's mandate also excluded extra-judicial executions and, in spite of the numbers of applications the IAP received, there were thousands more excluded because they missed the deadline to apply. It was also criticised for its lack of independence and transparency (hearings were not in public).[57]

In its place, the composition of the ERC set up in 2004 comprised less reserved seats for political parties and an increased presence of representatives from local NGOs. The ERC also included former detainees and political prisoners—which included even the President of the ERC itself, Driss Benzekri, who had spent seventeen years in jail as a political prisoner. All members, however, were still appointed by the State and did not include anyone who remained in active opposition to the king and his government.[58] The mandate of the ERC was also modified, removing any deadline for applying so as to close 'definitively the file of human rights abuses committed in the past'.

[56] Freeman, 'Transitional Justice in Morocco: A Progress Report', 11.

[57] Ibid., 11–12.

[58] See the research carried out by Najwa Belkziz, 'The Politics of Memory and Transitional Justice in Morocco', PhD Thesis, University of Melbourne, School of Historical and Philosophical Studies, 2018, https://minerva-access.unimelb.edu.au/bitstream/handle/11343/213791/Belkziz_Corrected%20thesis%2011%20June%202018.pdf?sequence=3&isAllowed=y, accessed 1 November 2019. She notes Islamists were excluded on grounds of the ongoing 'war on terror' and the Sahrawis from the disputed Western Sahara region because they were not sufficiently 'Moroccan' (ibid., 177).

The aims of the ERC were stated in Article 9 of the founding statute. These provided: establishing the scale of past violations; continuing with previous investigations; determining the responsibility of state *organs*; compensating victims for damage; making recommendations for the psychological and medical rehabilitation of victims; drafting a report to preserve memory of the events, and the development and promotion of a culture of dialogue. Notably, the notion of financial compensation from the IAP was replaced with the broader concept of 'reparation'.[59] Reparations also took on a collective dimension, with the possibility of addressing those communities, not just individuals, most affected by the human rights violations and marginalised in the country's national development. More importantly, perhaps, private hearings were replaced with public hearings—even though no express mention of this procedure was made in the mandate, it was interpreted through Article 9's aim to 'develop and promote a culture of dialogue'.

Public hearings were held in auditoriums and meetings halls across the country. These were widely attended, and all proceedings televised for local, national and international consumption. Hearings were 'victim-centred' and victims provided with emotional support. Each victim selected to appear reflected criteria of gender balance, regional representation, clarity, significance and diversity of their story, and psychological strength.[60] The ERC ended up taking 22,000 statements of victims and their families across the different regions of the country. Freeman notes 'hearings were well-received in their own locales and provoked significant debate and an outpouring of emotion'.[61]

It is important to note, however, that the hearings were not a criminal trial and the Commission was 'sanction free'. Indeed, the one thing none

[59] According to Freeman ('Transitional Justice in Morocco: A Progress Report', 13), the former financial package had been influenced by the Islamic concept of *diya*, which, commonly understood as 'blood money', restricted the package of restorative benefits to victims. With respect, the Islamic notion of *diya* always sat alongside forgiveness, which was the preferred option to help bring about rehabilitation of the offender and closure to the victim. It was one of the features of Islamic criminal justice that distinguished it from both Judaic punishments found in the Torah and Roman *lex talionis*. See further: Susan Hascall, 'Restorative Justice in Islam: Should Qisas be Considered a Form of Restorative Justice?' The fact that the IAP did not incorporate the dimension of forgiveness was more indicative of the secular and hybrid nature of Morocco's justice system.

[60] Freeman, 'Transitional Justice in Morocco: A Progress Report', 18.

[61] Ibid.

of the country got to hear was the personal identity of any of the perpetrators. Not one agent of repression, torturer or killer testified before the ERC in any of the seven public hearings held around the country because the Commission had no powers to compel attendance. Not only that, the victims who were allowed to present their testimonies in public had to sign a personal declaration that they would not name or identify their assailants.[62] Also the general public only got to hear a selection of testimony: more than 3000 Islamists detained as part of the 'War on terror', and therefore outside the ERC's timeframe, were specifically excluded.[63] Their stories were told outside the Commission, via parallel untelevised hearings held under the auspices of a human rights NGO.[64] For some, remains of bodies have not been returned and families have been denied the opportunities to visit the graves of their loved ones. To this day, sixty-six cases remain unresolved, some of which involved opposing the monarchy.[65]

While supporters of this procedure argued the absence of sanctions and identification facilitated the transition from an authoritarian dictatorship to a constitutional monarchy and a gentler, more liberal government, opponents of the process argued that it conferred continuing impunity for the commission of serious crimes and human rights violations. Many of those thought to have committed crimes under the *ancièn régime* continued to hold their posts within the police force, military and government and so amounted to a political whitewash.[66]

In many respects, the entire ERC, like the Human Rights Chamber for Bosnia Herzegovina, was not constructed on Islamic lines. Islamic law was not mentioned at all in the founding statute and some commentators even affirmed the ERC broke with Islamic tradition through its system of reparation. Previous compensation schemes had been based on traditional

[62] Pierre Hazan, 'The Nature of Sanctions: The Case of Morocco's Equity and Reconciliation Commission', 403.

[63] Ibid.

[64] See Belkziz, 'The Politics of Memory and Transitional Justice in Morocco', 57.

[65] ICTJ, 'Morocco still a Model for Justice in MENA, but Questions Remain', 2 August 2016, https://www.ictj.org/news/morocco-still-model-justice-mena-questions-remain, accessed 1 November 2019.

[66] Pierre Hazan, 'The Nature of Sanctions: The Case of Morocco's Equity and Reconciliation Commission', 404–405. See also: Belkziz, 'The Politics of Memory and Transitional Justice in Morocco', 57.

inheritance laws which had prevented wives of those disappeared from obtaining compensation. The ERC broke new ground by awarding wives reparations in their capacity as spouses of the victims.[67] And yet, on closer inspection, Islamic values and some Islamic ethics ran through it. In the preamble to the ERC, the king referred to the absence of punishment as 'pardoning' and not 'amnesty'. In a speech concluding the Commission's work, he even invoked the spirit of the traditional role of Imam:

> I am sure that the sincere work of reconciliation we have accomplished … is, in fact, a response to the divine injunction "Forgive with a gracious forgiveness". It is a gracious gesture of collective pardon.[68]

In their public testimonies before the ERC, Najwa Belkziz noted witnesses quoted Qur'anic verses of forgiveness and reconciliation, but also of vengeance.[69] The substance of the vast majority of testimonies, however, sought forgiveness and tolerance. Belkziz quotes the following statements of one of the witnesses as typical:

> Personally, I don't want the prosecution of those miserable torturers because I simply consider that I have won. I beat them when I preserved my sanity and some of my psychological balance, thanks to the support of my family and thanks to the support of my older brother. I don't want them prosecuted because I am convinced our country does not need to get into these mazes. I only want to know the truth.[70]

From an Islamic angle, the king may have seen the ERC as filling a gap. Justice demanded recognition of the harm done to individual victims and allowing victims to publicly explain what had happened to them and their suffering and give them the opportunity to offer their forgiveness to their abusers allowed them a degree of closure. The need for Goodness (*al-Ihsan*) required an emphasis on pardoning, forgiveness and reparation. None of the latter he was required to do but arguably was an ethical response, Islamically, to promote a culture of 'Islah'—of restitution. It is

[67] ICTJ, 'Morocco Still a Model for Justice in MENA, but Questions Remain', 53.

[68] Pierre Hazan, 'The Nature of Sanctions: The Case of Morocco's Equity and Reconciliation Commission', 406.

[69] Belkziz, 'The Politics of Memory and Transitional Justice in Morocco', 136–137.

[70] Ibid., 144.

possible, as Pierre Hazan noted, that justice would not have been served had individual perpetrators been identified, arrested, charged and then brought before a court. Judges were suspected of not being independent and important evidence substantiating crimes may have been missing. Moreover, as part of the Siyasa al-Shari'a, the king may have thought the continuation and loyalty of his government, policy and military relied upon his forgiveness of their faults, and better their rehabilitation than their punishment. Punishment may have let to instability, and instability to chaos.

Nevertheless, there was also a negative side. Justice was not complete; it was selective. Not all those who had suffered oppression and tyranny under the previous regime, nor indeed under the current regime, were given redress. Islamists and Sahrawis, and their families, were specifically excluded. On the one hand, the king might have legitimated their exclusion on grounds of 'rebellion'—they were rebels and continued to be so—and that demanded a tough response. And yet, so had been many of the communists whom the king was now embracing. Also, while the emphasis on forgiveness and reparation appeared indicative of Islamic ethical principles, there was also an indication that the statements given by witnesses were possibly coached and doctored by the government. The fact that so many witnesses praised the government and offered their forgiveness, and also evidence that the government actually censored some witnesses who could not contain their distress and upset at their captors,[71] indicated the government had their own purposes to serve and an image they wanted to project.

The most recent TRC in the Muslim World is Indonesia's experiment in Aceh, officially established by the Aceh provincial government in 2013. The process is ongoing and is not due to submit its final report until 2021.[72] The Aceh TRC has its origins in the 2005 Helsinki Agreement that ended the insurgency between the GAM (Free Aceh Movement) and the Indonesian Government following the devastation caused by the 2004 Asian Tsunami. Although that environmental disaster took hundreds of thousands of lives, more than 130,000 in Aceh alone, it focused the minds of warring factions to make peace. The insurgency and

[71] Belkziz, 'The Politics of Memory and Transitional Justice in Morocco', 136–139.

[72] Aceh Truth and Reconciliation Commission, 'A Diplomatic Brief', 2 April 2019, https://kkr.acehprov.go.id/aceh-truth-and-reconciliation-commission-a-diplomatic-brief/, accessed 9 September 2019.

state counter-insurgency in Aceh took place between 1976 and 2005 and cost more than 10,000 lives, involving dozens of mass murders, hundreds of extra-judicial executions, forced disappearances, rapes and thousands of incidences of torture.

According to established narrative, the MOU between the GAM and the Indonesian Government originally envisaged a Human Rights Court to try these past abuses and a *national* Truth and Reconciliation Commission to allow victims of the Suharto regime, including in Aceh, to speak and receive reparations. Under the MOU, central to the operation of the TRC and the awards of reparations was the granting of amnesty to combatants on both sides, a necessary fig leaf to guarantee the peace. In 2006, however, ignoring its practical importance, the Indonesian Constitutional Court struck down the agreement on the ground that a compensation regime could not be established without legal accountability.[73] The Aceh TRC as it existed only under the national umbrella was left in legal limbo, denying victims of the conflict their voice as well as continuing impunity for the perpetrators of human rights violations.

The failure to implement the TRC was regarded as a failure of both the GAM and the Indonesian Government. According to Elizabeth Drexler, the Indonesian Government and the GAM leaders had never been serious in the first place,[74] but the failure prompted NGOs and civil society groups in Aceh to push for a more victim-centred process (and with an Islamic focus[75]). It took a decade of their pushing for legislative reform in the Aceh state legislative chamber and using Aceh's own autonomy regulations (the Qanun), before an independent Aceh TRC was eventually passed in 2013.[76]

In some respects, the Aceh TRC is a more important case study for our purposes than Morocco's ERC as the provincial government of Aceh

[73] 'Indonesia Court Overturns Law Creating Truth and Reconciliation Commission', *Jurist*, 8 December 2006, https://www.jurist.org/news/2006/12/indonesia-court-overturns-law-creating/, accessed 9 September 2019.

[74] Elizabeth Drexler, *Aceh, Indonesia: Securing the Insecure State* (Philadelphia: University of Pennsylvania Press, 2008), 210.

[75] John Braithwaite, Valerie Braithwaite, Michael Cookson, and Leah Dunn, *Anomie and Violence: Non-Truth and Reconciliation in Indonesian Peacebuilding* (Canberra: ANU Press, 2010), 389–390.

[76] Ahmady, 'Eagerly Anticipating Aceh Truth Commission', *The Jakarta Post*, 18 December 2013, https://www.thejakartapost.com/news/2013/12/18/eagerly-anticipating-aceh-truth-commission.html, accessed 9 September 2019.

is democratic *and* set up on explicitly Islamic lines.[77] The Aceh TRC was established by Qanun No. 17/2013[78] within the legal framework of Aceh's special autonomy law No. 44/1999 and the Aceh Law No. 11/2006. The latter obligates the Aceh government to implement Islamic law,[79] including its criminal laws, and provides that 'every person of the Islamic faith must adhere to and practice Islamic law'.[80] In addition, it establishes a system of Shari'a courts to try cases (including involving non-Muslims where they submit to the jurisdiction)[81] and a Clerics Deliberation Council (MPU) to issue fatwas when considering formulation of policies. These provincial laws have not been superseded or replaced by national legislation.[82] Given the Aceh TRC was not implemented until 2013, this clearly implies, at least from the perspective of the Aceh government and people, that Aceh's TRC *is* an Islamic process, consistent with the laws and precepts of the Shari'a.

On reviewing the text of the Aceh TRC,[83] however, apart from a singular reference to working principles of the Commission to be based on Islam[84] and for its Commissioners to be 'spiritual', of good moral character and able to read the Qur'an,[85] the majority references are to

[77] Religious revival, religious symbolism and promotion of religious values were integral elements of post-*reformasi* Indonesia and particularly in Aceh in the wake of the Tsunami disaster; see R. Michael Feener, *Shari'a and Social Engineering: the Implementation of Islamic Law in Contemporary Aceh* (Oxford: OUP, 2013), 57. Islamic values and implementation of Islamic law were embodied in its 'special' identity. See the Crisis Management Initiative, 'Aceh Peace Process Follow-Up Project: Final Report', 2012, http://cmi.fi/wp-content/uploads/2016/04/aceh_report5_web.pdf, accessed 9 September 2019.

However, it should not be thought from this that all of the GAM (Free Aceh Movement) were in favour of implementing Shari'a. Indeed, the GAM governor of Aceh refused to sign the 2009 Islamic Criminal Code (though this remains Aceh law); see Feener, ibid.

[78] To view the original Indonesian text, see https://advokasi.elsam.or.id/assets/2015/09/2013_KKR_QanunAceh-Nomor-17-Tahun-2013-TentangKKR.pdf, accessed 9 September 2019.

[79] Aceh Law No. 11/2006, Chapter XVII, Article 125.

[80] Ibid., Article 126.

[81] Ibid., Chapter XVIII.

[82] Ibid., Chapter XVIX.

[83] See note 77, above.

[84] Qanun 17/2013, Chapter 2, section 2(a).

[85] Ibid., Chapter 3, section 11(b), (c) and (g).

international Human Rights and to international Human Rights instruments.[86] The source materials, reference points, value positions, content, procedures and even the mandates are derived from previous TRCs, not from the historical tomes of Islamic law or the juridical opinions of Shari'a scholars.[87] The TRC's three mandates: (1) to strengthen peace though revealing the truth about past human rights violations; (2) to help achieve reconciliation between perpetrators of human rights violations, both institutions and individuals, with victims; and (3) to recommend comprehensive reparations for victims of human rights violations in compliance with ***universal standards***, indicate an outward-looking, inclusive and universal frame of reference. There is no Islamic objection, here, to sourcing the values found in International Human Rights and to utilising procedural innovations applied in non-Muslim countries. The flexibility and ambiguity implicit within Islamic justice affords the universal norm an Islamic label. As Michael Feener notes, 'the very abstraction of the concept of Shari'a serves to facilitate diverse and dynamic conversations on social issues in contexts of contested political control'.[88]

Since 2013, the Aceh TRC has continued to document cases, collect information and organise public hearings. As part of its priority plan for 2019–2021, it also envisages developing reconciliation methods with religious actors and public actors. We have already noted the text of the Qanun 17/2013 is a blank slate when it comes to Islamic content and approach, so our future analysis and understandings of the workings of the Aceh TRC will have to come through observation of Islamic *praxis*.

[86] The text references the Convention Against Torture and Other Cruel, Inhuman or Degrading Treatment or Punishment (Preamble, para 4), the International Covenant on Economic, Social and Cultural Rights (ibid., para 9) and the International Covenant on Civil and Political Rights (ibid., para 10).

[87] This points to tensions within the Aceh TRC as it appears more a human rights body (especially given the lack of active involvement, until recently, of any Muslim scholars or local 'ulema'. I am grateful to Jess Melvin—a researcher on the Aceh TRC—for this point). The purported Islamic legitimation is not, however, to Islamic *fiqh*, but to Islamic principles (*al-maqasid*), indicating a broader and more political frame of reference. For a closer comparison of these human rights treaties with Islamic law and the *fiqh*, see Mashood A. Baderin, *International Human Rights and Islamic Law* (Oxford: OUP, 2005).

[88] Feener, *Shari'a and Social Engineering: the Implementation of Islamic Law in Contemporary Aceh*, 58.

Conclusion

The Aceh TRC and the Moroccan ERC were established on the back of a change of government, concerns over historical violations of human rights, on the need for reconciliation between opposition groups and government forces and for some reparation to be given to innocent civilian victims in the conflicts. In this respect, these TRCs were little different from the experiments elsewhere in the world. Yet Morocco's ERC was unique in that the procedure took place in a non-democratic context and a country that remained governed by a monarchy and whose cultural and societal values continued to reflect Islamic traditions. Similarly, in Aceh its TRC was and is being developed explicitly in the context of the region's 'Islamic identity'. Islam, then, is seen as part of the solution rather than the problem. Through the procedural device of the TRC, Islamic law and ethics combine with universal norms and standards to produce locally acceptable and viable mechanisms to help resolve conflicts and achieve a sustainable peace. They may also be fashioned to address ecological disasters.

The essential problem in the Muslim World remains the crisis and fragmentation of authority and the post-colonial 'state'. Zoli, Bassiouni and Khan rightly acknowledge a general 'ambivalence on the part of governments about acting in an open, fair and transparent way to establish the truth and to hold accountable those who committed domestic and international human rights crimes'.[89] State governments have also dragged their heels and failed to do what is necessary to combat climate change. While strong global leadership through the OIC and further reforms in its processes and collaborations with the UN and regional bodies might pay future dividends, the Aceh experiment is instructive. It was through education, active engagement and the coming together of civil society groups that the State governments were forced to effect real change. From West Papua to Lebanon, and even perhaps to war-torn regions of Syria, a suitably designed and properly directed TRC could provide the mechanism to bring about meaningful, just and lasting change. It is said in the West that we now live in an age of populism; much the same may be said of the Muslim world. Let us hope that we can harness it to promote

[89] Zoli, Bassiouni, and Khan, 'Justice in Post-Conflict Settings: Islamic Law and Muslim Communities as Stakeholders in Transition', 38.

the processes of peace and the sustainability of our environment. Failure is too hard to contemplate.

Acknowledgements I would like to thank Centre for Asian and Pacific Law, University of Sydney intern, Gang Yang, for his research assistance in writing this chapter.

Bibliography

Abdul Salam, Al-'Izz ibn. 2000. *Qawa'id al-Ahkam*. Damascus: Dar al-Qalam.

Aceh Truth and Reconciliation Commission. 2019. 'A Diplomatic Brief', 2 April. https://kkr.acehprov.go.id/aceh-truth-and-reconciliation-commission-a-diplomatic-brief/. Accessed 9 September 2019.

Ahmady, 2013. 'Eagerly Anticipating Aceh Truth Commission'. *The Jakarta Post*, 18 December. https://www.thejakartapost.com/news/2013/12/18/eagerly-anticipating-aceh-truth-commission.html. Accessed 9 September 2019.

Al-Ahsan, Abdullah and Young, Stephen (Eds.) 2017. *Qur'anic Guidance for Good Governance*. Kuala Lumpur: Palgrave-Macmillan.

Al-Bayhaqi, Abu Bakr, Ahmad Ibn Husayn. 2003. *Al-Sunan al-Kubra*. Beirut: Dar al-Kutub al-'Ilmiyyah (DKI).

Al-Mawardi, Abu'l Hasan 'Ali Ibn Muhammad Ibn Habib al-Basri al-Baghdadi. N.d. *Kitab Al-Ahkam Al-Sultaniyyah*. Beirut: Dar al-Fikr (English trans. Asadullah Yate. 1996. *Al-Ahkam as-Sultaniyyah: The Laws of Islamic Governance – Abu'l Hasan al-Mawardi*). London: Ta-ha.

Al-Nasafi, 'Abdullah Ibn Ahmad. 2014. *Madarik Al-Tanzil wa Haqa'iq al-Ta'wil*. Beirut: Dar al-Nafaes.

Al-Nawawi, Muyhideen Abu Zakariyya. 2002. *Al-Arba'uwna Al-Nawawiyyah fi al-Hadith al-Nabawiyyah*. Beirut: Dar al-Mashari'a.

Al-Razi, Muhammad Ibn Abi Bakr ibn 'Abd al-Qadir. 2001. *Mukhtar al-Sihah*. Beirut: Dar al-Fikr.

Al-Tabarani, Abu al-Qasim, Suleyman. 1989. *Makarim al-Akhlaq*. Beirut: Dar al-Kutub 'Ilmiyyah (DKI).

Al-Tabarani, Abu al-Qasim, Suleyman. 2007. *Mu'jam al-Kabeer*. Beirut: Dar al-Kutub al-'Ilmiyyah (DKI).

Anishina Bek News. 2015. 'Call to Action Report: Truth and Reconciliation Commission Recommends Environmental Stewardship and Protection of Our Natural Resources'. http://anishinabeknews.ca/2015/12/20/truth-and-reconciliation-commission-recommends-environmental-stewardship-and-protection-of-our-natural-resources/. Accessed 2 September 2019.

Baderin, Mashood A. 2005. *International Human Rights and Islamic Law*. Oxford: OUP.

Belkziz, Najwa. 2018. 'The Politics of Memory and Transitional Justice in Morocco'. PhD Thesis, University of Melbourne, School of Historical and Philosophical Studies. https://minerva-access.unimelb.edu.au/bitstream/handle/11343/213791/Belkziz_Corrected%20thesis%2011%20June%202018.pdf?sequence=3&isAllowed=y. Accessed 1 November 2019.

Braithwaite, John, Braithwaite, Valerie, Cookson, Michael and Dunn, Leah. 2010. *Anomie and Violence: Non-truth and Reconciliation in Indonesian Peacebuilding*. Canberra: ANU Press.

Crisis Management Initiative. 2012. 'Aceh Peace Process Follow-Up Project: Final Report'. http://cmi.fi/wp-content/uploads/2016/04/aceh_report5_web.pdf. Accessed 9 September 2019.

Drexler, Elizabeth. 2008. *Aceh, Indonesia: Securing the Insecure State*. Philadelphia: University of Pennsylvania Press.

Dugard, John. 1999. 'Dealing with Crimes of a Past Regime: Is Amnesty Still an Option?' *Leiden Journal of International Law* 12(4): 1001–1015.

Farrar, S.A. 2003. 'Islamic Jurisprudence and the Role of the Accused'. *Legal Studies* 23(4): 587–604.

Farrar, Salim. 2014. 'The Organisation of Islamic Cooperation: Forever on the Periphery of Public International Law?' *Chinese Journal of International Law* 13(4): 787–817.

Feener, R. Michael. 2013. *Shari'a and Social Engineering: The Implementation of Islamic Law in Contemporary Aceh*. Oxford: OUP.

Freeman, Mark. 2005. 'Transitional Justice in Morocco: A Progress Report'. ICTJ. https://www.ictj.org/sites/default/files/ICTJ-Morocco-Progress-Report-2005-English.pdf. Accessed 27 June 2019.

Hascall, Susan C. 2011. 'Restorative Justice in Islam: Should Qisas Be Considered a Form of Restorative Justice?' *Berkeley Journal of Middle Eastern & Islamic Law* 4: 35.

Hazan, Pierre. 2008. 'The Nature of Sanctions: The Case of Morocco's Equity and Reconciliation Commission'. *International Review of the Red Cross* 90(870): 399–407.

Hirst, Megan. 2009. 'An Unfinished Truth: An Analysis of the Commission of Truth and Friendship's Final Report on the 1999 Atrocities in East Timor'. https://www.ictj.org/sites/default/files/ICTJ-TimorLeste-Unfinished-Truth-2009-English.pdf. Accessed 3 November 2019.

Hodal, Kate. 2019. 'Arab World Turns Its Back on Religion—And Its Ire on the US'. *The Guardian*, June 24.

Ibn Abi Jamrat al-Azdi, 'Abdullah Ibn Sa'd. 1993. *Mukhtasar Sahih al-Bukhari*. Beirut: Dar Ibn al-Hazm.

Ibn Hibban, al-Busti, Muhammad. 2012. *Al-Saheeh Ibn Hibban*. Beirut: Dar Ibn Hazm.
ICTJ. 2016. 'Morocco Still a Model for Justice in MENA, but Questions Remain', 2 August. https://www.ictj.org/news/morocco-still-model-justice-mena-questions-remain. Accessed 1 November 2019.
Inwood, Joshua. 2017. 'Dealing with Hate: Can America's Truth and Reconciliation Commissions Help?' *The Conversation*, 1 March.
Jurist. 2006. 'Indonesia Court Overturns Law Creating Truth and Reconciliation Commission'. 8 December. https://www.jurist.org/news/2006/12/indonesia-court-overturns-law-creating/. Accessed 9 September 2019.
Lerner, S. 2006. 'Interview with Saloman Lerner.' *International Review of the Red Cross* 88(862): 225–233.
Liaw, Yock Fang. 1976. *Undang-undang Melaka: The Laws of Melaka*. The Hague: M. Nijhoff.
Liu, Zhongmin and Fan, Peng. 2018. 'Islamic Factors in Inter-State Cooperation of the OIC Members'. *Asian Journal of Middle Eastern and Islamic Studies* 12:1, 1–15. https://doi.org/10.1080/25765949.2018.1439616.
Masud, Muhammad Khalid. 2001. 'The Doctrine of Siyasa in Islamic Law'. *Recht van de Islam* 18: 1–29.
Muslim, Ibn Hajjaj. 2006. *Sahih Muslim*. Dar-Toyyibah.
OIC, SESRIC. 2019. 'Achieving Peace and Security in a World of Turmoil: An Arduous Challenge for the OIC'. www.sesric.org/files/article/659.pdf. Accessed 8 April 2019.
Pal, Leslie A. and Tok, M. Evren (Eds.). 2018. *Global Governance and Muslim Organisations*. Cham: Palgrave-Macmillan.
Pew Research Center. 'Religion and Public Life'. http://www.pewforum.org. Accessed 21 July 2019.
Pityana, N. Barney. 2018. 'The Truth and Reconciliation Commission in South Africa: Perspectives and Prospects'. *Journal of Global Ethics* 14(2): 194–207.
Quraishi-Landes, Asifa. 2008. 'Who Says Shari'a Demands Stoning of Women? A Description of Islamic Law and Constitutionalism'. *Berkeley Journal of Middle Eastern & Islamic Law* 1: 163–177.
Robinson, Francis. 2009. 'Crisis of Authority, Crisis of Islam?' *JRAS*, Series 3, 19(3): 339–354.
Scholtz, Werner and Ferreira, Gerrit. 2015. 'Climate Change Negotiations and Transitional Justice: The Advent of a Carbon Truth and Reconciliation Commission?' *Comparative and International Law Journal of South Africa* 48 (1): 42–58.
Sooka, Y. 2006. 'Dealing with the Past and Transitional Justice: Building Peace Through Accountability'. *International Review of the Red Cross* 88(862): 311–325.

Stan, L. and N. Nedelsky 2013. *Encyclopedia of Transnational Justice*. Cambridge: CUP.

Tolba, Mostafa and Saab, Najib (Eds.). 2008. 'Arab Environment, Future Challenges. Report of the Arab Forum for Environment and Development (AFED)'. http://www.fao.org/fileadmin/user_upload/rome2007/docs/Arab_Environment_Future_Challenges.pdf. Accessed 2 September 2019.

Yeager, J. David. 2004. 'The Human Rights Chamber for Bosnia and Herzegovina—A Case Study in Transitional Justice'. *International Legal Perspectives* 14(1): 44.

Yusuf, Hakeem O. 2010. *Transitional Justice, Accountability and the Rule of Law*. Abingdon and New York: Routledge, Glasshouse Books.

Zoli, Corri, Bassiouni, M. Cherif and Khan, Hamid. 2017. 'Justice in Post-Conflict Settings: Islamic Law and Muslim Communities as Stakeholders in Transition'. *Utrecht Journal of International and European Law* 33(85): 38–61.

Dr. Salim Farrar is Associate Professor and Associate Director of the Centre for Asian and Pacific Law within the Sydney Law School at the University of Sydney. He teaches and researches in Islamic Law, Islamic Ethics, Malaysian Law, International Law, Criminal Justice and Comparative Law. With Ghena Krayem, he is author of *Accommodating Muslims Under Common Law: A Comparative Analysis* (2016, 2018), Routledge Press.

CHAPTER 8

Innate Wisdom—Peace on/in/with Earth

Norman Habel

Introduction

In this chapter, I will return to a largely forgotten Biblical tradition, namely the Wisdom School belief in Wisdom as an innate force of nature that is an integral part of the cosmic design involving peace in/with Earth. I shall begin by identifying several Biblical traditions which did not promote peace in/with Earth, traditions which we have long recognised as normative while ignoring the Wisdom tradition of innate eco-Wisdom. Subsequently, I explore key texts where the various dimensions of innate Wisdom are embedded and articulated, texts such as Job 28, Proverbs 8 and Job 38–39. I also discuss how innate Wisdom is related to ecology, climate trauma and peace consciousness and how this awareness, following the journey of Job, enables us to discern a God of cosmic *shalom*.

N. Habel (✉)
Flinders University, Bedford Park, SA, Australia

J. Camilleri and D. Guess (eds.), *Towards a Just and Ecologically Sustainable Peace*, https://doi.org/10.1007/978-981-15-5021-8_8

Past Relationships with Earth/Nature

(a) *Imago Dei*. The *imago Dei* tradition, which many of us inherited, was based on the ancient text of Genesis 1.26–28. Humans were formed by a heavenly Deity to 'rule' over other living creatures and 'subdue' the Earth in accord with the image of the celestial Creator of the universe. The terms 'rule' (*rada)* and 'subdue' (*kabash*) are forceful expressions of domination and ought not to be harmonised in terms of later theology. Humans were identified as superior beings who had the divine right to tame the planet and control all its inhabitants. There was no image of peace with Earth or in Earth. Earth/nature was the servant of humanity.[1]

To interpret this text as a message to be 'good stewards' is to ignore the blatant empire orientation of the terminology: 'rule' means 'dominate' and 'subdue' often means 'rape'. A profound critical response to this tradition, reiterated in Psalm 8, is found in Job 7. For Job, the *imago Dei* tradition is a false dream. The God who 'visits' Job belittles rather than exalts, humiliates rather than glorifies, induces trauma rather than potential.[2]

(b) Mother Earth. Alongside the religious tradition of the *imago Dei* lies the folk tradition of Mother Earth, a tradition that some have connected with the Bible (Psalm 139.13–15). Mother Earth was not only our distant ancestor in legendary or evolutionary terms, but also the generous donor of all the necessities of life, the gifts of life and lifelines that surrounded us as children of Earth. The emphasis was on the richness of our heritage rather than peace-making relationships. We were privileged, it seems, to take advantage of our Mother's generosity.

In many Christian traditions, however, it was considered heretical or at least unwise to designate Earth as our Mother. Earth was a creation of God, a natural resource that we, as the superior beings on Earth, had a right to explore, exploit and consume. With the so-called progress of humanity in agricultural, industrial and economic terms, the diverse domains of nature became legitimate opportunities to take advantage of the rich resources embedded

[1] Norman Habel, *An Inconvenient Text* (Adelaide: ATF Press, 2009), 1–10.

[2] Norman Habel, *Finding Wisdom in Nature: An Ecological Reading of the Book of Job* (Sheffield: Sheffield Phoenix Press, 2014), Chapter 8.

in the planet. Earth was an object to be exploited, not a Mother with whom to negotiate peace and within whom peace might be discerned.

(c) The Promised Land. Another powerful tradition we have inherited is the Promised Land tradition. Most of our ancestors assumed that just as Moses and Joshua are portrayed as believing that the land of Canaan is theirs to conquer by divine right, they could follow suit in colonial lands. As a result, they, like the Israelites, were forbidden to make a covenant/peace treaty with the peoples of the land or the land itself. In the case of Jericho, for example, 'the city and all that is within it shall be devoted to YHWH for destruction' (Jos. 6.17).

In the colonial era, many countries, including Australia, were viewed as the Promised Land which could be invaded and possessed without any covenant or treaty, totally ignoring the precedent of Abraham and the land of Canaan.

(d) The Abraham Tradition. If we return to the original Abraham tradition, especially in Genesis 14, we discover a radically different understanding of a settler's relationship to the land and the peoples of the land. A recent volume of mine entitled, *Acknowledgement of the Land and faith of Aboriginal Custodians After Following the Abraham Trail*, retrieves the ancient tradition of Abraham in Canaan. The essentials of that tradition are:

- Canaan is a host country that welcomes Abraham and his community;
- Abraham makes allies with indigenous Canaanite communities and lives at peace with them and the land;
- Abraham defends the land rights of the Canaanites by routing Eastern invaders;
- Abraham receives and accepts the blessing of the deity El Elyon, the Creator Spirit of Canaan;
- Abraham swears by El Elyon that he will take none of the possessions of the Canaanites retrieved in battle; and
- Abraham makes a formal treaty/covenant with one Canaanite community, swearing by El Olam, a treaty embracing loyalty to the people and to the land (Gen. 21).

To a large extent, we have ignored the preceding Abraham tradition and appropriated the Old Testament traditions about Earth as the land we are permitted to rule and whose resources we are free to appropriate regardless of the consequences. The Wisdom tradition of the ancient Near East, however, reflects a radically different orientation that provides a valuable alternative in the current ecological context.

The Wisdom Traditions

When we mention Wisdom in general, we tend to understand Wisdom in one of the following ways:

- Inherent Divine Wisdom—the inherent capacity of God to comprehend and sustain the numerous domains in the design of the cosmos, everything from ants to asteroids, black holes to wise humans (Job 11.6; 12.13).
- Accumulated Human Wisdom—the profound knowledge accumulated by the elders/fathers of the Wisdom School of Thought. This profound knowledge was expected to enable the wise to live a successful and righteous life. This is the Wisdom we expect to find in the Book of Proverbs.
- Acquired Wisdom—the specific insight acquired by close observation of natural phenomena or human behaviour, an insight that is tantamount to a 'scientific' discernment of reality, a procedure of the ancient Wisdom School (e.g. Prov. 6.6).
- Primordial Wisdom—the primordial blueprint that determines the design of the cosmos and its component parts, the Wisdom 'acquired' by God prior to the creation of the cosmos (see Prov. 8.22).

There is, however, another understanding of Wisdom that is related to the above but is especially relevant in the current context, namely Innate Wisdom, the innate cosmic driving force in nature.[3]

[3] Norman C. Habel, *Discerning Wisdom in God's Creation: Following the Way of Ancient Scientists* (Melbourne: Morning Star, 2015), Chapters 1 and 2.

The Wisdom Manifesto

A key text for appreciating innate Wisdom is Job 28, a text I have designated The Wisdom Manifesto. The significance of this Wisdom text was recognised in 2002 when an International Conference was convened in Amsterdam with the focus explicitly on Job 28.[4] Further research on this text, in the context of the *Book of Job* and the Wisdom School tradition, has enabled a fuller appreciation of the 'innate' dimension of Wisdom in the scientific thinking of the Wisdom School.[5]

According to this classic Wisdom text, everything has its 'place' (*maqom*), its designated locus in the design of Earth and indeed of the entire cosmos. There is a designated 'place' on Earth for metals, precious stones, volcanoes, food supplies and so on. Everything has its 'place' thereby establishing order and just peace. The key question posed by the narrator of this Manifesto is 'Where can Wisdom be found?' Where is the 'place' of Wisdom, the most precious element in the design of the cosmos? (Job 28.12).

The narrator then introduces us to God, portrayed as a primordial Wisdom scientist. In accord with the practice of ancient Wisdom scientists, this God 'discerns' (*bin*) the 'way' (*derek*) and comes to know the 'place' (*maqom*) of Wisdom. At this crucial point, we are introduced to two other technical Wisdom terms, namely 'discern' (*bin*) and 'way' (*derek*). While 'place' (*maqom*) refers to the designated locus of an entity in the design of the cosmos, 'way' refers to its designated innate character, its innate Wisdom, and 'discern' refers to the 'scientific' process of identifying the truth by 'observing' and researching an entity.

How did God, the Wisdom scientist, achieve this goal?

> *He looked to the ends of the Earth*
> *and 'observed' everything under the skies.* (28.24)

To 'observe' (*bin*) is another technical Wisdom School term for researching an entity to determine its true nature. As the ancient proverb advises, 'observe the way (*derek*) of the ant and gain Wisdom' (Prov. 6.6).

[4] Ellen van Wolde (ed.), *Job 28: Cognition in Context* (Biblical Interpretation Series, vol. 64) (Leiden/Boston: Brill, 2003).

[5] Habel, *Discerning Wisdom in God's Creation.*

And where does God, the Wisdom scientist, discern Wisdom? In his head? No! In Nature! The key insight into this mystery is revealed at the end of the Wisdom Manifesto:

> *When he fixed the 'force' of the wind*
> *and meted out the waters by 'measure',*
> *when he made a 'rule/law' for the rain*
> *and a 'way' for the thunderstorm,*
> *then he 'observed' her and appraised her,*
> *established her and probed her.* (28.25–27)

We may speak today about the laws of nature as given phenomena, but for the ancient Wisdom scientist the force of the wind, the laws for the rain, and the 'way' of the thunderstorm are what God 'observed' as the innate Wisdom that governs the function of these domains of nature. God, however, not only 'observes' this innate dimension but also double checks and establishes innate Wisdom as a permanent factor in the design of planet Earth. The very act of establishing highlights the role of innate Wisdom as the underlying order or innate peace that governs the habitats of Earth.

Innate Wisdom is an integral dimension of the design of the cosmos—its character, its dimensions, its integral operations, its natural forces and its ecology on/in Earth. Innate Wisdom may be designated eco-Wisdom, an innate integrating force that works for 'peace' and authenticity in nature.

Innate Life Force

If we turn to Proverbs, we discover that the 'way' (*derek*) is prevalent in living things as well as in the inanimate forces of nature. The Wisdom School has taught me that there is an innate life-force programmed into all living creatures: a life-force called Wisdom, a capacity for living beings to be true to their nature.

The 'way' of things in the Wisdom Literature, especially in the Book of Proverbs, refers to the characteristic behaviour of living beings, including humans. The novice scientist is encouraged by wise mentors to 'watch' the ant, for example, and 'observe' its 'way', its distinctive code of behaviour, its innate Wisdom, and so gain some personal Wisdom:

Go to the ant, you lazybones!
Observe its 'way' and be wise.
Without having any chief
Or officer or ruler,
It prepares food in summer
And gathers its sustenance in harvest. (Prov. 6.6–8)

The Wisdom in an ant colony, according to these ancient scientists, is an inner capacity to function as a corporate body without any hierarchy, a mystery that modern scientists still find fascinating.

The author of Proverbs 30 explores a range of innate mysteries both in society and nature. He claims that he lacks the intellectual capacity to understand the mysteries in nature that surround him. There are four things (innate 'ways') in nature, he declares, that are too wonderful to comprehend:

The way of an eagle in the sky,
The way of a snake on a rock,
The way of a ship on high seas,
The way of a man with a girl. (Prov. 30.18–19)

In relation to the process/theory of evolution, modern scientists may well explore the role of DNA, genetics and other factors to determine the discrete behaviour of living creatures. The Wisdom School would argue that each species has its own unique 'way', its innate Wisdom. Today, I believe we can not only learn from scientists about the driving factors in each living being, but also from the Wisdom School who 'discerns' innate Wisdom by observation and not by theory or by revelation. Innate Wisdom is that distinctive life force in each living entity that enables it to be true to its character, its place in nature and its relationship with all other domains of nature.

Celia Deane-Drummond identifies this dimension of innate Wisdom as 'natural wisdom'. In her discussion of natural Wisdom in evolutionary biology, she returns to the example of the ant and asks whether we are not better informed than the 'way' of the ant. She distinguishes between natural Wisdom and cosmic design, which she claims implies a deliberate designer. She concludes:

> Natural Wisdom by contrast puts more emphasis on the process of the evolutionary searchlight scanning forms of biological diversity. Convergence across all traits simultaneously has not been found, for the way/wisdom of each species is unique. (2005, 70)

The Wisdom School would agree that each species has its own innate Wisdom, its 'way' to be true to its species. It would seem, then, that we can not only learn from the School of Wisdom but also from the evolutionary scientist to appreciate and sustain the unique Wisdom of each species in Planet Earth.

Primordial Wisdom

If we listen again to the famous voice emanating from the whirlwind (Job 38) to extend Wisdom therapy to Job, we discover that God, the Wisdom scientist who is introduced in the Wisdom Manifesto, now takes Job back to the primordial to explore the original design of the various domains of the cosmos.

To appreciate the relationship between this primordial design and Wisdom, it is helpful to reconsider Wisdom's bold proclamation in Proverbs 8. She begins by declaring,

> *YHWH acquired (qana) me first,*
> *His way (derek) before his works.*
> *From of old, from antiquity I was established,*
> *from the first, from the beginnings of Earth.* (Prov. 8.22–23)

While many translations render *qana* as 'create', its basic meaning throughout Proverbs is to 'acquire' (e.g. 4.3, 7). God 'acquires' Wisdom, the primal blueprint for the design of the cosmos. The primordial character of Wisdom is announced at several key points in her address in Proverbs 8. She not only existed before the beginnings of Earth (8.23), but she emerged or was 'brought forth' when there were no depths or mountains (8.24–25). Wisdom was 'there' in the primordial before there were fertile fields or celestial realms (8.26–28). In short, Wisdom is identified as the primordial design that existed from the beginning.

When God challenges Job to explain the mysteries of the 'design' of the cosmos, the primordial blueprint of Wisdom hovers in the background. And God, the primordial scientist who acquires and discerns Wisdom in the design of the cosmos, changes roles when he addresses Job and is tantamount to a therapist who takes Job on a journey to explore the Wisdom dimensions of the cosmos as a pathway to healing and to a new understanding of the nature of God.

To undertake this journey and undergo divine therapy, Job is forced to face the fact that the 'design' of the cosmos has been darkened, presumably by Job's traumatic screams against God. If he is to undertake this cosmic journey, Job needs to 'gird his loins' and 'act like a man' or 'a hero' as some translations read. Job dispenses with his dust and ashes and joins God in a primordial tour of the cosmos.

Innate Wisdom and Ecology

In his discourse from the whirlwind, the Wisdom scientist discloses an integral relationship between innate Wisdom and ecology as well as the integrating role of Wisdom to establish stable 'order' or, in terms of our current context, just peace.

The Wisdom therapist's opening challenge is for Job to no longer cloud the design of the cosmos with ignorance, but to explore and illuminate the mysteries of the universe, integral to which is the role of innate Wisdom in the primordial domains of nature.

Light and darkness, for example, are identified here as discrete realms that have hidden abodes like the *locus* of Wisdom herself (Job 38.20–21). Each realm has its 'place' and 'way', evidence of innate Wisdom in the primordial blueprint of the cosmos. To discern their locations, the Wisdom scientist would also need to learn cosmic geography. Here, God seems to be challenging Job to play the role of a Wisdom scientist alongside the Creator and 'observe' the locus of these domains in the primordial design of the cosmos.

Included also is a series of mysteries associated with the skies and the weather (38, 22–30). Job is challenged to 'see/observe', as a true Wisdom scientist, the 'places' of snow and hail in the skies. Snow and hail are not only viewed as forces of nature, but also as arsenals in times of adversity (Deut. 28.12; Jer. 10.13).

The thunderstorm (in vv. 26–27) may at first appear to be an uncontrolled meteorological happening. The use of the Wisdom term 'way' in relation to both lightning and thunder, however, emphasises that both of these weather phenomena are governed by innate Wisdom, innate forces of nature, just as rain is controlled by an innate 'rule' (28.26).

The significance of the rainstorm here is the beneficial role it plays across Earth, even in uninhabited wastelands. Thunderstorms are the source of a divine blessing for fields and forests, a blessing Job did not appreciate in the misery of his trauma.

God also challenges Job to explain how the constellations of the heavens/skies function. Pleiades, Orion and Mazzaroth are celestial wonders that function in the space of the skies above. Then comes a classic ecological question:

> *Do you know the 'laws' of space?*
> *Can you establish their 'order' on Earth?* (38.33)

How the laws of space, such as gravity or antigravity, influence the order of things on Earth is a profound ecological question that challenges contemporary ecologists. As we noted in the Wisdom Manifesto, the 'laws' of nature, whether in the sky or on Earth, are the innate Wisdom that governs/guides the operation of a given phenomenon, whether it be a constellation or a lightning flash. Innate Wisdom not only activates the forces of nature but integrates and sustains their 'place' in the design of the cosmos. The innate Wisdom may well be designated eco-Wisdom, a pivotal insight for interpreting texts like the *Book of Job*, or contemporary ecology.

Another question posed by the divine therapist is found in the famous query:

> *Who put Wisdom in the cloud canopy*
> *and gave Discernment to my pavilion?* (v. 36)

The Wisdom located in the cloud canopy is to be distinguished from the Wisdom skills of Wisdom students who attempt to 'count the clouds'. The initial response of many readers is to assume that God placed Wisdom in the clouds. But, as the Wisdom science of Job 28 has revealed, Wisdom is an innate force of nature, including the clouds. Innate Wisdom is tantamount to the ecology that guides the clouds, and all domains of nature, to operate as designed within their 'place', their eco-locus in the cosmos. Innate Wisdom is a traditional understanding of what we today could call eco-Wisdom.

Eco-Wisdom and Eco-Peace

The innate Wisdom or eco-Wisdom integral to every domain of the cosmos has not only a specific 'locus' and a discrete character or 'way', but also a profound function to maintain 'peace', an operation that is true to its character.

The portrayal of the Sea as a baby (in Job. 38.8–11), often interpreted as but an expression of divine irony or humour, is a colourful case in point. Listen again to the voice:

Who hedged in the Sea with doors
when he gushed forth from the womb,
when I wrapped him in robes of cloud
and swaddled him in dense cloud,
when I prescribed my limit for him
by fixing bars and doors
and saying, "Thus far you may come and no farther!
Here your proud waves break!" (vv. 8–11)

In the ancient Biblical world, Sea (*yam*) was often identified with chaos, the turbulent primordial waters. But in this portrayal, Sea is transformed from chaos to order, from being a violent force to being a restrained child, from being at odds with God to being at peace with the world. In the eco-Wisdom design of the cosmos, the Sea has 'limits' that govern the locus and character of the Sea, God's 'peaceful' cosmic child.

Job is also taken into the world of the wild where all the untamed beings know their place and live at peace within their domain. Lions know how to take care of their whelps, the ibex have an innate Wisdom that enables them to know when and where to deliver their young and the wild ass has a designated home in the wilderness (Job 38.39–40.8).

Especially pointed is the portrayal of the wild ox (39.9–12). Again this passage has a touch of humour that sometimes leads interpreters to dismiss its significance. The passage begins:

Is the wild ox willing to 'serve' you?
Will he spend the night beside your crib?
Can you hold the wild ox in the furrow with ropes?
Will he harrow the valleys behind you?
Can you rely on his great strength
and leave your toil to him?

> *Can you trust him to harvest your grain*
> *and gather it in from the threshing floor?* (39.9–12)

The use of the term 'serve' immediately evokes a connection with the mandate to 'rule' found in the mandate for humans to dominate over living creatures and Earth in Gen. 1.26–28, a mandate discussed above.

In this passage (Job 39.9–12), God asks whether the wild ox is willing to be Job's servant, to treat Job as master, to consider Job his ruler and to reap his harvests for him. The answer is obviously, 'No!' It is not the norm for wild animals to be subjugated by humans; the mandate to dominate is here exposed as contrary to the innate impulses of natural order.

The ecology of the wild is for creatures to be free from human domination, to live at peace together within the Wisdom laws that govern their domains of the cosmos. In this passage from the *Book of Job*, we detect how the Wisdom School of thought was willing to explore a world where domination and servitude were replaced by a world of interrelated domains at peace with each other and governed by discrete innate Wisdom impulses.

Another relevant passage is the portrayal of the eagle who has the innate Wisdom to soar on high, make a nest on a cliff, dwell on a rock and yet have the capacity to discern food on the ground far below. The eagle is depicted as 'wise', being at peace with the mountains (39.27–30).

Throughout the divine discourse from the whirlwind, each of the many domains of the cosmos, each governed by innate Wisdom that determines its character, exists side by side as part of the overall design of the universe. There is no primordial conflict between these domains, no hierarchy, no inner tensions. A Wisdom-inspired cosmos is a cosmos at peace.

Wisdom Consciousness and Peace Consciousness

Pivotal to Job's therapy via the divine therapist is the acute cosmic consciousness and Wisdom consciousness aroused in Job. Ultimately, Job responds to the questions from the therapist: Who put Wisdom in the clouds? Who put discernment in the eagle? Job becomes acutely aware of the Wisdom factor inherent in all domains of nature, whether animate or inanimate.

I must now confess that the experience of Job as portrayed by the narrator, a mentor of the Wisdom School, has stirred within me a similar Wisdom consciousness. I now reflect on the natural creatures around me and wonder, for example, about the innate Wisdom of the blue wren that enables it to be true to its nature as a blue wren, regardless of its evolutionary history. Or to view a lightning flash and wonder about its 'way', its innate Wisdom, the force of nature that governs its modus operandi.

Wisdom consciousness makes me reflect further on the nature of Earth as a living organism, a planet whose many domains are governed, guided and integrated by innate Wisdom. Ultimately, that innate Wisdom is part of my being, the spirit that leads me to be true to myself.

If we view again the final chapters of the *Book of Job*, we recognise the role of the divine therapist and the response of Job. Via the therapeutic voice from the whirlwind, Job is guided to gain both a cosmic consciousness and an innate Wisdom consciousness that overcome his prior trauma about his God and his world. He responds in a way that I would call a 'peace consciousness' (42.1–6). Job's 'peace consciousness' consists in being:

- At peace with the cosmos, its primordial design and its domains which he has explored with the guidance of God, the Wisdom scientist cum Wisdom therapist;
- At peace with the innate Wisdom that governs the realms and creatures of the cosmos and informs him of their true character;
- At peace with the God with whom he was in violent conflict and whom he now 'sees' within the cosmos;
- At peace with his community who berated him and for whom he intercedes; and
- At peace with himself by dispensing with the 'dust and ashes' of his trauma and getting on with his life.

I believe that a profound understanding of innate Wisdom leads us to discern a Peace consciousness, a cosmic *shalom*, a deep awareness of our pivotal relationships with the cosmos, the planet, the natural world, the community and our own psyche.

Climate Consciousness

Job reaches a level of peace consciousness that represents an ideal found in the Wisdom traditions of Israel. But, given the climate crisis, can we readily speak of being at peace with any of the domains reflected in Job's response to the God who took him on a whirlwind tour of the cosmos? Are we at peace with the innate Wisdom that governs the various domains of nature? Or are we experiencing climate trauma?

An eco-psychologist named Zhiwa Woodbury explores the field of climate trauma. He writes:

> The disarmingly innocuous term 'climate change' expresses a psychosocial defence mechanism that prompts us to recoil when we consider the implications of climate science. When viewed honestly through the lens of traumatology, this deepening existential crisis prompts an entirely new, unprecedented and higher order category of trauma: Climate Trauma. What is unique about this category of trauma is that it is an ever-present threat to the biosphere, one that calls into question our shared identity: What does it mean to be human in the Anthropocene?[6]

According to scholars in this field, we are in trauma denial when we simply use the term climate change. We need to come to terms with the existential reality that we are experiencing climate trauma. Where do we start to face the crisis—the crises of our planet, our own consciousness and our relationship with Earth?

Deep Wisdom and Just Peace

Recent efforts to link peace and justice are not new, as Joseph Camilleri argues in his chapter in this volume. The background to the discussion of just peace is frequently the issue of whether just war can be justified. My exploration of the role of Wisdom, however, would suggest that when relating peace to planet Earth, the insights of the Wisdom School offer an alternative understanding of justice, namely an integrated harmonious order, better known as *shalom*, that is effected by the role of deep Wisdom. This just peace dimension is governed by Wisdom in at least three deep 'ways'; the cosmos, natural forces and living creatures.

(a) Cosmic Wisdom is the primordial blueprint that governs the design of the cosmos, the primordial order that integrates the diverse realms of the universe, the deep peace that creates a cosmos filled

[6] Zhiwa Woodbury, 'Climate Trauma: Toward a New Taxonomy of Trauma', *Ecopsychology* 11, 31 January 2019, 1, https://doi.org/10.1089/eco.2018.0021.

with wonder, mystery and harmonious relationships.[7] It is not insignificant that after her address in Proverbs 8, she declares that she was beside God like a 'little child' and gave him great delight, rejoicing in his inhabited world (8.30–31). Wisdom is personified as being at peace with God and with the cosmos in the primordial, a truth Job also discovers.

(b) The Wisdom discerned by God in Job 28 focuses on Wisdom as a force innate in the laws of nature. God, the Creator Spirit in the Wisdom Literature, is not only the Primordial Sage but also the Primordial Scientist who investigates and establishes the role of Wisdom in the forces of nature.

In the Wisdom School, order or just peace in the domains of nature is determined by Wisdom, an inner code or intelligent force that the 'laws' of nature obey. When we are dealing with laws such as gravity or antigravity, the Wisdom School discerns an innate Wisdom that establishes a harmonious order in nature.[8]

(c) The scientists of the Wisdom School are not only expected, like Job on his cosmic tour, to discern the presence of innate Wisdom operating in the laws of nature, but also to 'observe' the living phenomena around them to discern the 'way', that is innate Wisdom, functioning as a life force.

Fundamental to the mystery of Wisdom in nature is that every living phenomenon has an innate code (*derek*) that governs its characteristic behaviour as an integral part of a harmonious ecosystem. Even in the wild, each living creature is guided by Wisdom to maintain its place and establish a peaceful order.[9]

[7] Habel, *Discerning Wisdom in God's Creation*, Chapter 4.

[8] Job 28.17; Habel, *Discerning Wisdom in God's Creation*, Chapter 3.

[9] Habel, *Discerning Wisdom in God's Creation*, Chapter 2.

A God of Justice or Cosmic Shalom?

Recent research into the concept of divine justice articulated in the *Book of Job* has associated it with a trauma hermeneutic that highlights the portrayal of the deity that Job claims to have 'punished' him unjustly and who reveals Wisdom as a governing/healing presence in the cosmos.[10]

Job's friends claim to know the truth about a God of retributive justice as espoused in Deuteronomistic theology, a jealous God of reward and retribution. They see Job's condition as clear evidence of divine punishment for crimes that he must have committed. One friend even claims that Job, in spite of his devastating trauma, is suffering less than he deserves from their God of justice.

Job, who is tempted to commit suicide, knows the depth of despair as he faces meaningless existence because his 'way' in life has been hidden by God (3.23). He experiences this God of justice as a 'watcher of humans', an 'eternal eye' waiting to discover what humans have done wrong so that he can crush them. Job, however, is innocent of any sin and cannot tolerate a cruel God of reward and retribution. Ultimately, then, he is ready to 'scream bloody murder', with the hope that Earth, his mother and friend, will hear his cry, as she did the cry of Abel, and have Job's case heard in a celestial court.

What the narrator achieves, via this portrayal of Job's trauma, is to expose the folly of viewing God as a celestial power that is 'just' as he rewards and punishes humans on Earth. The narrator then reveals an alternative portrayal of God who is a God of Wisdom, a revelation introduced at the pivotal point in the *Book of Job*, namely the Wisdom Manifesto of Job 28. Instead of a deity who searches the Earth to find guilty mortals, the Wisdom School affirms a God who searches all of Earth to discern Wisdom. The Wisdom God of the narrator is radically different from the righteous God of the friends and the terrifying God Job has experienced.

The Wisdom God then takes Job on a cosmic journey to discern the role of innate Wisdom operating as the source of integrated relationships and cosmic order. This is the God of *shalom*, of positive peace relationships, not the jealous God of Moses or the warrior God of Joshua. Innate

[10] David Janzen, *The Violent Gift: Trauma's Subversion of the Deuteronomistic History's Narrative* (New York: T & T Clark, 2012); Andre Lacocque, 'The Deconstruction of Job's Fundamentalism', *JBL*, 2007, 126, 83–97.

Wisdom reveals not only an eco-Wisdom portrayal of the cosmos, but also an understanding of God as a Peace Presence permeating the cosmos, nature and life.

If we return to Job's response in Job 42, he claims that in spite of all he has heard about God from his friends and community, he now claims that he 'sees' God. How? Not in the acts of reward and punishment by a righteous God, but in the presence of innate Wisdom that permeates, orders and animates the cosmos, in the God of cosmic *shalom*.

Conclusion

Innate Wisdom is one of the understandings of Wisdom in the biblical world that has been ignored. Innate Wisdom is that vibrant force in all domains of life and nature that enables them to be true to their character and place in the cosmic order. A close examination of the key texts in which this concept is embedded reveals that within each of the domains of Earth there is an innate impulse that not only governs these domains but is equivalent to eco-Wisdom, an impulse that integrates the various inner forces or laws of nature to establish order, or cosmic *shalom*.

In response, we humans, as Earth beings, ought to relate to Earth in a personal way so that we promote peace with Earth, and in Earth, by cooperating with eco-Wisdom and gaining a genuine Wisdom consciousness which also embraces a deeper Peace consciousness, a cosmic *shalom*. We are invited by the Wisdom School to gain the capacity to 'see' God, as Job did, in the presence of Wisdom in, with and under all our surroundings.

Bibliography

Buxton, Graham & Habel, Norman (2016). *The Nature of Things: Rediscovering the Spiritual in God's Creation* (Eugene, OR: Pickwick).

Deane-Drummond, Celia (2006). *Wisdom and Wonder: Conversations in Science, Spirituality and Theology* (Philadelphia: Templeton Foundation Press).

Habel, Norman (2001). 'Is the Wild Ox Willing to Serve You? Challenging the Mandate to Dominate'. In *The Earth Bible: Volume 3* (Sheffield: Sheffield Academic Press), 179–89.

Habel, Norman (2009). *An Inconvenient Text* (Adelaide: ATF Press).

Habel, Norman (2014). *Finding Wisdom in Nature: An Ecological Reading of the Book of Job* (Sheffield: Sheffield Phoenix Press).

Habel, Norman C. (2015). *Discerning Wisdom in God's Creation: Following the Way of Ancient Scientists* (Melbourne: Morning Star Publishing).

Habel, Norman C. (2017). *The Earth-Care Charter and 95 Eco-Theses* (Adelaide: Lutheran Education, Australia).
Habel, Norman (2018). *Acknowledgement of the Land and Faith of Aboriginal Custodians After Following the Abraham Trail* (Melbourne: Morning Star).
Janzen, David (2012). *The Violent Gift: Trauma's Subversion of the Deuteronomistic History's Narrative* (New York: T & T Clark).
Lacocque, Andre (2007). 'The Deconstruction of Job's Fundamentalism'. *Journal of Biblical Literature* 126, 83–97.
van Wolde, Ellen (ed.) (2003). *Job 28: Cognition in Context*. Biblical Interpretation Series, vol. 64 (Leiden/Boston: Brill).
Woodbury, Zhiwa (2019). *Climate Trauma: Toward a New Taxonomy of Trauma* (Published on line 31 January 2019, https://doi.org/10.1089/eco.2018.0021).

Norman Habel is a Professorial Fellow at Flinders University. He is a biblical scholar who has specialised in the *Book of Job*, ecology and the Bible, and Wisdom Literature. His initiatives include *The Earth Bible Series*, *The Season of Creation* and *Ecological Hermeneutics*. His recent works include *Discerning Wisdom in God's Creation* and *The Earth Care Charter.* He has also been involved with building positive relationships with the Aboriginal Peoples of Australia. He worked with the Rainbow Spirit elders in Queensland to produce *Rainbow Spirit Theology*. He is the author of *Reconciliation, Searching for Australia's Soul.* A recent work entitled *Acknowledgement of the Land and Faith of the Indigenous Custodians After Following the Abraham Trail* is a challenge for the churches of Australia and was the background for his presentation at the Common Dreams Conference in July 2019. His most recent work is *Making Peace with the Land of Australia*, the Charles Strong Trust Lecture delivered in November 2019 in Newcastle.

CHAPTER 9

Pope Francis's Moral Compass for Climate Change and Global Justice

Bruce Duncan

Pope Francis has emerged Mandela-like as one of the most prominent advocates for social justice and thorough-going reform of international economic systems. In his strong critiques of the astonishing maldistribution of wealth and power, he has endeavoured to help mobilise world opinion for social reform and to address the unprecedented threat from climate change. He has invited cooperation in this response with other churches, religious traditions and all socially concerned people.

In his message for the World Day of Peace for 2019, Francis urged all to support even 'radical change' for the common good, especially 'all politicians, whatever their culture or religion ... to work together for the good of the human family and to practise those human virtues that sustain all sound political activity: justice, equality, mutual respect, sincerity, honesty, fidelity'.[1]

[1] Pope Francis, *2019 World Day of Peace*, 8 December 2018, http://w2.vatican.va/content/francesco/en/messages/peace/documents/papa-francesco_20181208_messaggio-52giornatamondiale-pace2019.html, accessed 10 December 2018.

B. Duncan (✉)
University of Divinity, Melbourne, VIC, Australia
e-mail: b.duncan@ytu.edu.au

J. Camilleri and D. Guess (eds.), *Towards a Just and Ecologically Sustainable Peace*, https://doi.org/10.1007/978-981-15-5021-8_9

This chapter considers the social thinking of Pope Francis, his role in debates about unfair patterns of globalisation as well as climate change, and his resolute rejection of 'savage capitalism', forms of neoliberal economics and politics that he blames for extreme inequality, exacerbating poverty and hunger. He insists that charity is not enough, but that the structural causes of inequality must be addressed. The chapter also considers efforts by Francis to collaborate with leaders of world religions, particularly Muslim leaders, to oppose instrumentalising religion to justify violence; it also considers the significance of Indigenous peoples in preserving natural resources and helping restore our sense of care for the Earth.

Francis's strong social views have been warmly welcomed by many, though some others consider them exaggerated and too sweeping.[2] The most determined opposition to Pope Francis comes from powerful political elites and financial interests. Francis speaks as a voice for marginalised masses, invoking the spirit of Jesus and the prophets, and insists that the cry of the poor is 'drowned out by the din of the rich few, who grow ever fewer and more rich'. The poor are weeping 'while the wealthy few feast on what, in justice, belongs to all. Injustice is the perverse root of poverty'.[3] He is echoing Pope John Paul II who in 2003 warned: 'The war of the powerful against the weak has today, more than ever, created profound divisions between rich and poor'.[4]

Bergoglio's Background

Francis did not spring on to the world stage as a novice leader. In Argentina, Bergoglio lived through revolutionary violence, military dictatorship, and the so-called dirty war which claimed the lives of more than

[2] Michael D'Emic, 'The Crisis of Ethics in the Financial System and Pope Francis', *Evangelium Gaudii [sic]', World Capital Advisory Partners LLC*, 22 April 2013, https://www.google.com/search?client=firefox-b-d&q=Michael+D%27Emic%7C%2C+%E2%80%98The+Crisis+of+Ethics+in+the+Financial+System+and+Pope+Francis%E2%80%99+Evangelium+Gaudii.

[3] 'Pope: The Din of "Ever More Rich" Drown Out Cries of Poor', *Crux Online*, 18 November 2018, https://cruxnow.com/vatican/2018/11/18/pope-the-din-of-ever-more-rich-drown-out-cries-of-poor/, accessed 24 November 2018.

[4] John Paul II, 'Post-Synod Apostolic Exhortation', *Pastores Gregis*, Vatican, 16 October 2003, http://w2.vatican.va/content/john-paul-ii/en/apost_exhortations/documents/hf_jp-ii_exh_20031016_pastores-gregis.html, accessed 29 October 2003.

eight thousand people, the ill-fated Falklands war, and economic collapse in 2002 which plunged millions of Argentinians into desperate poverty.[5] It was the largest sovereign default in history till that time. As Archbishop of Buenos Aires Bergoglio helped mobilise Church and civil networks to take urgent action to address the crisis, Thomas R. Rourke writes:

> Bergoglio worked with the government to create 'Argentine Dialogue', a seven-month period of extensive civic dialogue, during which the people rose up in a whirlwind of participation, organizing neighbourhood initiatives, mutual-aid societies, transportation, child care and bartering systems.[6]

Bergoglio developed his views on social and economic matters over many decades.[7] He became closely involved in shaping the *Aparecida* assembly of the bishops' conferences of Latin America in Brazil in 2007 and supervised the writing of *The Aparecida Document*, summarising the continental-wide consultations. This 160-page document also highlighted the plight of the Indigenous and other marginalised people, in the context of 'a process of concentration of power and wealth in the hands of a few'.[8]

Prefigured in The *Aparecida Document* are the central themes that Bergoglio expanded as Pope: the incisive critique of extreme inequality; failures in the global economic systems and neoliberal ideology; concerns about the environmental crisis; climate change and ecological sustainability; the need to promote the active participation of women in the Church and society; the 'option for the poor', individually and as societies; and promoting collaboration with all people of good will.[9]

[5] Austen Ivereigh, *The Great Reformer: The Making of a Radical Pope* (New York: Henry Holt and Company, 2014) offers a detailed portrait of Bergoglio's life and background. See also Paul Vallely, *Pope Francis: Untying the Knots* (London: Bloomsbury, 2013).

[6] Thomas R. Rourke, *The Roots of Pope Francis's Social and Political Thought: From Argentina to the Vatican* (Lanham, MD: Rowman & Littlefield, 2018), 105.

[7] See Massimo Borgesi, *The Mind of Pope Francis: Jorge Mario Bergoglio's Intellectual Journey* (Collegeville, MN: Liturgical Press, 2017).

[8] *The Aparecida Document: V General Conference of the Bishops of Latin America and the Caribbean* (Lexington, KY: CreateSpace Independent Publishing, 2007), 39.

[9] See Bruce Duncan, 'Pope Francis's Call for Social Justice in the Global Economy', *The Australasian Catholic Record* 91, 3 (July 2014).

Joy of the Gospel and *Laudato Si'*

Pope Francis in November 2013 expanded themes from The *Aparecida Document* and elsewhere into a 48,000-word document, *The Joy of the Gospel*.[10] It included an uncompromising attack on neoliberal economics and the power of special interests, and demonstrated that Francis was putting the full moral authority of the Church behind his critique of injustice and inequality in the current pattern of globalisation. He forcefully rejected 'an economy of exclusion and inequality. Such an economy kills'. 'Today everything comes under the laws of competition and the survival of the fittest, where the powerful feed upon the powerless' (#53). He blamed 'trickle-down' economic theories that 'assume that economic growth, encouraged by a free market, will inevitably succeed in bringing about greater justice and inclusiveness' (#54). He said 'the idolatry of money' had replaced the ancient golden calf, resulting in the dictatorship of an impersonal and inhumane economy (#55). 'A new tyranny is born', leading to 'absolute autonomy of the marketplace and financial speculation' (#56). He said the problems could not be solved without 'attacking the structural causes of inequality' (#173).

Francis expounded his vision more fully in his 42,000-word social encyclical, *Laudato Si': On Care for Our Common Home* in May 2015, arguing that the major social, economic and political issues of our time are closely intertwined. Unless we resolved these urgent threats, the results could be 'catastrophic', he warned. He was following closely in the footsteps of his predecessors, with Pope Paul VI warning of 'ecological catastrophe' as long ago as 1970.[11]

Laudato Si' was timed to bring maximum moral support behind international efforts to secure the common well-being, particularly through the UN Sustainable Development Goals (SDGs) and the UN Climate

[10] Pope Francis, *The Joy of the Gospel* (*Evangelii Gaudium*), 24 November 2013, http://w2.vatican.va/content/francesco/en/apost_exhortations/documents/papa-francesco_esortazione-ap_20131124_evangelii-gaudium.html, accessed 26 November 2013.

[11] Pope Francis, *Laudato Si': On Care for Our Common Home*, 24 May 2015, #4, http://w2.vatican.va/content/francesco/en/encyclicals/documents/papa-francesco_20150524_enciclica-laudato-si.html, accessed 25 May 2015.

Change Conference COP21 in Paris in the following December.[12] Immediately after Pope Francis addressed the UN General Assembly on 25 September 2015, delegates from more than 190 nations voted to approve the resolution 'Transforming our World: the 2030 Agenda for Sustainable Development'.

The encyclical received extensive media coverage. In the view of the prominent Vatican commentator, Austen Ivereigh, 'It wasn't just the activists and the experts who were impressed. Laudato si' [sic] was by a long shot the most widely read papal document in history'.[13] The scientific community and many commentators welcomed it warmly, recognising its importance in helping mobilise public opinion on the moral basis of preserving life and sustaining the planet for coming generations. It gave the lie to the climate deniers. Lord Nicholas Stern, former chief economist at the World Bank and chair of the Grantham Institute in London, said that this 'extraordinarily important and original' encyclical was 'perfectly timed' in the six-month lead-up to the Paris climate conference, since poverty and climate change were 'the two defining challenges of our generation'.[14]

An adviser to UN secretary-generals in devising the Millennium Development Goals and the Sustainable Development Goals, University Professor at Columbia University Jeffrey Sachs regarded the document as 'absolutely magnificent', based on the best science, and that it had been well received by scholars in the United States, though he noted that conservative sectors in the Church opposed the Pope making such statements on social issues.[15] Sachs said the Pope had played a 'huge role' in making the Paris conference a success.[16]

[12] See Bruce Duncan, 'The Economics Behind the Social Thought of Pope Francis', *Australasian Catholic Record* 94, 2 (April 2017), 157.

[13] Austen Ivereigh, 'Pope Francis's Critique of Modern Life', *Commonweal Magazine* 146, 15 (October 2019), https://www.commonwealmagazine.org/why-call-it-progress, accessed 6 October 2019.

[14] Megan Cornwell, 'Lord Stern: Pope's Encyclical Was 'Perfectly Time' for UN Climate Summit', *The Tablet*, 11 March 2016, https://www.thetablet.co.uk/news/5282/lord-stern-pope-s-encyclical-was-perfectly-timed-for-un-climate-summit, accessed 16 March 2016.

[15] Christopher White, 'Economist Sachs acts as Pope's Cheerleader on "*Laudato Si*"'. *Crux*, 30 June 2018, https://cruxnow.com/vatican/2018/06/30/economist-sachs-acts-as-popes-cheerleader-on-laudato-si/.

[16] Ivereigh, 'Pope Francis's Critique of Modern Life'.

The prominent US environmentalist Bill McKibben in the *New York Review of Books* felt 'great relief' reading the encyclical. 'This marks the first time that a person of great authority in our global culture has fully recognised the scale and depth of our crisis'. It 'does indeed accomplish all the things that the extensive news coverage highlighted'.[17]

Australian political commentator Robert Manne concurred with McKibben's comments, adding that *Laudato Si'* was 'one of the most important documents of our era', and 'the most consequential intervention in the discussion of climate change since Al Gore's film *An Inconvenient Truth*'. Manne thought it 'preposterous that some commentators rejected it as "Marxist"'.[18]

The eminent economist Herman Daly also welcomed the encyclical, though he noted that it had little to say about world population.[19] Francis was firm in opposing abortion but emphasised the role of conscience for Catholics in making decisions about the number of their children; he affirms 'responsible parenthood' according to their circumstances and what is reasonable, even if it means limiting the number of children, taking account also of environmental constraints.[20]

A number of leading economists disagreed with the encyclical's criticisms of the free market and particularly the confusing paragraph 171 on carbon credits, which reads:

> The strategy of buying and selling 'carbon credits' *can* [my italics] lead to a new form of speculation which would not help reduce the emissions of polluting gases worldwide… [It] may simply become a ploy which permits maintaining the excessive consumption of some countries and sectors.

[17] Bill McKibben, 'The Pope and the Planet', *New York Review of Books*, 13 August 2015, https://www.nybooks.com/articles/2015/08/13/pope-and-planet/, accessed 20 August 2015.

[18] Robert Manne, 'Laudato Si': A Political Reading', *The Monthly*, 1 July 2015, https://www.themonthly.com.au/blog/robert-manne/2015/01/2015/1435708320/laudato-si-political-reading, accessed 7 July 2015.

[19] Herman Daly, 'Thoughts on Pope Francis' Laudato Si'', 23 June 2015, *The Daly News* blog, https://steadystate.org/thoughts-on-pope-francis-laudato-si/, accessed 28 June 2015.

[20] See Bruce Duncan, 'Did Pope Francis Fudge the Population Issue?', *Social Policy Connections News*, 23 September 2015, https://www.socialpolicyconnections.com.au/?p=9286, accessed 23 September 2015.

This paragraph became a 'lightening rod' for criticism by some economists who understood it to mean that the Pope was opposed to market-based approaches to reducing carbon emissions. Robert Stavins, director of Harvard's Environmental Economics Program, wrote that the Pope's 'unfortunate attack on the use of the market to address climate change' was out of step with expert opinion. Carbon taxes or cap-and-trade systems were the best instruments to use in addressing climate change.[21]

Yale economist, William D. Nordhaus, wrote that 'the encyclical overlooks the central part that markets, particularly... carbon pricing, must play'. 'For climate change, the major need is to raise the price of CO2 emissions sufficiently high that they are reduced sharply. This can be done either by taxing emissions or by a system of cap-and-trade'.[22]

People familiar with Catholic social teaching would know that the Church was not opposed to free markets as long as they operate justly and that the benefits are spread reasonably equitably. What *Laudato Si'* highlighted was that many markets were not working equitably, and the benefits were going overwhelmingly to top income groups, leaving millions of others struggling.

In relation to scepticism about carbon markets, Steffen Boehm, director of the Essex Sustainability Institute at the University of Essex, wrote that carbon markets in Europe had worked poorly and were riddled with 'financial fraud, phantom emissions reductions and polluter subsidies ... and should be scrapped'.[23] The authors of *Laudato Si'* could have avoided this misunderstanding about the role of markets if they had clearly restricted their criticism specifically to the initial mismanagement of carbon credits in Europe and elsewhere. Speaking with top executives from global energy companies on 14 June 2019 Pope Francis

[21] Eduardo M. Penalver, 'Carbon Trading and the Morality of Markets in Laudato Si', Legal Studies Research Paper No. 17-03, Cornell Law School, 31 January 2017, https://papers.ssrn.com/sol3/papers.cfm?abstract_id=2908996, accessed 19 February 2017.

[22] William D. Nordhaus, 'The Pope & the Market', *New York Review of Books*, 8 October 2015, https://www.nybooks.com/articles/2015/10/08/pope-and-market/, accessed 15 October 2015.

[23] Steffen Boehm, Gareth Bryant and Siddhartha Dabhi, 'Even the Pope Gets It—Carbon Markets Won't Fix the Climate', *The Conversation*, 23 June 2015, https://theconversation.com/even-the-pope-gets-it-carbon-markets-wont-fix-the-climate-38950, accessed 23 June 2015.

firmly endorsed carbon pricing as 'essential' in order to quickly reduce greenhouse emissions.[24]

One of the most vehement critics of *Laudato Si'* was the veteran journalist Paul Kelly, editor-at-large of *The Australian*, who on 24 June 2015 declared the Pope's language 'almost hysterical. Profound intellectual ignorance is dressed up as honouring God'. Francis and his advisers were 'environmental populists and economic ideologues of a quasi-Marxist bent'. The editorial in *The Weekend Australian* of 27–28 June continued that the Pope 'appears to have swallowed a new pernicious dogma' and denounced free market principles in the guise of 'religious instruction'. Kelly vehemently denounced warnings of catastrophic climate change. A series of other writers in *The Australian* joined the attack on the encyclical.[25]

Also in Australia, a Queensland university lecturer Merv Bendle in *Quadrant Online* in an article 'An Apostate Pope?' contended that the Pope was 'economically illiterate' and *Laudato Si'* 'a re-hash of the familiar Marxist-Leninist diatribes against Western imperialism'.[26]

Such intemperate opposition to Church social teaching is often driven by commercial or financial interests, especially in the United States. A group of Catholic writers and advocates coalesced around right-wing think tanks and networks from the 1970s, determined to interpret Catholic social teaching to be more acceptable to their wealthy patrons. These Catholic neoconservatives were employed to contest the social justice direction of the papacy and the US bishops, and became like court theologians for the Republican Party, including even its libertarian wing. They included the late Michael Novak, the late Richard John Neuhaus and George Weigel who frequently rejected church teaching on social and distributive justice, and especially the critique of popes since Paul VI of what today we could call neoliberalism, with its exaggerated emphasis on

[24] 'Pope Francis Declares "Climate Emergency" and Urges Action', *The Guardian*, 15 June 2019, https://www.theguardian.com/environment/2019/jun/14/pope-francis-declares-climate-emergency-and-urges-action, accessed 15 June 2019.

[25] Bruce Duncan, 'The Economic Credibility of Pope Francis: The Australian Newspaper and Laudato Si'', in John Capper (ed.), *Economic and Ecological Responses to Laudato Si'* (Hindmarsh, SA: ATF Theology, 2015), 55–74.

[26] Merv Bendle, 'An Apostate Pope?', *Quadrant Online*, 15 July 2015, https://quadrant.org.au/opinion/doomed-planet/2015/07/apostate-pope/, accessed 19 July 2015.

free markets and capital flows with minimal regulation.[27] Sharon Beder in *Free Market Missionaries: The Corporate Manipulation of Community Values* showed how the neoconservative networks like the American Enterprise Institute, with enormous funding from wealthy patrons, propagated their neoliberal philosophy and politics through the United States and elsewhere.[28]

Pope Francis is facing more determined opposition from the US right-wing, including Catholics like Steve Bannon, former editor of the far-right online magazine *Breitbart* and later strategic adviser to Donald Trump. The US Cardinal Raymond Burke had invited Bannon to speak by video link to a conference in the Vatican in 2014. In an apocalyptic speech, Bannon said civilisation was 'at the beginning stages of a global war against Islamic fascism'.[29]

After leaving the administration of Donald Trump in August 2017, Bannon went to Europe and hoped in 2019 to transform the Dignitatis Humanae Institute south of Rome into a centre to mobilise populist movements in Europe, but the Italian government revoked rights to use the building in late May. Bannon was closely allied to Matteo Salvini, leader of the Northern League and minister of the Interior after the March 2018 elections. Salvini is a leading populist and strongly opposes Pope Francis's views, particularly on migration issues. Bannon advised Salvini to attack the Pope, using migration issues.[30] Burke is one of the leading conservative cardinals opposed to the policies of Pope Francis. Burke fell out with Bannon in June 2019.[31]

[27] See Bruce Duncan, 'Tackling Capitalism: What Vatican II Achieved and What Still Needs to Be Done', *Pacifica* 26, 2 (2013), 207–214.

[28] Sharon Beder, *Free Market Missionaries: The Corporate Manipulation of Community Values* (London: Earthscan, 2006).

[29] Andrew Brown, 'The War against Pope Francis', *The Guardian*, 27 October 2017, https://www.theguardian.com/news/2017/oct/27/the-war-against-pope-francis, accessed 28 October 2017.

[30] 'Steve Bannon Told Italy's Populist Leader: Pope Francis Is the Enemy', *The Guardian*, 13 April 2019, https://www.theguardian.com/world/2019/apr/13/steve-bannon-matteo-salvini-pope-francis-is-the-enemyhttps://www.theguardian.com/world/2019/apr/13/steve-bannon-matteo-salvini-pope-francis-is-the-enemy, accessed 14 April 2019.

[31] Nicolas Seneze, '"How America Wanted to Change the Pope": Chapter Four: Facing the Power of Money', 12 August 2019, https://international.la-croix.com/news/how-america-wanted-to-change-the-pope-introduction/10670, accessed 15 August 2019.

According to Antonio Spadaro SJ, a close collaborator of Pope Francis, 'documents like the encyclical Laudato Si' have led the church on a collision course against those powers that are imprinting a nationalistic, populist and fundamentalist direction on the political dynamics of the world'.[32]

Francis strongly opposes populist movements, saying on 9 August 2019 he was frightened about 'speeches that resembled those by Hitler in 1934'. Responding to concerns about large numbers of migrants, he said if countries could not reasonably accept more migrants, the flow of migrants would only stop if source countries could solve their problems with international help.[33]

Building Alliances for Action on Climate and Poverty

Francis laid out his moral agenda in *Laudato Si'* in collaboration with leading scientists and organisations, including advisers involved in developing the UN Sustainable Development Goals, such as Professors Joseph Stiglitz of Columbia University (former chief economist of the World Bank and Nobel laureate), Jeffrey Sachs and Hans Joachim Schellnhuber, founding director of the Potsdam Institute for Climate Impact Research and a member of the Intergovernmental Panel on Climate Change.[34] Francis warned of an unprecedented global 'catastrophe' unless we reversed global warming. He also warned how economic systems were failing, with extreme inequality and poverty undermining social cohesion in many countries, resulting in civil conflict and wars.

Francis not only used Church resources and networks, especially the prestigious Pontifical Academy of Sciences and the Pontifical Academy of Social Sciences, but he constantly tried to establish personal contact

[32] Quoted in Michael Sean Winters, 'Sixth Year May Go Down as the Most Decisive in Francis' Papacy', *National Catholic Reporter*, 13 May 2019, https://www.ncronline.org/news/opinion/distinctly-catholic/sixth-year-may-go-down-most-decisive-francis-papacy, accessed 16 May 2019.

[33] Carol Glatz, 'In New Interview, Pope Explains Aim of Synod, Warns Against Nationalism', *National Catholic Reporter*, 9 August 2019, https://www.ncronline.org/news/vatican/new-interview-pope-explains-aim-synod-warns-against-nationalism, accessed 11 August 2019.

[34] See Duncan, 'The Economics Behind the Social Thought of Pope Francis', 148–166.

and rapport with key political leaders, scientists and scholars, economists and business leaders. He did not claim technical or scientific expertise but relying on expert advice, he was highlighting the moral dimensions in these issues.

For instance, in an address to the president of the UN General Assembly and finance ministers from various nations, Francis on 27 May 2019 urged them 'to help prevent a crisis that is leading the world towards disaster'. Investments in fossil fuels were still increasing 'even though scientists tell us that fossil fuels should remain underground'. 'We need to act decisively to put an end to all emissions of greenhouse gases by mid-century', and instead switch to renewable energy sources. He said the 'ledger of life itself, of human dignity and survival' were at stake, as he appealed for them to help 'save the world from indifference and from the idolatry of money' along with 'the corruption of vested interests. We still reckon as profit what threatens our very survival'.[35]

On 9 June he lamented that progress on achieving the SDGs 'has in some cases been slow and even non-existent', when inadequate development models have failed to be inclusive or sustainable, and were instead exacerbating poverty and undermining social equity. He called for a renewed ethical vision to overcome 'hunger and food insecurity, persistent social and economic distress, the degradation of ecosystems, and a "culture of waste"'. Francis urged his listeners to help all become 'stewards of the earth'.[36]

Then at a two-day conference on 14 June 2019 with executives from the energy sectors, including the CEOs of Exxon, Mobil, BP, Royal Dutch Shell, Total, Chevron and Eni, plus Black Rock, BNP Paribas and Hermes Investment, Francis appealed to them to use their expertise to avert 'a climate emergency'. He noted that the Intergovernmental Panel on Climate Change warned that 'effects on the climate will be catastrophic if we cross the threshold of 1.5°C' of warming. Yet only a decade remained to confine global warming to that level. Francis encouraged the

[35] Pope Francis, Address to Finance Ministers from Various Nations, on 'Climate Change and New Evidence from Science, Engineering, and Policy', Vatican, 27 May 2019, http://w2.vatican.va/content/francesco/en/speeches/2019/may/documents/papa-francesco_20190527_climate-change.html, accessed 29 May 2019.

[36] Jim Fair, 'Pope Pushes Progress on Sustainable Development', *Zenit*, 9 June 2019, https://zenit.org/articles/pope-pushes-progress-on-sustainable-development/, accessed 15 June 2019.

energy executives to ensure a just transition to a low-carbon economy while protecting jobs and living standards as well as reducing inequality. He endorsed carbon pricing as 'essential'. Finally he called for 'transparency in reporting climate risk' so capital could do the most good. He concluded that 'time is running out! … Today a radical energy transition is needed to save our common home'.[37]

In response the meeting issued a joint statement agreeing to increase investment and innovation in sustainable energy, and introduce new carbon pricing mechanisms, even carbon taxes. But different views in the group meant they did not endorse the Pope's call for a 1.5 °C cap on temperature rise, and instead aimed to keep global warming below 2 °C.[38]

Fostering Collaboration of World Religions

Pope Francis has repeatedly urged all religious traditions to support the Sustainable Development Goals, and 'to open up the treasures of our best traditions' to dialogue 'on how to build the future of our planet'. Speaking at a conference on 'Religions and the Sustainable Development Goals' in March 2019, he commended the 2030 Agenda approved in September 2015, which integrated the '"five Ps": people, planet, prosperity, peace and partnership'. He stressed the need to listen to all voices, especially the poor, migrants, Indigenous people and the young. Reiterating *Laudato Si'*, he called for a powerful 'ecological conversion' to guide economic and political objectives to build a fair and sustainable future.[39]

[37] Pope Francis to a meeting on 'The Energy Transition and Care for Our Common Home', Vatican, 14 June 2019, http://w2.vatican.va/content/francesco/en/speeches/2019/june/documents/papa-francesco_20190614_compagnie-petrolifere.html, accessed 17 June 2019.

[38] Bernhardt Warner, 'Pope Francis Just Convinced These Big Oil CEOs to Alter Their Message on Climate Change', *Fortune*, 17 June 2019, https://www.google.com/search?client=firefox-b-d&q=Bernhardt+Warner%2C+%E2%80%98Pope+Francis+just+convinced+these+big+oil+CEOs, accessed 19 June 2019.

[39] Pope Francis to the 'Conference on Religions and the Sustainable Development Goals (SDGs): Listening to the Cry of the Earth and of the Poor', Vatican City, 8 March 2019, http://w2.vatican.va/content/francesco/en/speeches/2019/march/documents/papa-francesco_20190308_religioni-sviluppsostenibile.html, accessed 11 March 2019.

Francis has significantly advanced diplomacy and theological dialogue with sections of the Muslim world, opening new paths of collaboration on many social issues.[40] In 2014, he had instructed that encyclicals of the Holy See be also available in Arabic translation. As a gesture of respect and openness, *Laudato Si'* was released on the first day of the Muslim holy month of Ramadan.[41]

Shortly after *Laudato Si'*, Muslim scholars and leaders published a significant document dealing almost exclusively with climate change. At the conclusion of the International Islamic Climate Change Symposium in Istanbul, involving Islamic leaders and scholars from some twenty countries, *The Islamic Declaration on Global Climate Change* was released in August 2015 to add Muslim moral support for the December Paris Climate Summit. The document endorsed global efforts to avert 'catastrophic' disasters. The authors linked the moral obligations to protect the poor and the planet with key passages from the Qur'an, assuring it strong religious legitimation.[42]

An accompanying statement said the Islamic Declaration was in 'harmony with the Papal Encyclical and has won the support of the Pontifical Council on Justice and Peace of the Holy See'. Its president, Cardinal Peter Turkson, greeted the document with 'great joy' and 'in a spirit of solidarity' prayed for its success and continued collaboration to protect our common home.[43]

Similar declarations about climate change from Buddhist, Jewish, Hindu and other religious groups followed. A Buddhist Declaration on Climate Change was adopted in Paris by fifteen senior world Buddhist leaders, including the Dalai Lama, on 28 November 2015, just days

[40] See Damien Howard SJ 'An Islamic Declaration on Climate Change', *Thinking Faith*, 29 October 2015, https://www.thinkingfaith.org/articles/islamic-declaration-climate-change, accessed 15 November 2015.

[41] Russell Powell, '*Laudato' Si'*: Engaging Islamic Tradition and Implications for Legal Thought', Seattle University Law Report, Vol. 40 (October 2017), 1330, https://digitalcommons.law.seattleu.edu/sulr/vol40/iss4/8/, accessed 20 November 2017.

[42] *The Islamic Declaration on Global Climate Change*, August 2015, http://www.ifees.org.uk/wp-content/uploads/2016/10/climate_declarationmMWB.pdf, accessed 25 August 2015.

[43] Kevin Clarke, 'A Francis Effect? Islamic Leaders Issue Statement on Climate Change', *America Magazine*, 18 August 2015, https://www.americamagazine.org/content/dispatches/francis-effect-islamic-leaders-issue-statement-climate-change, accessed 25 August 2015.

before the start of the Climate Summit. The Declaration updated a 2009 statement, warning of the consequences of climate change for all living species. It was reportedly the first time so many Buddhist leaders had spoken with one voice on global issues.[44] Other major religious traditions also issued statements warning of the dangers of climate change, including those by Hindu[45] and Jewish[46] leaders.

Francis and his staff have long interacted with leaders of other Christian groups, particularly the Anglican and Lutheran Churches, the World Council of Churches and evangelical networks.[47]

Pope Francis followed Popes John Paul II and Benedict XVI in inviting leaders of the world's religions to meet in Assisi. On the thirtieth anniversary of the first gathering in 1986, more than 400 religious leaders met in Assisi for three days over 18–20 September 2016 for the World Day of Prayer for Peace. Religious leaders came from Buddhist, Jewish, Christian, Hindu and Muslim traditions, as well as from traditional religions. A close friend of Francis, the Ecumenical Patriarch of Constantinople, Bartholomew, as well as the Archbishop of Canterbury, Justin Welby, and the general secretary of the World Council of Churches were among those present. Pope Francis said 'we are all children of God. And God is the God of peace. There is no god of war. He who makes war is evil; it is the devil who wants to kill everyone'.[48]

In the light of the resurgence of anti-Semitism in Europe and elsewhere, Francis insistently reaffirms his Church's firm opposition to

[44] 'The Time to Act Is Now. A Buddhist Declaration on Climate Change', 28 November 2015, https://oneearthsangha.org/statements/the-time-to-act-is-now/, accessed 30 May 2019.

[45] 'A Hindu Declaration on Climate Change', 2015, http://www.hinduclimatedeclaration2015.org/english, accessed 30 May 2019.

[46] 'To the Jewish People, to All Communities of Spirit, and to the World: A Rabbinic Letter on the Climate Crisis', 11 May 2015, https://theshalomcenter.org/250-rabbis-sign-rabbinic-letter-climate-crisis; The Forum on Religion and Ecology at Yale has resources on statements by world religions on climate change, http://fore.yale.edu/climate-change/statements-from-world-religions/christianity-protestant-denominations-and-organizations/, accessed 30 May 2015.

[47] See Austen Ivereigh, 'Is Francis Our First Charismatic Pope?' *America Magazine*, 14 June 2019, https://www.americamagazine.org/faith/2019/06/14/francis-our-first-charismatic-pope, accessed 17 June 2019.

[48] 'Religious Leaders of Many Faiths Talk Peace in Assisi—World Council of Churches', September 2016, https://www.oikoumene.org/en/press-centre/news/religious-leaders-of-many-faiths-talk-peace-in-assisi, accessed 16 June 2019.

anti-Semitism,[49] and endeavours to deepen relationships with Jewish communities and leaders.[50] This is complicated by Israel's policies of displacement of Palestinians and its treatment of people in the Occupied Territories.

Dialogue with Islamic Leaders

Alarmed by the outbreak of Islamist violence and terrorism, Francis sees deepening friendships between Christians and Muslims as vital for world peace, the eradication of poverty and developing a sustainable ecology. In April 2017, he met the Grand Imam of the al-Azhar mosque in Cairo, Sheikh Ahmad Al-Tayyib, in the most eminent centre of Sunni Islamic learning, encouraging cooperation and dialogue. Welcoming the Pope, the Grand Imam said: 'Islam is not a religion of terrorism because a group of followers carelessly manipulates Islamic texts and misinterprets them ignorantly'. All religious leaders had to 'liberate religions from false concepts'.[51]

In his address in Cairo on 28 April 2017 to the Al-Azhar and Muslim Council of Elders' Global Peace Conference, Francis said:

> We have an obligation to denounce violations of human dignity and human rights, to expose attempts to justify every form of hatred in the name of religion, and to condemn these attempts as idolatrous caricatures of God: Holy is his name, he is the God of peace, God *salaam*. Peace alone, therefore, is holy and no act of violence can be perpetrated in the name of God, for it would profane his Name.[52]

[49] ADL Los Angeles, 'ADL Leaders and Pope Francis Join to Reaffirm Jewish-Catholic Relations and Denounce Anti-Semitism; LA ADL Leaders Among Group', 10 February 2017, https://la.adl.org/adl-leaders-and-pope-francis-join-to-reaffirm-jewish-catholic-relations-and-denounce-anti-semitism-la-adl-leaders-among-group/, accessed 15 February 2017.

[50] Carol Glatz, 'Anti-Semitism Must Be Banned from Society, Pope Says', *National Catholic Reporter*, 5 November 2018, https://www.ncronline.org/news/vatican/francis-chronicles/anti-semitism-must-be-banned-society-pope-says, accessed 7 November 2018.

[51] Joshua McElwee, 'In Egypt, Francis Exhorts Religions to Condemn Violence Masquerading as Sanctity', *National Catholic Reporter*, 28 April 2017, https://www.ncronline.org/news/vatican/egypt-francis-exhorts-religions-condemn-violence-masquerading-sanctity, accessed 30 April 2017.

[52] Pope Francis, 'Address to the Participants in the International Peace Conference', Cairo, 28 April 2017, http://w2.vatican.va/content/francesco/en/speeches/2017/april/documents/papa-francesco_20170428_egitto-conferenza-pace.html#_ftn5, accessed 30 April 2017.

The Ecumenical Patriarch Bartholomew also spoke, insisting that 'terrorism is a stranger to any religion', including Islam. 'The credibility of religions today depends on their attitude towards the protection of human freedom and dignity, as well as on their contribution to peace. This is the presupposition not only of peaceful coexistence, but even the sheer survival of humanity'.[53]

Pope Francis met again with the Grand Imam, Sheikh Ahmad al-Tayyib, in Abu Dhabi in February 2019. Sheikh Al-Tayyib welcomed the Pope, saying that 'the name of God must not be used to justify violence. God did not create us to cause suffering'. People who used religious texts to justify violence should be denounced as 'murderers'.[54] Francis reiterated that 'No violence can be justified in the name of religion'. 'There is no alternative: we will either build the future together or there will not be a future'. Standing 'on the side of the poor', they 'must keep watch as sentinels of fraternity in the night of conflict', and specifically oppose wars and increasing armaments.[55]

In what may be one of the most significant documents of our time, Francis signed a joint declaration with the Grand Imam on 4 February 2019: 'Human Fraternity for World Peace and Living Together'. Francis had compared this meeting with that of St Francis and Sultan al-Malik al-Kamil of Egypt at Damietta, precisely eight centuries earlier. The document drew from both Christian and Muslim writings and traditions, reiterating that all 'believers are called to express this human fraternity by safeguarding creation and the entire universe and supporting all persons, especially the poorest and those most in need'. The document insisted that true moral bases of religion were 'tolerance and fraternity'. 'God has

[53] Ecumenical Patriarch Bartholomew, 'Religions and Peace', *Zenit*, 28 April, 2017, https://zenit.org/articles/address-of-ecumenical-patriarch-bartholomew-religions-and-peace-at-al-azhar-and-muslim-council-of-elders-global-peace-conference/, accessed 30 April 2017.

[54] 'Pope in UAE: Oppose War with Sweet Prayer; Abu Dhabi Declaration Signed', *Gulf News*, 4 February 2019, https://gulfnews.com/uae/government/pope-in-uae-oppose-war-with-sweet-prayer-abu-dhabi-declaration-signed-1.1549298647331, accessed 6 February 2019.

[55] Pope Francis, 'Address to Interreligious Meeting at Founder's Memorial', Abu Dhabi, 4 February 2019, http://w2.vatican.va/content/francesco/en/speeches/2019/february/documents/papa-francesco_20190204_emiratiarabi-incontrointerreligioso.html, accessed 6 February 2019.

created us to understand one another, cooperate with one another and live as brothers and sisters who love one another'.[56]

As the director of Vatican communications, Andrea Tornielli commented: 'for the first time, there was a strong commitment by an Islamic authority to walk in a certain direction'. Not all the Islamic world accepted that authority, he said, but this was a new beginning: 'there is an Islam that dialogues and with which a peaceful coexistence can be built'. The real battle, Tornielli commented, was within Islam itself.[57]

Indigenous Peoples and the Environment

Indigenous peoples are extremely important in Pope Francis's mind, having much to teach our world about 'the sacredness of our earth', respect for nature and the environment, their values and how to live sustainably. 'This means that their voice and their concerns should be at the centre of the implementation of the 2030 Agenda', he said in an address on the Sustainable Development Goals in March 2019. 'Although they represent only five percent of the world's population, they look after about twenty-two percent of the earth's landmass. Living in areas such as the Amazon and the Arctic, they help protect approximately eighty percent of the planet's biodiversity'.[58]

In Peru, on 19 January 2018 Pope Francis had forcefully denounced business malpractice and exploitation in the Amazon, the 'lungs' of the earth, and defended its peoples against oppression and marginalisation. Speaking to 4000 Indigenous people near the border with Bolivia, Francis warned about global warming and the devastation of illegal mining, with slave labour and human trafficking. 'Great business interests want to lay

[56] Document on 'Human Fraternity for World Peace and Living Together' signed by His Holiness Pope Francis and the Grand Imam of Al-Azhar al-Tayyib (Abu Dhabi), 4 February 2019, http://w2.vatican.va/content/francesco/en/events/event.dir.html/content/vaticanevents/en/2019/2/4/fratellanza-umana.html, accessed 6 February 2019.

[57] Deborah Castellano Lubov, 'Andrea Tornielli on the Islam That Wants to Dialogue', *Zenit*, 20 June 2019, https://zenit.org/articles/interview-andrea-tornielli-on-the-islam-that-wants-to-dialogue-never-respond-to-hatred-with-hatred/, accessed 22 June 2019.

[58] Pope Francis, at the Conference on 'Religions and the Sustainable Development Goals (SDGs): Listening to the Cry of the Earth and of the Poor', 8 March 2019, http://w2.vatican.va/content/francesco/en/speeches/2019/march/documents/papa-francesco_20190308_religioni-svilupposostenibile.html, accessed 10 March 2019.

hands on petroleum, gas, lumber and gold' from the Amazon. 'The native Amazonian peoples have probably never been so threatened on their own lands as they are at present'.[59]

To help mobilise support for the Indigenous peoples and to protect the Amazon, Pope Francis called a special Synod of Bishops for the Pan-Amazon region entailing wide participation: by April 2019, 80,000 people were engaged in parish or community events.[60] Francis hoped the Synod in Rome in October 2019, a child of *Laudato Si'*, would drive home the urgency of tackling the climate crisis. An indigenous leader, Sr Laura Vicuna, who had received death threats by 'invaders' of their lands, said this was the first time in history that the Amazon and its original villages 'are at the center of church reflection. Pope Francis ... has made possible this unique moment with the Amazon synod'.[61]

Conclusion

Pope Francis has vigorously promoted a global agenda to avert the unprecedented threats to human life especially from global warming, and to advocate the well-being of all Earth's creatures. Scarcely a week goes by without him making a major address on the concerns of *Laudato Si'*. His call for renewed international efforts to guide policies and action by the moral compass of solidarity and fraternity may once have seemed utopian, but many now endorse his plea as a matter of survival, including leading scientists, economists and international organisations.

Francis sees religious beliefs as reinforcing commitment to this common moral compass. He has tried to deepen dialogue and cooperation among religious leaders and traditions, focussing on universal human

[59] 'Pope's Address to Indigenous People of Amazon Region in Peru', *Zenit*, 19 January 2018, https://zenit.org/articles/popes-address-to-indigenous-people-of-amazon-region-in-peru/, accessed 21 January 2018.

[60] Linda Bordoni, 'Pope Asks Amazon Bishops to Open Up New Paths for the Church and for Humanity', *Vatican News*, 14 March 2018, https://www.vaticannews.va/en/church/news/2018-03/synod-bishops-pan-amazon-region-repam-mauricio-lopez.html, accessed 16 March 2018.

[61] Soli Salgado, 'Indigenous from Brazil Bring Plea to Rome: "We Need More Protection"', *National Catholic Reporter*, 8 October 2019, https://www.google.com/search?client=firefox-b-d&q=Soli+Salgado%2C+%E2%80%98Indigenous+from+Brazil+bring++plea+to+Rome%3A+%E2%80%9CWe+need+more+protection, accessed 8 October 2019.

well-being, the core of all religions and respectable philosophical beliefs despite major differences and interpretations over history.

Pope Francis is endeavouring to mobilise all people of good will for urgent action in the face of current crises. It is not hyperbole to suggest millions of lives will depend on the outcome of this global agenda.

Bibliography

ADL Los Angeles. 'ADL Leaders and Pope Francis Join to Reaffirm Jewish-Catholic Relations and Denounce Anti-Semitism; LA ADL Leaders Among Group', 10 February 2017. https://la.adl.org/adl-leaders-and-pope-francis-join-to-reaffirm-jewish-catholic-relations-and-denounce-anti-semitism-la-adl-leaders-among-group/. Accessed 15 February 2017.

The Aparecida Document: V General Conference of the Bishops of Latin America and the Caribbean. Lexington, KY: CreateSpace Independent Publishing, 2007.

Bartholomew, Ecumenical Patriarch. 'Religions and Peace'. *Zenit*, 28 April 2017. https://zenit.org/articles/address-of-ecumenical-patriarch-bartholomew-religions-and-peace-at-al-azhar-and-muslim-council-of-elders-global-peace-conference/. Accessed 30 April 2017.

Beder, Sharon. *Free Market Missionaries: The Corporate Manipulation of Community Values*. London: Earthscan, 2006.

Bendle, Merv. 'An Apostate Pope?'. *Quadrant Online*, 15 July 2015. https://quadrant.org.au/opinion/doomed-planet/2015/07/apostate-pope/. Accessed 19 July 2015.

Boehm, Steffen, Gareth Bryant, and Siddhartha Dabhi. 'Even the Pope Gets It—Carbon Markets Won't Fix the Climate'. *The Conversation*, 23 June 2015. https://theconversation.com/even-the-pope-gets-it-carbon-markets-wont-fix-the-climate-38950. Accessed 23 June 2015.

Bordoni, Linda. 'Pope asks Amazon Bishops to Open Up New Paths for the Church and for Humanity'. *Vatican News*, 14 March 2018. https://www.vaticannews.va/en/church/news/2018-03/synod-bishops-pan-amazon-region-repam-mauricio-lopez.html. Accessed 16 March 2018.

Borgesi, Massimo. *The Mind of Pope Francis: Jorge Mario Bergoglio's Intellectual Journey*. Collegeville, MN: Liturgical Press, 2017.

Brown, Andrew. 'The War against Pope Francis'. *The Guardian*, 27 October 2017. https://www.theguardian.com/news/2017/oct/27/the-war-against-pope-francis. Accessed 28 October 2017.

Clarke, Kevin. 'A Francis Effect? Islamic Leaders Issue Statement on Climate Change'. *America Magazine*, 18 August 2015. https://www.americamagazine.org/content/dispatches/francis-effect-islamic-leaders-issue-statement-climate-change. Accessed 25 August 2015.

Cornwell, Mega. 'Lord Stern: Pope's Encyclical Was 'Perfectly Time' for UN Climate Summit'. *The Tablet*, 11 March 2016. https://www.thetablet.co.uk/news/5282/lord-stern-pope-s-encyclical-was-perfectly-timed-for-un-climate-summit. Accessed 16 March 2016.

Daly, Herman. 'Thoughts on Pope Francis' Laudato Si''. *The Daly News*, 23 June 2015. https://steadystate.org/thoughts-on-pope-francis-laudato-si/. Accessed 28 June 2015.

D'Emic, Michael. 'The Crisis of Ethics in the Financial System and Pope Francis' *Evangelium Gaudii [sic]', World Capital Advisory Partners LLC*, 22 April 2013. https://www.google.com/search?client=firefox-b-d&q=Michael+D%27Emic%7C%2C+%E2%80%98The+Crisis+of+Ethics+in+the+Financial+System+and+Pope+Francis%E2%80%99+Evangelium+Gaudii+.

'Document on "Human Fraternity for World Peace and Living Together" Signed by His Holiness Pope Francis and the Grand Imam of Al-Azhar al-Tayyib' [Abu Dhabi]. *Vatican News*, 4 February 2019. http://w2.vatican.va/content/francesco/en/events/event.dir.html/content/vaticanevents/en/2019/2/4/fratellanza-umana.html. Accessed 6 February 2019.

Duncan, Bruce. 'Did Pope Francis Fudge the Population Issue?' *Social Policy Connections News*, 23 September 2015. https://www.socialpolicyconnections.com.au/?p=9286. Accessed 23 September 2015.

Duncan, Bruce. 'Pope Francis's Call for Social Justice in the Global Economy'. *The Australasian Catholic Record* 91, 3 (July 2014): 178–193.

Duncan, Bruce. 'Tackling Capitalism: What Vatican II Achieved and What Still Needs to Be Done'. *Pacifica* 26, 2 (2013): 207–214.

Duncan, Bruce. 'The Economic Credibility of Pope Francis: The Australian Newspaper and *Laudato Si'*', in John Capper (ed.), *Economic and Ecological Responses to* Laudato Si'. Hindmarsh, SA: ATF Theology, 2015, 55–74.

Duncan, Bruce. 'The Economics Behind the Social Thought of Pope Francis'. *Australasian Catholic Record* 94, 2 (April 2017): 148–166.

Fair, Jim. 'Pope Pushes Progress on Sustainable Development'. *Zenit*, 9 June 2019. https://zenit.org/articles/pope-pushes-progress-on-sustainable-development/. Accessed 15 June 2019.

Glatz, Carol. 'Anti-Semitism Must Be Banned from Society, Pope Says'. *National Catholic Reporter*, 5 November 2018. https://www.ncronline.org/news/vatican/francis-chronicles/anti-semitism-must-be-banned-society-pope-says. Accessed 7 November 2018.

Glatz, Carol. 'In new interview, Pope Explains Aim of Synod, Warns Against Nationalism'. *National Catholic Reporter*, 9 August 2019. https://www.ncronline.org/news/vatican/new-interview-pope-explains-aim-synod-warns-against-nationalism. Accessed 11 August 2019.

Harvey, Fiona, and Jillian Ambrose. 'Pope Francis Declares "Climate Emergency" and Urges Action'. *The Guardian*, 15 June 2019. https://www.theguardian.com/environment/2019/jun/14/pope-francis-declares-climate-emergency-and-urges-action. Accessed 15 June 2019.

'A Hindu Declaration on Climate Change'. 2015. http://www.hinduclimatedeclaration2015.org/english. Accessed 30 May 2019.

Howard, Damien SJ. 'An Islamic Declaration on Climate Change'. *Thinking Faith*, 29 October 2015. https://www.thinkingfaith.org/articles/islamic-declaration-climate-change. Accessed 15 November 2015.

The Islamic Declaration on Global Climate Change. August 2015. http://www.ifees.org.uk/wp-content/uploads/2016/10/climate_declarationmMWB.pdf. Accessed 25 August 2015.

Ivereigh, Austen. 'Is Francis Our First Charismatic Pope?' *America Magazine*, 14 June 2019. https://www.americamagazine.org/faith/2019/06/14/francis-our-first-charismatic-pope. Accessed 17 June 2019.

Ivereigh, Austen. 'Pope Francis's Critique of Modern Life'. *Commonweal Magazine* 146, 15 (October 2019). https://www.commonwealmagazine.org/why-call-it-progress. Accessed 6 October 2019.

Ivereigh, Austen. *The Great Reformer: The Making of a Radical Pope*. New York: Henry Holt & Company, 2014.

Lubov, Deborah Castellano. 'Andrea Tornielli on the Islam That Wants to Dialogue'. *Zenit*, 20 June 2019. https://zenit.org/articles/interview-andrea-tornielli-on-the-islam-that-wants-to-dialogue-never-respond-to-hatred-with-hatred/. Accessed 22 June 2019.

Manne, Robert. 'Laudato Si': A Political Reading'. *The Monthly*, 1 July 2015. http://www.themonthly.com.au/blog/robert-manne/2015/01/2015/1435708320/laudato-si-political-reading.

McElwee, Joshua. 'In Egypt, Francis Exhorts Religions to Condemn Violence Masquerading as Sanctity'. *National Catholic Reporter*, 28 April 2017. https://www.ncronline.org/news/vatican/egypt-francis-exhorts-religions-condemn-violence-masquerading-sanctity. Accessed 30 April 2017.

McKibben, Bill. 'The Pope and the Planet'. *New York Review of Books*, 13 August 2015. https://www.nybooks.com/articles/2015/08/13/pope-and-planet/. Accessed 20 August 2015.

Nordhaus, William D. 'The Pope & the Market'. *New York Review of Books*, 8 October 2015. https://www.nybooks.com/articles/2015/10/08/pope-and-market/. Accessed 15 October 2015.

Penalver, Eduardo M. 'Carbon Trading and the Morality of Markets in Laudato Si'', 31 January 2017, Legal Studies Research Paper No. 17-03, Cornell Law School. https://papers.ssrn.com/sol3/papers.cfm?abstract_id=2908996. Accessed 19 February 2017.

Pope Francis. 'Address to Interreligious Meeting at Founder's Memorial', Abu Dhabi, 4 February 2019. http://w2.vatican.va/content/francesco/en/speeches/2019/february/documents/papa-francesco_20190204_emiratiarabi-incontrointerreligioso.html. Accessed 6 February 2019.

Pope Francis. Address to the 'Conference on Religions and the Sustainable Development Goals (SDGs): Listening to the Cry of the Earth and of the Poor'. Vatican City, 8 March 2019. http://w2.vatican.va/content/francesco/en/speeches/2019/march/documents/papa-francesco_20190308_religioni-svilupposostenibile.html. Accessed 11 March 2019.

Pope Francis. Address 'To the Participants in the International Peace Conference', Cairo, 28 April 2017. http://w2.vatican.va/content/francesco/en/speeches/2017/april/documents/papa-francesco_20170428_egitto-conferenza-pace.html#_ftn5. Accessed 30 April 2017.

Pope Francis. 'Climate Change and New Evidence from Science, Engineering, and Policy'. Address to Finance Ministers from Various Nations. Vatican City, 27 May 2019. http://w2.vatican.va/content/francesco/en/speeches/2019/may/documents/papa-francesco_20190527_climate-change.html. Accessed 29 May 2019.

Pope Francis. *Laudato Si': On Care for Our Common Home*. Vatican City, 24 May 2015. http://w2.vatican.va/content/francesco/en/encyclicals/documents/papa-francesco_20150524_enciclica-laudato-si.html. Accessed 25 May 2015.

Pope Francis. 'The Energy Transition and Care for Our Common Home'. Vatican City, 14 June 2019. http://w2.vatican.va/content/francesco/en/speeches/2019/june/documents/papa-francesco_20190614_compagnie-petrolifere.html. Accessed 17 June 2019.

Pope Francis. *The Joy of the Gospel* (*Evangelii Gaudium*). Vatican City, 24 November 2013. http://w2.vatican.va/content/francesco/en/apost_exhortations/documents/papa-francesco_esortazione-ap_20131124_evangelii-gaudium.html. Accessed 26 November 2013.

Pope Francis. *2019 World Day of Peace*. Vatican City, 8 December 2018. http://w2.vatican.va/content/francesco/en/messages/peace/documents/papa-francesco_20181208_messaggio-52giornatamondiale-pace2019.html. Accessed 10 December 2018.

'Pope in UAE: Oppose War with Sweet Prayer; Abu Dhabi Declaration Signed'. *Gulf News*, 4 February 2019. https://gulfnews.com/uae/government/pope-in-uae-oppose-war-with-sweet-prayer-abu-dhabi-declaration-signed-1.1549298647331. Accessed 6 February 2019.

Pope John Paul II. 'Post-Synod Apostolic Exhortation', *Pastores Gregis*, 16 October 2003. http://w2.vatican.va/content/john-paul-ii/en/apost_exhortations/documents/hf_jp-ii_exh_20031016_pastores-gregis.html. Accessed 29 October 2003.

'Pope: The Din of "Ever More Rich" Drown Out Cries of Poor', Crux Online, 18 November 2018. https://cruxnow.com/vatican/2018/11/18/pope-the-din-of-ever-more-rich-drown-out-cries-of-poor/. Accessed 24 November 2018.

'Pope's Address to Indigenous People of Amazon Region in Peru'. *Zenit*, 19 January 2018. https://zenit.org/articles/popes-address-to-indigenous-people-of-amazon-region-in-peru/. Accessed 21 January 2018.

Powell, Russell. '*Laudato Si'*: Engaging Islamic Tradition and Implications for Legal Thought'. *Seattle University Law Report*, Vol. 40 (October 2017): 1326–1342. https://digitalcommons.law.seattleu.edu/sulr/vol40/iss4/8/. Accessed 20 November 2017.

'Religious Leaders of Many Faiths Talk Peace in Assisi'. *World Council of Churches* blog, 21 September 2016. https://www.oikoumene.org/en/press-centre/news/religious-leaders-of-many-faiths-talk-peace-in-assisi. Accessed 16 June 2019.

Rourke, Thomas R. *The Roots of Pope Francis's Social and Political Thought: From Argentina to the Vatican*. Lanham, MD: Rowman & Littlefield, 2018.

Salgado, Soli. 'Indigenous from Brazil Bring Plea to Rome: "We Need More Protection"'. *National Catholic Reporter*, 8 October 2019. https://www.google.com/search?client=firefox-b-d&q=Soli+Salgado%2C+%E2%80%98Indigenous+from+Brazil+bring++plea+to+Rome%3A+%E2%80%9CWe+need+more+protection. Accessed 8 October 2019.

Seneze, Nicolas. '"How America Wanted to Change the Pope": Chapter Four: Facing the Power of Money'. *La Croix*, 12 August 2019. https://international.la-croix.com/news/how-america-wanted-to-change-the-pope-introduction/10670. Accessed 15 August 2019.

'The Time to Act Is Now. A Buddhist Declaration on Climate Change', 28 November 2015. https://oneearthsangha.org/statements/the-time-to-act-is-now/. Accessed 30 May 2019.

'To the Jewish People, to All Communities of Spirit, and to the World: A Rabbinic Letter on the Climate Crisis', 11 May 2015. https://theshalomcenter.org/250-rabbis-sign-rabbinic-letter-climate-crisis.

Townsend, Mark. 'Steve Bannon Told Italy's Populist Leader: Pope Francis is the Enemy'. *The Guardian*, 13 April 2019. https://www.theguardian.com/world/2019/apr/13/steve-bannon-matteo-salvini-pope-francis-is-the-enemyhttps://www.theguardian.com/world/2019/apr/13/steve-bannon-matteo-salvini-pope-francis-is-the-enemy. Accessed 14 April 2019.

Vallely, Paul. *Pope Francis: Untying the Knots*. London: Bloomsbury, 2013.

Warner, Bernhardt. 'Pope Francis Just Convinced These Big Oil CEOs to Alter Their Message on Climate Change'. *Fortune*, 17 June 2019, https://www.google.com/search?client=firefox-b-d&q=Bernhardt+Warner%2C+%E2%80%98Pope+Francis+just+convinced+these+big+oil+CEOs. Accessed 19 June 2019.

White, 'Christopher. 'Economist Sachs Acts as Pope's Cheerleader on "*Laudato Si'*"'. *Crux*, 30 June 2018. https://cruxnow.com/vatican/2018/06/30/economist-sachs-acts-as-popes-cheerleader-on-laudato-si/.

Winters, Michael Sean. 'Sixth Year May Go Down as the Most Decisive in Francis' Papacy'. *National Catholic Reporter*, 13 May 2019. https://www.ncronline.org/news/opinion/distinctly-catholic/sixth-year-may-go-down-most-decisive-francis-papacy. Accessed 16 May 2019.

Dr. Bruce Duncan is a Catholic priest of the Redemptorist Congregation. He studied economics and politics at the University of Sydney and is a senior lecturer in history and social justice studies at Yarra Theological Union, a college of the University of Divinity in Melbourne. He is also the director of the advocacy organisation, Social Policy Connections. His published books include: *Social Justice: fuller life in a fairer world*; *Crusade or conspiracy?: Catholics and the anti-communist struggle in Australia*; *Catholic Social Teaching: From Rerum Novarum to 1931.*

CHAPTER 10

Restoring Our Interconnected Spiritual and Ecological Integrity: Imperative for a Just and Ecologically Sustainable Peace

Shelini Harris

Introduction

A just and ecologically sustainable peace means positive peace, the existence of justice and a holistic sense of well-being and goodwill among all species. This has been difficult to achieve due to the complex interrelationship between the multiple human and non-human forms of life. At the core of this crisis is a fundamental misunderstanding of the nature and higher goals of human beings, and this misunderstanding distorts relationships with other humans and with other species, causing both the gross injustice we see within human societies and ongoing ecocide. This misconception is informed by epistemologies based on the dominant constructions of rationality/reason and the concomitant dualistic separation from nature, resulting in the tendency to perceive the relationship between humans and the rest of nature as one of conflict. This tendency is

S. Harris (✉)
Australian National University, ACT Canberra, Australia
e-mail: shelini.harris@anu.edu.au

J. Camilleri and D. Guess (eds.), *Towards a Just and Ecologically Sustainable Peace*, https://doi.org/10.1007/978-981-15-5021-8_10

apparent even among environmentalists and conservationists in the solutions they proffer for ecological problems. Drawing on spiritual traditions, I describe non-dualistic notions of the human and their concomitant epistemology, illustrating (through the example of the Sri Lankan Buddhist soil/water conservation system) how maintaining this integrity of the human is inseparable from an understanding of our interrelatedness with other humans and other species, thereby ensuring ecological well-being.

Economic Technical Misconception of the Human Being

There has been much debate about the origins, timeline and causes of the erroneous path that has put us in the current predicament of gross intra-human and inter-species injustice and depredation. In his well-known article Lynn White attributed the environmental crisis to the Judeo-Christian tradition and its teaching regarding dominion over nature (Genesis 1), arguing that during the middle ages in Europe various mechanical innovations in farming led to an ethos of aggressive domination of nature.[1] He claimed that Christian beliefs supported these developments and were responsible for the later marriage of science and technology, which has led to catastrophe. However, more recent research on the medieval period challenges White's account of a monolithic ethos dominated by Christian belief, especially one that had a unified interpretation of Christianity.[2] Furthermore, even in his account, he acknowledges detrimental ecological practices by the Greeks and Romans in earlier periods, and endorses the very different approach taken even within Latin (Western) Christianity by St. Francis of Assisi, and approaches within the Eastern Orthodox tradition. Although versions of Latin Christianity did lend support to more aggressive attitudes to nature, a range of other

[1] Lynn White Jr, 'The Historical Roots of Our Ecologic Crisis,' in *Science*, New Series, Vol. 155, No. 3767 (March 1967), 1203–1207.

[2] For a good analysis of White's legacy, errors, exaggerations and relevance, see Elspeth Whitney, 'The Lynn White Thesis: Reception and Legacy,' in *Environmental Ethics*, Vol. 35 (Fall 2013), 313–331. Eric Ashby, in *Reconciling Man with the Environment* (Stanford: Stanford University Press, 1978), 4, notes that John Passmore, while agreeing that Christianity has encouraged the view of humans as nature's absolute master, notes that it was Bacon and Descartes who really set in motion this idea: John Passmore, *Man's Responsibility for Nature* (Duckworth: London, 1974). Galileo's trouble with the church is well known.

factors were significant, including the major contributions of Francis Bacon and René Descartes (in the 1600s) to the dominance of particular notions of reason, rationality and scientific methodology.[3]

Science and rationality, as observed by many including ecofeminists such as Vandana Shiva, Val Plumwood and Sandra Harding, entail a particular way of thinking that has privileged white, male, middle-class Europeans.[4] As Shiva notes, Bacon's method is characterized by 'a dichotomising between male and female, mind and matter, objective and subjective, rational and emotional, and the conjunction of masculine and scientific domination over nature, women and the non-west'.[5] In the scientific turn, knowledge came to be associated with power over nature and often over women and non-western peoples who were considered incapable of rational, dispassionate, neutral and universal ways of thinking. Being objective meant a denial of any relationship, and of emotional or moral connection. As Plumwood notes, what eventually passed for a universal and neutral understanding was the particular way of thinking of this dominant group, shaped by the taken-for-granted facts and circumstances of the status quo.[6] Any challenge to these underlying assumptions appeared to be relative or 'special interest'. This led to a fundamental change in 'humanity's epistemic relations with nature',[7] in other words, in the way we know about and understand nature. Animals, trees, etc., were objects to be studied in mechanistic ways by an almost disembodied reason: knowledge apart from the knower. The knowledge that was gained about them was purely by way of mental perception, reason being in absolute control throughout the process. This was considered crucial for progress and for what is known as modern civilization. Thus, ecological epistemologies were classified as superstitious and dismissed.

[3] Samuel Ijsseling, *Rhetoric and Philosophy in Conflict: An Historical Survey* (The Hague: Martinus Nijhoff, 1976), 60.

[4] See discussion in Sandra Harding (ed.), *The 'Racial' Economy of Science: Toward a Democratic Future* (Bloomington: Indiana University Press, 1993), especially her introductory chapter.

[5] Vandana Shiva, *Staying Alive: Women, Ecology and Survival in India* (New Delhi: Kali for Women, 1988), 15.

[6] Val Plumwood, *Environmental Culture: The Ecological Crisis of Reason* (London: Routledge, 2002), 43–44.

[7] David Delaney, *Law and Nature* (Cambridge: Cambridge University Press, 2003), 59.

Human societies were also considered to function almost through mechanistic processes similar to evolution in nature, culminating in modernization and the rationalization of all aspects of society. These notions guided the construction of the dominant notion of development expressed by Samuel Huntington and Walt Rostow, who were very influential in shaping development policies that impacted much of the world. Rostow differentiates between modern Western and traditional societies based on whether they hold what he refers to as a Newtonian or pre-Newtonian worldview.[8] By Newtonian, he means a mechanistic understanding of nature, using nature to increase material production.[9] Huntington points out that the difference between traditional societies and modern ones is that in the former people were 'passive and acquiescent' conceiving themselves as continuous with nature, not as exerting control over their environment, especially for production processes.[10] James Scott analyses the historical and scientific processes of separating and classifying human societies and other entities such as forests in order to subject them to state and legal control, ignoring the complex interrelationships between humans and other species, and between all other species.[11] Epistemologies that located humans as continuous with the rest of nature were deemed ignorant and an impediment to modernization, progress and economic interest.

For example, the assessment by former Harvard professor Stanley Tambiah, of the cherished ancient tank agroforest soil/water conservation ecosystem of Sri Lanka, which has been marvelled at by engineers around the world[12] and cited at the International Court of Justice in the

[8] Walt W. Rostow, 'Stages of Economic Growth: A Non-Communist Manifesto (1960),' in J. Timmons Roberts and Amy Hite (eds), *From Modernization to Globalization: Perspectives on Development and Social Change* (Oxford: Blackwell, 2000), 100.

[9] Ibid., 101.

[10] Samuel Huntington, 'The Change to Change: Modernization, Development, and Politics,' in *Comparative Politics*, Vol 3, No. 3 (1971), 283–322, 287.

[11] James C. Scott, *Seeing Like a State: How Certain Schemes to Improve the Human Condition Have Failed* (New Haven: Yale University Press, 1998), 13.

[12] M. Kravcik, J. Pokorny, J. Kohutiar, M. Kovac, and E. Toth, *Water for the Recovery of the Climate: A New Water Paradigm* (Kosice: Typo-Press, 2008).

Hague as an example of a healthy human-nature ecosystem,[13] illustrates this point quite clearly:

> This devaluation, if not rejection of urban existence as a necessary contemporary fact of life, may impede the literati and the ideologues from formulating and envisioning a "plausible" satisfying and creative urban form of life for Sri Lankans. And the fixation on an idealized mode of rural, "egalitarian," temple-focused community life may also act as a brake on thinking innovatively about agro-industrial forms of life that transcend the limits of imagined village republics.[14]

This tank system was created by ancient kings to capture water during the monsoon rains to hydrate the landscape during the dry seasons, based on the dictate that not a drop of water should be wasted without benefiting humans and other species, as the land belonged to all of them (according to Buddhist beliefs). This system of landscape management is very much along the lines of permaculture ideas of complex biodiverse systems which function through the interactions of multiple species, with humans as one of them.[15] Because these small cascades of tanks were quite contrary to the modernist rational notion of large-scale water catchment and orderly, neat landscapes geared towards mass production, they were described as useless and disorderly by the British when they colonized the country. Relying on multispecies epistemology, the peasants took their cues for when to fish, when not to fish, when to harvest and so on from the surrounding plants and animals, which is consistent with teachings and beliefs within the Hindu, Buddhist and Jain systems of epistemology and sense of self that includes the rest of nature as subjects and teachers.[16] It directly violates the dominant rationalist notion that requires control

[13] Opinion of Vice President of the International Court of Justice on the decision based on the 1997 case concerning the Gabcikovo-Nagymaros (Hungary/Slovakia) Project. Christopher G. Weeramantry, *Environmental Aspects of Sri Lanka's Ancient Irrigation System* (Ratmalana, Sri Lanka: Sarvodaya Vishva Lekha, 2000).

[14] Stanley Tambiah, *Buddhism Betrayed? Religion, Politics, and Violence in Sri Lanka* (Chicago: University of Chicago Press, 1992), 109–110.

[15] Bill Mollison, *Permaculture: A Designers' Manual* (Tyalgum, Australia: Tagari, 1988), ix. The manual as a whole is a guide to setting up such systems.

[16] Cited from *Sun Newspaper*, 2 April 1983, by D.L.O. Mendis in *Eppawala—Destruction of Cultural Heritage in the Name of Development* (Ratmalana, Sri Lanka: Vishva Lekha, 1999), ii. Also, by the same author, *Alternative Development: An Anthology: Volume 1, Part II, Remember Your Humanity: If You Want Peace, Prepare For Peace* (Ratmalana: Vishva Lekha, 2005).

by the human mind. Instead of being in control of nature and manipulating it to provide for human needs, lifestyles and fulfilment of needs are coordinated to align with the functions and needs of everything else in the landscape. Human intervention in the form of constructing the little tanks was in accord with nature's patterns. Recent modern mega development projects that constructed huge tanks (e.g. Lunugamvehera and Uda Walawe), disregarding the advice of holistic ecological engineers such as the well-known D.L.O. Mendis who understood the workings of this ancient ecosystem, failed miserably to provide any water, having disrupted large areas of the ecosystems and displaced many species. This contributed to the region becoming a major source of human-human (Marxist uprising) and human-elephant conflict.

This is unsurprising given that rational control in the form of power over nature and others easily translates, as Plumwood argues, into self-interested egoism and the economic rationalism so central to acquisitive capitalism.[17] Ecological economists and sociologists observe that in social and political policies, humans are theorized primarily as self-interest maximizers; for this is considered the only rational motivation. Consequentially, needs and behaviour are determined by statistics, in isolation from any regard of the other, whether human or non-human.[18] Critical economist Stephen Marglin says:

> … in adopting a particularly extreme form of individualism, in abstracting knowledge from context, in limiting community to the nation, and in positing boundless consumption as the goal of life, economics offers us no way of thinking about the human relationships that are the heart and soul of community other than as instrumental to the individual pursuit of happiness.[19]

In the process of extracting 'raw material' from 'inert' nature in such an instrumental fashion, only a narrow set of empirically observable circumstances, consequences and other factors are considered. Thus, as Marglin

[17] Plumwood, *Environmental Culture*, 22–25.

[18] Mary Douglas and Steven Ney, *Missing Persons: A Critique of Personhood in the Social Sciences* (Berkeley: University of California Press, 1998), 22–3; Robert Costanza, 'Social Goals and the Valuation of Ecosystem Services,' in *Ecosystems*, Vol. 3, No. 1 (Jan–Feb. 2000), 4–10, 6.

[19] Stephen A. Marglin, *The Dismal Science: How Thinking Like an Economist Undermines Community* (Cambridge: Harvard University Press, 2008), 9.

observes, what cannot be measured is considered not to exist.[20] It is not difficult to imagine how this easily led to the destruction of marginalized peoples and nature, whose interests and knowledge neither qualified as 'reason' or as 'empirically demonstrable' according to this narrow rationality, as illustrated in the example above. Because human happiness is associated with increased consumption, which necessitates perpetual growth in the economy, entailing the continual production of material goods and infrastructure to facilitate growth which is dependent on resources and energy extracted from nature, the result is the displacement of other species and marginalized humans.[21] In the above example, the government built the mega-tanks and provided chemicals for the peasants to raise crops; however, the tanks failed to produce water, so the whole project failed. Plumwood observes that the majority of scientists work for corporations, resulting in instrumental goals that are defined by a narrow set of interests aimed at increasing productivity.[22]

The conflicts that ensued were almost inevitable. Psychologists observe that the values and goals promoted by the dominant capitalist economic structures (in the above case, by the International Financial Institutions via the local government) end up shaping people's perceptions of themselves and their concomitant behaviour.[23] As they argue, 'broad socio-cultural practices and ideologies often have an enormous influence on people's self-concepts, motivation, behaviour, and interpersonal relationships'.[24] Thus, the more communal and cooperative values and goals are dismissed or discouraged in order to foster materialist desires, goals and notions of self-interest which are always perceived as being in competition with others. Social structures and institutions are constructed on the assumption that humans are motivated almost exclusively by self-interest,

[20] Marglin, *The Dismal Science*, 6.

[21] Richard Heinberg, *The End of Growth: Adapting to Our New Economic Reality* (Gabriola Island, Canada: New Society Publishers, 2011), 6.

[22] Plumwood, *Environmental Culture*, 38–39.

[23] Tim Kasser, Steve Cohn, Allen D. Kanner, and Richard M. Ryan, 'Some Costs of American Corporate Capitalism: A Psychological Exploration of Value and Goal Conflict,' in *Psychological Inquiry*, Vol. 18, No. 1 (2007), 2, 3, 6.

[24] Kasser et al., 'Some Costs of American Corporate Capitalism', 2.

with the expectation of self-interest being the most 'normal' accepted kind of behaviour, thus making it a self-fulfilling prophecy.[25]

This is apparent for instance in how some environmental/conservation groups try to protect certain forests and wildlife by forming reserves separating them from the local humans. For instance, in the situation described above and in similar other contexts, due to the conflicts that erupted, the wildlife experts (trained in Western science) and other officials constructed electric fences (a hard tech approach, requiring a lot of material resources that are unsustainable in the long run) separating humans and wildlife. These demarcations paid attention neither to the migratory paths etched in the long memories of the elephants nor the historic crop raising practices of the peasants. Consequently, neither party stayed on its side of the fence, resulting in casualties on both sides. The authorities blame both of the victims—'rogue' elephants and 'illegal' peasant farmers. Historically, despite instances of exploitation and violence towards elephants,[26] local zoologists have noted a rather synergistic relationship. They describe elephants as synanthropic, whereby, especially in the past, humans and elephants occupied the same landscapes; elephants enjoying the remnants of the shifting agriculture practiced by the peasants.[27] Because of the shifting agriculture, conflict was avoided (This is not to say all human-elephant interactions and religious use of elephants were not exploitative). Therefore, these local scientists argue for a more synergistic, low tech, organic solution:

> ... it is not really necessary to resort to the so called high-tech "Action Plans" which are in most cases drawn at great financial cost, by the so called foreign consultants and experts, who have little knowledge and understanding of local ecological conditions and processes, as well as of the socio-economic conditions and the interactions with wildlife and their habitat by the local human population.[28]

[25] Dale T. Miller, 'The Norm of Self-Interest,' in *American Psychologist*, Vol. 54, No. 12 (1999).

[26] Martha Chaiklin, 'Ivory in Early Modern Ceylon: A Case Study in What Documents Don't Reveal', in *International Journal of Asian Studies*, Vol. 6, No. 1 (2009), 37–63

[27] Mangala De Silva and Padma Kumari De Silva, *The Sri Lankan Elephant: Its Evolution, Ecology and Conservation* (Colombo: WHT Publications, 2007).

[28] H. I.E. Katugaha, Mangala De Silva, and Charles Santiapillai, 'A Long-Term Study on the Dynamics of the Elephant (*Elephas Maximus*) Population in Ruhuna National Park, Sri Lanka,' in *Biological Conservation*, Vol. 89, No. 1 (1999), 51–59, 58.

Often, dominant solutions remain trapped within the same frameworks and end up reinforcing them or achieving partial liberation, arriving at some form of negative peace where there is cessation of open conflict, but structures and practices of injustice/oppression remain intact. The two major challenges facing the globe today are the ecological crisis and stark inequality, the ill effects of each impinging on the other. For example, human rights lawyer and UN rapporteur Philip Alston has recently issued a scathing report with his findings of a 'climate apartheid' where the rich and powerful, who are primarily responsible for the climate catastrophe, are able to cushion themselves from the worst effects of it, while the poor face the brunt of it; driven to such levels of deprivation that he says even basic human rights could become impossible goals.[29] His concern is that the dominant approaches, including human rights efforts, are far from adequate. For instance, the Millennium Development Goals and the subsequent Sustainable Development Goals continue to stress economic growth (largely dependent on environmental depletion) despite their acknowledgement of environmental concerns and inequalities within and between countries (the opposite of the 'trickle down' promised by neoliberal economic growth).[30] As noted above, the mastery of nature in order to increase productivity is central to the rationality of capitalist modernity, a rationality that privileges the status quo, resulting in further deprivation for marginalized human beings, making Alston's claim of climate apartheid inevitable.

Scholars in Subaltern Studies (e.g. Ramachandra Guha and Joan Martinez-Alier) speak of 'full-stomach' and 'empty-belly' environmentalism.[31] As noted in the example above, many conservation efforts problematize the livelihoods and actions of the indigenous and other

[29] UN Human Rights Office of the High Commissioner, 'UN Expert Condemns Failure to Address Impact of Climate Change on Poverty,' https://www.ohchr.org/EN/NewsEvents/Pages/DisplayNews.aspx?NewsID=24735&LangID=E, accessed 21 June 2019.

[30] Thomas Pogge and Mitu Sengupta, 'Assessing the Sustainable Development Goals from a Human Rights Perspective,' in *Journal of International and Comparative Social Policy*, Vol. 32, No. 2, (2016), 83–97; Viktoria Spaiser, Shyam Ranganathan, Ranjula Bali Swain and David J. T. Sumpter, 'The Sustainable Development Oxymoron: Quantifying and Modelling the Incompatibility of Sustainable Development Goals,' in the *International Journal of Sustainable Development and World Ecology*, Vol. 24, No. 6 (2017), 457–470.

[31] Ramachandra Guha and Joan Martinez-Alier, *Varieties of Environmentalism: Essays North and South* (London: Earthscan, 1997), 12.

marginalized people who happen to live on the fringes of forests and other areas significant for conservation.[32] This is evident in the way problems are framed. Conflicts over forest resources or crop raiding by wildlife are framed as an issue of balancing the rights and well-being of the human population vs that of wildlife, with separate 'experts' attempting to find solutions in isolation from each other. For example, on the one hand, development workers try to help villagers; on the other, conservationists try to protect flora or fauna threatened by villagers who rely on forests for their needs, occasionally killing crop-raiding wildlife in desperation. Not surprisingly, the solutions proffered tend to facilitate artificial means of keeping them separate from each other to avoid conflict, expecting them to always act according to 'self-interest'.[33] What is missing is an understanding of how humans, other animals, and the landscape, have survived together for centuries, often millennia, and a willingness to be open to epistemologies and knowledges, not only of other humans but of non-humans as well.

Solutions continue to be embedded within and to perpetuate the dominant neoliberal framework emphasizing self-interest and consumption, and focus on controlling the lives and behaviour of the wildlife and local people rather than the dominant structures or elites.[34] Again, this has to do with the resilience of the idea of economic rationality, of faith in the 'objective' science of economic methods to solve problems. Often, ecotourism and other such projects of the global neoliberal market are offered as the solution. Rosaleen Duffy notes that 'In essence, nature-based tourism allows neoliberalism to turn the very crises it has

[32] Ramachandra Guha, 'Radical American Environmentalism and Wilderness Preservation: A Third World Critique,' in *Environmental Ethics*, Vol. 11, No. 1 (1989).

[33] Subhabrata Bobby Banerjee, 'Who Sustains Whose Development? Sustainable Development and Reinvention of Nature,' in *Organization Studies*, Vol. 24, No. 1 (2003), 158.

[34] Paul J. Funston, Rosemary J. Groom, Peter A. Lindsey, 'Insights into the Management of Large Carnivores for Profitable Wildlife-Based Land Uses in African Savannas,' in *PLoS One*, Vol. 8, No. 3 (2013). For discussions problematizing these approaches, see Rosaleen Duffy, 'Nature-Based Tourism and Neoliberalism: Concealing Contradictions,' in *Tourism Geographies*, Vol. 17, No. 4 (2015), 529–543. Maria Magdalena Tancau, 'Crafting Wild Nature in South-West Uganda: Rich Communities, Happy Tourists,?' *Journal of Eastern African Studies*, Vol. 5, No. 3 (2011), 465–481.

created into new sources of accumulation ... allowing it to extend neoliberal logics to a greater range of non-human biophysical phenomena'.[35] These projects require that peasants and forest dwellers are torn away from their lifestyles that are integrated with the forests and wildlife, to turn them into commercial actors (serving global capitalism), instead of challenging the large-scale commercial practices that create the conditions for these immediate local level conflicts between the marginalized humans and non-humans. This entails epistemic, spiritual and structural violence which would be classified as slow violence in Peace Studies, '...a violence that occurs gradually and out of sight, a violence of delayed destruction that is dispersed across time and space, an attritional violence that is typically not viewed as violence at all'.[36] This persistent misconstrual in multiple attempts to 'solve' problems indicates that the error is deep and fundamental—in our epistemology and sense of self. Several indigenous and other spiritual/religious systems offer alternative worldviews, epistemologies and sense of self. I would caution against an uncritical idealistic romanticizing of indigenous/native cultures and note Helen Kopnina's concerns regarding the misuse of indigenous/native interests in a way that is not consistent with ecologically sound lifestyles.[37] Furthermore, none of the cultures I discuss can be said to have embodied or upheld consistently and ubiquitously the teachings, beliefs and practices I highlight. The spread of modernist/development values and worldviews has almost displaced holistic worldviews, especially in the urban areas. Nevertheless, they are of sufficient salience within these cultures to have invited the assessments noted above by development theorists, and are thus of use in providing resources and ideas as we try to construct just and ecological lifestyles, and most importantly, alternative understandings of what it means to be human. Within these cultures as well, there are movements that reject the dominant paradigm, aligning themselves to some of these traditional insights. Due to space, I limit my discussion to beliefs found within some Buddhist and Hindu traditions.

[35] Duffy, 'Nature-Based Tourism,' 529.

[36] Rob Nixon, *Slow Violence and the Environmentalism of the Poor* (Cambridge: Harvard University Press, 2011), 2.

[37] Helen Kopnina, 'Toward Conservational Anthropology: Addressing Anthropocentric Bias in Anthropology,' in *Dialectical Anthropology*, Vol. 36, Nos. 1 and 2 (2012), 127–146.

The Holistic Ecological Human Being

While the contribution of Mahatma Gandhi to non-violence is well known, what is less known is his major goal, which was to rid India of a form of civilization (much of what has been described above) that he believed violated and kept captive the truer sense of the self. 'I bear no enmity towards the English, but I do towards their civilization. … But there is no end to the victims destroyed in the fire of civilization. Its deadly effect is that people come under its scorching flames believing it to be all good'.[38] He was concerned that by creating a false sense of self that relied on external material power, often entailing violence to humans and the rest of nature, the true spiritually empowered self was enslaved. Gandhi's notion of *swaraj* is control over one's own mind and self and is the crucial step towards any other kind of freedom.[39] Gandhi pronounced that 'strength lies in the absence of fear', meaning that one exhibited greater strength in being able to coexist with that which is fearsome, without being fearful. He pointed out that the British claimed that they were protecting the Indians from the dangerous wild animals, but the Indian villagers slept peacefully at night, quite vulnerable to the wild animals, long before the British arrived.[40] This attitude is a reflection of a sense of self that is continuous with the rest of nature: one lives with and cooperates with all other entities, accepting the concomitant contingencies. Modern rationalistic culture has placed an emphasis on controlling nature itself as a threat and source of fear, as well as being a source of raw material. As the modernizers Rostow and Huntington argued, it is precisely this attitude of the 'traditional' people which separated them from modern culture—a sign of their fatalism and backwardness.[41]

The source of this seeming fatalism is predicated on an elaborate belief system incorporating a kinship with the rest of nature, wherein the material and spiritual are inseparable. In the Hindu and Buddhist worldviews of reincarnation, animals are part of the cosmic lifecycle of humans on the

[38] Mohandas K. Gandhi, *Hind Swaraj: And Other Writings*, Anthony J. Parel (ed.) (Cambridge: Cambridge University Press, 1997), 39, 43, 74, 119. Gandhi wrote Hind Swaraj in 1909.

[39] Gandhi, *Hind Swaraj*, 94.

[40] Gandhi, *Hind Swaraj*, 38–39.

[41] Rostow, *'Stages of Economic Growth,'* 100. Huntington, *'The Change to Change'*, 145.

path to escaping the wheel of *samsara*, and towards *nirvana*. As Christopher Key Chapple points out, this belief is at the basis of the strong sense of kinship and *ahimsa* (non-violence), which also relativizes the human birth within the broader cosmic journey. He notes that the teachings of the sages, yoga and so on reference animals and other aspects of nature as teachers and kin meriting respect and protection. He gives the example from Hindu scriptures, where a spiritual seeker, Satyakama, goes into the forest to learn, and on being instructed by a bull, other animals and elements of nature, 'discovered the inner working of the human being, living in a body through the sense and the mind'.[42] Chapple notes that yoga manuals instruct how to 'mimic the stance and mood of specific animals',[43] because, contra Descartes' mechanistic view of them, animals in this worldview are a great mystery, only understood through a recognition of their agency and power with a sense of intimate awe. Religious practices such as the narratives of past lives as inseparable from animals, dietary and agricultural practices that respect other lives, yoga movements and meditations which invite one to view and experience the 'world through the body and senses of an animal', incorporate one into a very different epistemology and sense of self. The conceptual tools we use to understand our surroundings shape us and our environment accordingly. Through this kind of engagement with other species, we form a different kind of subjecthood. The founder of one of the biggest and most successful movements for an alternative to the dominant model of development, inspired by the Gandhian notion of *Sarvodaya* (the awakening of all) and Buddhism, A. T. Ariyaratne, points out that humans have tried to study and subjugate nature with technology as if they were not a part of nature.[44] Gandhi states emphatically that ethics and economics cannot be separated, that the use of machines and other tools of industrialization are not to be at the cost of human beings or the environment.[45] Economics has to be moral. If it hurts people or causes some to prey

[42] Christopher Key Chapple, 'Animal Ethics', in *Sophia*, Vol. 57 (2018), 76.

[43] Chapple, 'Animal Ethics,' 77, 82.

[44] A.T. Ariyaratne, *Buddhism and Sarvodaya: Sri Lankan Experience* (Delhi: Sri Satguru Publications, 1996), 56.

[45] Gandhi, *Hind Swaraj*, 67, 94.

on others, it is sinful. He says: 'That economics is untrue that ignores or disregards moral values'.[46]

Ariyaratne says that one needs to recognize the dominant rationality and the falsehoods it promotes as defilements and cleanse them from one's mind and then proceed to 'recognize that there are unjust and immoral socio-economic chains which keep the vast majority of people enslaved, and that these chains have to be removed if the human being is to experience true freedom'.[47] Buddhists say: 'Only the mind thus cleansed and nonfragmentary will recognise that one's own welfare is inseparable from that of all life and will struggle against unjust and immoral socio-economic structures that keep other lives enslaved and suffering'.[48] Gandhi said:

> Identification with everything that lives is impossible without self-purification; without self-purification the observance of the law of *ahimsa* must remain an empty dream; God can never be realized by one who is not pure of heart. Self-purification therefore must mean purification in all the walks of life. A purification being highly infectious, purification of oneself necessarily leads to the purification of one's surroundings.[49]

That means that spiritual purification and awareness will have implications for every aspect of life. The venerable Pategama Gnanarama says that trees are living organisms, silently serving us, just as dogs, cows or other creatures do, and that it is the forest of passions that leads to the cutting down of trees in the forest, so one ought to cut down desires, not trees, citing the teaching from the Dhammapada: 'Cut down the forest of passions (*vana*) but not real trees. From the forests of passions springs fear'.[50] Therefore, cutting down a tree is a serious offence for a higher ordained monk. Ariyaratne and Nandasena Ratnapala emphasize the inseparability of the well-being of animals from any notion of justice

[46] Gandhi, *All Men Are Brothers: Autobiographical Reflections* (New York: Continuum, 1995), 113–118.

[47] Ariyaratne, *Buddhism and Sarvodaya*, 4.

[48] Ariyaratne, *Buddhism and Sarvodaya*, 4. 'All lives' means human and non-human.

[49] Gandhi, *An Autobiography: The Story of My Experiments with Truth* (Boston: Beacon Press, 1993), 753.

[50] Venerable Pategama Gnanarama, *An Approach to Buddhist Social Philosophy* (Singapore: Ti-Sarana Buddhist Association, 1996), 145.

or democracy in the Buddhist worldview.[51] Such consideration of other species is primarily perceived as *dhamma*, or the eternal way of reality (it can also be understood as law, morality or truth). Therefore, Ariyaratne's Sarvodaya societies (in over 15,000 villages across Sri Lanka) begin with spiritual practices, recognizing that no alternative to the current models of development is possible without a transformation of one's epistemology and concomitant sense of self.[52]

These spiritual teachings point to spiritual cleansing and the openness it creates to other epistemologies, and thus, the restoration of a healthier, truer sense of who we are. It is these kinds of selves that can facilitate cultures of peace which are just and ecologically sustainable. Peace scholar and Quaker sociologist Elise Boulding asserts:

> ... peace culture includes ways of life, patterns of belief, values, behaviour, and accompanying institutional arrangements that promote mutual caring and well-being as well as an equality that includes an equitable sharing of the earth's resources among its human members and with all living beings. It offers mutual security for humankind in all its diversity through a profound sense of species identity as well as kinship with the living earth.[53]

The tank system described above fits well into this understanding. It is an agro-eco-forestry system that works in a soil and water conservation landscape. Due to the complex interactions and ecological cycles, including the dense forest surroundings and small size of the tanks, moisture is retained in the system, encouraging healthy microbial activity in the soil, minimizing the need even for organic fertilizers. The crucial lesson here is that if not for the respect given to the epistemology and needs of the large number of species of animals and plants, this system will not work. The spiritual beliefs uniting the material, non-material, mind, body and

[51] Ariyaratne, *Buddhism and Sarvodaya*, 67. Nandasena Ratnapala, *Buddhist Democratic Political Theory and Practice* (Ratmalana, Sri Lanka: Sarvodaya Vishva Lekha Publications, 1997), vi, xi–xiii. For a discussion of the way the Buddhist emperor Asoka established universal health care for animals in the fourth century BCE, see Asoka King of Magadha, *The Edicts of Asoka*, N. A. Nikam and Richard McKeon (ed. and trans.) (Chicago: University of Chicago Press, 1959), especially 64–65.

[52] George Bond, *Buddhism at Work: Community Development, Social Empowerment, and the Sarvodaya Movement* (Bloomfield: Kumarian Press, 2004).

[53] Elise Boulding, *Cultures of Peace: The Hidden Side of History* (Syracuse: Syracuse University Press, 2000), 1.

spirit worlds are an integral and essential part of this world; temples and shrines are present as an inseparable part of the landscape and ethos. Many of the Sarvodaya villages are located in and participate in some of the few remaining such landscapes. Although much of it has been disrupted over the course of modern development and earlier neglect, it provides useful guidance. Rural sociologist Philip McMichael, who has contributed much to critical thought on development and peasant movements against it, observes:

> Subjecting the "environment" to management is accomplished either through some mediating relationship such as price, or through restorative relationships to more closely integrate human activity with natural cycles … An ecological calculus, then, is the obverse of the market calculus in privileging and protecting the environment, and reorienting human activity to restoring and valuing natural relations and processes on their own terms, rather than endowing them with market values and being subjected to a cost/benefit calculus.[54]

This is what the beliefs and practices previously discussed provide—a means for creating an ecological calculus.

Plumwood embraces what she calls a materialist spirituality of place and expresses concerns about calls for spirituality as providing deeper motivation, as she feels they arise from the duality and rationality we have criticized above, which perceives a split between the material and spiritual, with the former being reduced in significance and often rejected.[55] Most religious traditions have some form of criticism or other of 'materialism', and certainly, the forms of Christianity (and aspects of some other religions as well) that support the aggressive attitudes mentioned by White do justify her concerns. Plumwood cites John Passmore's account of how Christians saw paganism and the veneration of nature as their main enemy, with missionaries destroying such belief systems in many of the colonies. Many such instances in history caution us against an uncritical acceptance of spirituality. However, notions of spirituality and

[54] Philip McMichael, 'Food system Sustainability: Questions of Environmental Governance in the New World (Dis)Order', in *Global Environmental Change*, Vol. 21 (2011), 805.

[55] Plumwood, *Environmental Culture*, 219, 220–223.

materiality within various traditions are quite diverse. Even within Christianity (especially the Eastern Orthodox tradition) and in much of the Hindu and Buddhist traditions I have discussed above, the material and spiritual worlds are not seen as a duality. I cannot address the Christian strands here due to space. From the discussions above, it should be apparent that the material and non-material world are intertwined and inseparable within the belief structures, shared epistemology and formation of subjecthood.[56] Acknowledging that the visible material world is part of, and not all of, reality does not necessarily result in a rejection or devaluation of it. Criticism of materialism, even above, by Ariyaratne (and ultimately this is Gandhi's concern), is a rejection of the consumptive lifestyle that erodes our sense of self and freedom, ultimately rendering us incapable of perceiving our interrelatedness with other species and prone to destroying the material world.[57]

Though they may not have always been ecological, just and peaceful, cultural beliefs, epistemologies, and knowledge, similar to those I have described above, derived in communion with landscapes and other lives, offer some resources and guidance in our struggle to restore our sense of spiritual integrity inseparable from the rest of nature. It is a call to adopt practical lifestyles, and in doing so, think with the other animal—the elephants, the trees—in our deliberations over solving our current crises, instead of the dominant patterns of alienated monolithic thinking that reduces everything to a financial calculus.[58] As noted above, there

[56] For a discussion of this kind of approach within Christianity, see, for instance, Fred Bahnson and Norman Wirzba, *Making Peace with the Land: God's Call to Reconcile with Creation* (Westmont: InterVarsity Press, 2012).

[57] I have discussed how spirituality offers us an alternative sense of self and values, giving us a vantage point outside of the dominant ecocidal rationality and facilitating actual alternative lifestyles, in 'Moral and Spiritual Change as a Basis for a Shift to Ecological Integrity: The Amish and Buddhists,' in *Climate Change, Cultural Change: Religious Responses and Responsibilities*, Anne Elvey and David Gormley-O'Brien (eds), (Eugene, Oregon: Wipf and Stock, 2013).

[58] Numerous and multifarious efforts, including alternative forms of education, Transition Towns, movements like Via *Campesina*, permaculture/agro-ecoforestry, among others, are already taking place and more needs to be done. Due to space they can't be addressed here. I discuss some examples in 'Moral and Spiritual Change' (ibid.). See also Juliana Birnbaum and Louis Fox (eds), *Sustainable Revolution: Permaculture in Ecovillages, Urban Farms, and Communities Worldwide* (Berkeley: North Atlantic Books, 2014); Philip McMichael, 'Peasants Make Their Own History, But Not Just as They Please... ,' in *Journal of Agrarian Change*, Vol. 8, Nos. 2 and 3 (April and July 2008), 205–228.

is a growing body of knowledge demonstrating what traditional/spiritual peoples have known all along about our interrelatedness.[59] This understanding of our interrelatedness and mutuality contradicts the ones upheld by current dominant economical/technical systems and ideologies, thus offering us inspiration and resources to restore our spiritual, and consequently, ecological integrity.

Bibilography

Ariyaratne, A. T. 1996. *Buddhism and Sarvodaya: Sri Lankan Experience*. Delhi: Sri Satguru Publications.

Ashby, Eric. 1978. *Reconciling Man with the Environment*. Stanford, CA: Stanford University Press.

Aśoka, King of Magadha. 1959. *The Edicts of Asoka*, eds. and trans. Narayanrao Appurao Nikam, and Richard Peter McKeon. Chicago: University of Chicago Press.

Bahnson, Fred, Norman Wirzba, and Bill McKibben. 2012. *Making Peace with the Land: God's Call to Reconcile with Creation*. Westmont: InterVarsity Press.

Banerjee, Subhabrata Bobby. 2003. 'Who Sustains Whose Development? Sustainable Development and the Reinvention of Nature'. *Organization Studies* 24 (1): 143–180.

Birnbaum, Juliana, and Louis Fox. 2014. *Sustainable [R]Evolution: Permaculture in Ecovillages, Urban Farms, and Communities Worldwide*. Berkeley, CA: North Atlantic Books.

Bond, George Doherty. 2004. *Buddhism at Work: Community Development, Social Empowerment and the Sarvodaya Movement*. Bloomfield, CT: Kumarian Press.

Boulding, Elise. 2000. *Cultures of Peace: The Hidden Side of History*. Syracuse: Syracuse University Press.

Bradshaw, G. A. 2010. 'An Ape among Many: Animal Co-Authorship and Trans-Species Epistemic Authority'. *Configurations: A Journal of Literature, Science, and Technology* 18 (1/2): 15–30.

[59] Frans de Waal, *Good Natured: The Origins of Right and Wrong in Humans and Other Animals* (Cambridge: Harvard University Press, 1996), 50–2; G. A. Bradshaw, 'An Ape Among Many: Animal Co-Authorship and Trans-species Epistemic Authority,' in *Configurations*, Vol. 18, Issues 1 and 2 (Winter 2010), 15–30. I discuss some of these findings in the fields of biology, zoology and psychology in 'Our War with Nature and Each Other from a Buddhist Perspective: Insights from Buddhism and Sharon Welch,' in *Ecological Aspects of War: Religious and Theological Perspectives*, Anne Elvey, Deborah Guess, and Keith Dyer (eds), (Hindmarsh, South Australia: ATF Press, 2016).

Chaiklin, Martha. 2009. 'Ivory in Early Modern Ceylon: A Case Study in What Documents Don't Reveal'. *The International Journal of Asian Studies* 6 (1): 37–63.

Costanza, Robert. 2000. 'Social Goals and the Valuation of Ecosystem Services'. *Ecosystems* 3 (1): 4–10.

Delaney, David. 2003. *Law and Nature*. New York: Cambridge University Press.

De Silva, Mangala, and Padma Kumari De Silva. 2007. *The Sri Lankan Elephant: Its Evolution, Ecology and Conservation*. Colombo: WHT Publications.

De Waal, F. B. M. 1996. *Good Natured: The Origins of Right and Wrong in Humans and Other Animals*. Cambridge, MA: Harvard University Press.

Douglas, Mary, and Steven Ney. 1998. *Missing Persons: A Critique of Personhood in the Social Sciences*. Berkeley: University of California Press.

Duffy, Rosaleen. 2015. 'Nature-Based Tourism and Neoliberalism: Concealing Contradictions'. *Tourism Geographies* 17 (4): 529–543.

Funston, Paul J., Rosemary J. Groom, and Peter A. Lindsey. 2013. 'Insights into the Management of Large Carnivores for Profitable Wildlife-Based Land Uses in African Savannas'. *PloS One* 8 (3): e59044.

Gandhi, Mahatma. 1993. *An Autobiography: The Story of My Experiments with Truth*. Boston: Beacon Press.

Gandhi, Mahatma. 1995. *All Men Are Brothers: Autobiographical Reflections*. New York: Continuum.

Gandhi, Mahatma. 1997. *Hind Swaraj and Other Writings*, ed. Anthony J. Parel. Cambridge: Cambridge University Press.

Gnanarama, Venerable Pategama. 1996. *An Approach to Buddhist Social Philosophy*. Singapore: Ti-Sarana Buddhist Association.

Guha, Ramachandra. 1989. 'Radical American Environmentalism and Wilderness Preservation: A Third World Critique'. *Environmental Ethics* 11 (1): 71–83.

Guha, Ramachandra, and Juan Martínez Alier. 1997. *Varieties of Environmentalism: Essays North and South*. London: Earthscan Publications.

Harding, Sandra G. 1993. *The "Racial" Economy of Science: Toward a Democratic Future*. Bloomington: Indiana University Press.

Harris, Shelini. 2013. 'Moral and Spiritual Change as a Basis for a Shift to Ecological Integrity: The Amish and Buddhists'. In *Climate Change, Culture Change: Religious Responses and Responsibilities*, eds Anne Elvey and David Gormley-O'Brien, 133–144. Eugene, Oregon: Wipf and Stock Publishers.

Harris, Shelini. 2016. 'Our War with Nature and Each Other from a Buddhist Perspective: Insights from Buddhism and Sharon Welch'. In *The Ecological Aspects of War: Religious and Theological Perspectives*, eds Anne Elvey, Deborah Guess, and Keith Dyer, 47–71. Hindmarsh, South Australia: ATF Press.

Heinberg, Richard. 2011. *The End of Growth: Adapting to Our New Economic Reality*. Gabriola Island, Canada: New Society Publishers.

Huntington, Samuel P. 1971. 'The Change to Change: Modernization, Development, and Politics'. *Comparative Politics* 3 (3): 283–322.

Ijsseling, Samuel. 1976. *Rhetoric and Philosophy in Conflict: An Historical Survey*. The Hague: M. Nijhoff.

Kasser, Tim, Steve Cohn, Allen D. Kanner, and Richard M. Ryan. 2007. 'Some Costs of American Corporate Capitalism: A Psychological Exploration of Value and Goal Conflicts'. *Psychological Inquiry* 18 (1): 1–22.

Katugaha, H. I. E., Mangala de Silva, and Charles Santiapillai. 1999. 'A Long-Term Study on the Dynamics of the Elephant (Elephas Maximus) Population in Ruhuna National Park, Sri Lanka'. *Biological Conservation* 89 (1): 51–59.

Kopnina, Helen. 2012. 'Toward Conservational Anthropology: Addressing Anthropocentric Bias in Anthropology'. *Dialectical Anthropology* 36 (1/2): 127–146.

Kravcik, M., J. Pokorny, J. Kohutiar, M. Kovac, and E. Toth. 2008. *Water for the Recovery of the Climate: A New Water Paradigm*. Kosice: Typo-Press Publishing House.

Marglin, Stephen A. 2008. *The Dismal Science: How Thinking Like an Economist Undermines Community*. Cambridge, MA: Harvard University Press.

McMichael, Philip. 2008. 'Peasants Make Their Own History, but Not Just as They Please'. *Journal of Agrarian Change* 8 (2–3): 205–228.

McMichael, Philip. 2011. 'Food System Sustainability: Questions of Environmental Governance in the New World (Dis)Order'. *Global Environmental Change* 21 (3): 804–812.

Mendis, D.L.O. 1999. *Eppawala: Destruction of Cultural Heritage in the Name of Development*. Ratmalana, Sri Lanka: Vishva Lekha.

Mendis, D.L.O. 2005. *Alternative Development: An Anthology: Volume 1, Part II, Remember Your Humanity: If You Want Peace, Prepare for Peace*. Ratmalana: Vishva Lekha.

Miller, Dale T. 1999. 'The Norm of Self-Interest'. *American Psychologist* 54 (12): 1053–1060.

Mollison, Bill. 1988. *Permaculture: A Designers' Manual*. Tyalgum, Australia: Tagari Publications.

Nixon, Rob. 2011. *Slow Violence and the Environmentalism of the Poor*. Cambridge, MA: Harvard University Press.

Passmore, John Arthur and RWV Elliott collection. 1974. *Man's Responsibility for Nature: Ecological Problems and Western Traditions*. London: Duckworth.

Plumwood, Val. 2002. *Environmental Culture: The Ecological Crisis of Reason*. London, New York: Routledge.

Pogge, Thomas, and Mitu Sengupta. 2016. 'Assessing the Sustainable Development Goals from a Human Rights Perspective'. *Journal of International and Comparative Social Policy* 32 (2): 83–97.

Ratnapala, Nandasena. 1997. *Buddhist Democratic Political Theory and Practice.* Ratmalana, Sri Lanka: Sarvodaya Vishva Lekha Publications.

Rostow, Walt W. 2000. 'Stages of Economic Growth: A Non-Communist Manifesto (1960)'. In *From Modernization to Globalization: Perspectives on Development and Social Change*, eds J. Timmons Roberts and Amy Hite, 100–109. Oxford: Blackwell.

Scott, James C. 1998. *Seeing Like a State: How Certain Schemes to Improve the Human Condition Have Failed.* New Haven: Yale University Press.

Shiva, Vandana. 1988. *Staying Alive: Women, Ecology, and Development.* London, New Delhi: Zed Books: Kali for Women.

Spaiser, Viktoria, Shyam Ranganathan, Ranjula Bali Swain, David J. T. Sumpter, Institutionen för samhällsvetenskaper, Södertörns högskola, and Nationalekonomi. 2017. 'The Sustainable Development Oxymoron: Quantifying and Modelling the Incompatibility of Sustainable Development Goals'. *International Journal of Sustainable Development & World Ecology* 24 (6): 457–470.

Tancau, Maria Magdalena. 2011. 'Crafting Wild Nature in South-West Uganda: Rich Communities, Happy Tourists?' *Journal of Eastern African Studies* 5 (3): 465–481.

Tambiah, Stanley. 1992. *Buddhism Betrayed? Religion, Politics, and Violence in Sri Lanka.* Chicago: University of Chicago Press.

UN Human Rights Office of the High Commissioner. 2019. 'UN Expert Condemns Failure to Address Impact of Climate Change on Poverty'. https://www.ohchr.org/EN/NewsEvents/Pages/DisplayNews.aspx?NewsID=24735&LangID=E. Accessed 21 June 2019.

Weeramantry, Christopher G. 2000. *Environmental Aspects of Sri Lanka's Ancient Irrigation System.* Ratmalana, Sri Lanka: Sarvodaya Vishva Lekha Publishers.

White, Lynn. 1967. 'The Historical Roots of Our Ecologic Crisis'. *Science* 155 (3767): 1203–207. http://www.jstor.org.virtual.anu.edu.au/stable/1720120.

Whitney, Elspeth. 2015. 'Lynn White Jr.'s 'The Historical Roots of Our Ecologic Crisis' After 50 Years'. *History Compass*, 13: 396—410. https://doi.org/10.1111/hic3.12254.

Shelini Harris is currently at the Australian National University and is analysing the effects of the dominant human rights paradigm especially in terms of its entanglement with modernist neoliberal frameworks and concomitant understanding of religion and law. Formerly a lecturer in the fields of Peace and Conflict Studies and Religious Studies in the United States, her research includes studying the connections between spirituality, values and sustainable lifestyles, and the forms of religious beliefs that are either detrimental to or conducive to justice and peace between humans, as well as with the environment. Shelini Harris focuses especially on the contribution of Buddhist beliefs and practices and Russian spiritual/philosophical traditions and lives of saints, to ecological thinking and lifestyles. Her academic training includes Russian Studies, especially Russian religious thought, see her thesis *Nicolas Berdiaev's Concept of "Personality" as it Relates to His Understanding of Good and Evil, Ethics, and Politics*. She has also trained in Permaculture Design (PDC), and other ecological water and soil management systems, conducting field research in these areas, including human-animal conflict which she incorporates as essential elements in Peace Education.

CHAPTER 11

Breathing the Others, Seeing the Lives: A Reflection on Twenty-First-Century Nonviolence

Chaiwat Satha-Anand

Introduction: Pope Francis' Question

On 8 July 2013, Pope Francis visited Lampedusa, a small Italian island some 70 miles from Tunisia known as North Africa's 'gateway to Europe'. The island is the destination to the world's deadliest migration route: More than twenty thousand have lost their lives trying to cross the Mediterranean from Africa to Europe. Looking out over the Port of Lampedusa, standing at an altar which was assembled from remnants of wooden refugee boats while performing mass, the Pope asked the people in the congregation: 'Has anyone of us grieved for the death of these brothers and sisters?' The Pope then pointed out that if the answer is no, it is probably because their lives have become a mere abstraction which has occurred through the globalization of indifference. They have become nameless, voiceless, and shapeless. Their lives have no value and

C. Satha-Anand (✉)
Thammasat University, Bangkok, Thailand

J. Camilleri and D. Guess (eds.), *Towards a Just and Ecologically Sustainable Peace*, https://doi.org/10.1007/978-981-15-5021-8_11

their deaths have little or no consequence. As such, grief for their deaths is out of the question.[1]

Perhaps it is the victims' tales that best echo Pope Francis' reflection on this human tragedy. One Mexican migrant who crossed the US border, arrested and sent back dozens of times, expressed his experiences with painful eloquence: 'When one travels as an illegal, he is dragged like the tail of an animal, like trash. I imagined how the sea washes trash onto the shore, and I told myself, maybe here it's just like I'm in the ocean, being tossed out again and again'.[2]

The Pope's question is in fact quite difficult, especially if one looks at how countries around the world are coping with the migration crisis where in some cases politicized immigration rhetoric has cast them in the role of 'animals, gang members, rapists, or even aliens'.[3] But even among countries with best practices on the issue, the struggle has been immensely hard. In the Danish Social Democrat Party's effort to advance a 'realistic and fair' immigration policy as a way to unite a country torn apart by this issue like so many, the party leader Mette Frederiksen, now the country's prime minister, pointed out that the world is in a 'new' situation where record numbers of refugees are on the move. But unlike their predecessors, the majority of them fled their homes not because of wars or civil unrest, but because of climate change or other ecological disasters which force them to better their lives in places far from home. With the impending demographic shift—for example the population in Africa is expected to double by about 2050—a country such as Denmark has to search its soul for a realistic and fair immigration policy in an effort to retain its self-understanding as 'a compassionate country'.[4]

It could be asked whether Pope Francis' question was mere rhetoric and therefore should not be taken literally.[5] Rhetorical or not, I would argue that Pope Francis' question is very illuminating for two reasons. First, even if his question is considered 'rhetorical', it does not mean that

[1] Francisco Cantu, 'Has Any One of Us Wept?', *The New York Review of Books*, Vol. LXVI, No. 1 (January 17–February 6, 2019), 4–6 (6).

[2] Ibid.

[3] Ibid.

[4] Mette Frederiksen, 'Realist and Fair Immigration: A Policy to Unite Denmark', Inter Press Service, May 5, 2019, http://www.ipsnews.net/2019/05/realistic-fair-immigration-policy-unite-denmark/, accessed 29 August 2019.

[5] I thank my anonymous reader for his/her question on this point.

it is less significant. In fact, one could argue, with Aristotle, that traditionally rhetoric involves an attempt to persuade the audience to accept the speaker's position. In addition, contemporary study of political rhetoric and the ways it is linked to the social organization of responsibility has been central to anthropology's understanding of how human beings account for their predicaments and burdens.[6] But since it was raised in the form of a question, Pope Francis' rhetoric, if one chooses to believe it to be so, assumes the form of 'invitational rhetoric' where the audience is invited into the world of the speaker, to reflect on the question posed before them, and possibly arrive at their own answers.[7] In this sense, the Pope's 'rhetorical' question, again if this is presumed to be the case, could result in the possibility that the audience would emerge from the question feeling unsettlingly empowered because rhetoric also contains within it an ability to make everything seem strange, while revealing insights that have been hidden in the dominant culture one lives in.[8] Second, since the power of rhetoric cannot be overlooked (and I would say that those in all religious circles are well aware of this fact) it does not mean that one can separate the rhetorical from the literal. Here I would choose to believe that Pope Francis' question was both rhetorical and literal, especially since it was raised in the context of the holy Catholic mass he was performing. Consequently, his question is used here as an entry point into the discussion that will follow precisely because it serves as an invitation for his audience, as well as for those who read his question, to give themselves a chance to genuinely think about grieving for others. I suspect that beneath the symptom of 'globalization of indifference' that he suggests lies the way we 'think' about ourselves that make it difficult, if not impossible, to grieve for others.

[6] Michael Herzfeld, *Anthropology: Theoretical Practice in Culture and Society* (Malden, MA: Blackwell, 2001), 169.

[7] Carlo A. Pedrioli, 'Beyond Aristotle: Alternative Rhetorics and the Conflict Over the U.S. Law Professor Persona(e)', *Ohio Northern University Law Review*, Vol. 28 (2012), 919–955. See especially 923–924; SSRN: https://ssrn.com/abstract=2307908.

[8] Jack Lee Sammons, 'Legal Writing Scholarship, Making Strange, and the Aesthetics of Legal Rhetoric', *Mercer Law Review*, Vol. 61 (April 13, 2010). SSRN: https://ssrn.com/abstract=1591123.

The Thesis

Rephrasing the Pope's question to underscore the pain it delicately suggests, I am curious about why, when seeing the sights of human suffering, we do not grieve. Why can't we? In meditating on why it is so very difficult to answer Pope Francis' question, this chapter is an attempt to advance a philosophical assertion that the difficulty arises from a profound sense of human failure to realize how one's life is connected to the lives of others, and that to be able to do so, it is necessary to call into question the dominant role of 'thinking' as the defining quality of being human.

A popular work recounting the history of humankind highlights the human ability to think, and the genetic mutation that enhanced it, and with that the development of human language that made telling stories outside of oneself possible.[9] But the question I am interested in exploring here is that if thinking is the defining element of being human, as is commonly believed, what does one do when one thinks? Thinking is an act that could only be done in isolation. But the understanding that living beings are human because they think fails to recognize the human reality of living as connected beings in the world. This failure, I would argue, is responsible for the way one treats others and the living world that is the human home, and thereby engenders current ecological disaster and undermines peace, which has in turn led to so much human suffering.

If this is indeed the case, there is a need to let go of the dominant conceptual mode which renders humans blind to others' sufferings and deaf to the cries of those who are oppressed, seeing them, as well as the world we live in, only as objects that can be utilized at will. An alternative mode of being could, and in fact should, be conceptualized, not of the human being as one who thinks, but as one who breathes, an act that naturally connects one's life with others'. In short, this chapter is an attempt to demonstrate that it is difficult to engage the Pope's question precisely because of the dominant epistemic belief. It advances a critique of the dominant understanding of being human while proposing an alternative epistemic ground drawn from two very different religious inspirations, namely Buddhism and Islam. This is done not as an exercise in comparative religious studies, which would be far beyond the scope

[9] Yuval Noah Harari, *Sapiens: A Brief History of Humankind* (New York: HarperCollins, 2015).

of this study, but to show that the global problems such as migration crisis, as well as the issue of ecological disaster affect all people—with or without faith. But for those with religious beliefs in at least three major religions—Christianity, Islam, and Buddhism, the philosophical route to be taken can be traced back to their own beliefs. The differences between faith traditions can be traversed in a struggle against the 'globalization of indifference'. This chapter concludes with discussion of people who have nonviolently risked their lives to save the forest in India, an illustration of how people can move beyond the dominant mode of being and nonviolently fight for a more ecologically peaceful world.

Methodology

In order to unpack the way in which we 'think' about ourselves and the rest of lives in the world as an answer to why it is so very difficult to answer the Pope's question, this chapter proceeds through four related questions. First, what is the prevalent way of thinking about how ecological disaster could be avoided and a just and peaceful society created? By exploring public imagination as depicted in popular culture, I hope to find some answers. But popular culture, like other types of culture, is often inevitably, and perhaps unconsciously, under the influence of dominant modern philosophy. Therefore the second question is: What kind of foundational idea exists which makes it possible for such imagination to flourish? I will deal with this question by examining the most common and influential modern philosophy of being, one which has dominated the world since the sixteenth century. Third, is there an alternative mode of being human that could be considered a possible philosophical replacement of the dominant one? Inspired by religious teachings drawn from Buddhist and Islamic sources, I will elaborate on the possibility that this new mode of being human is conducive to the creation of a more ecologically peaceful world precisely because it refuses to disengage the self from other humans and the world, and instead seeks to profoundly reconnect with them. Fourth, apart from philosophical rumination, does the world of practice offer any real-life examples of how such a philosophical shift could possibly manifest itself? In this final section of the paper, the example of how Indian villagers risked their lives to protect their forests will be discussed as a possible course of nonviolent action to help save our world by preventing us from becoming the authors of destruction of our lives and the planet.

In line with these four related questions, this chapter is divided into four sections using four words as metaphorical tools: snapping, thinking, breathing, and hugging.

Snapping

The phenomenal success of the movies *Avengers: Infinity Wars/Endgame* in the Marvel Cinematic Universe marked the end of the second decade of the twenty-first century. The super-villain in this decade-long superhero franchise is Thanos, an alien (a Titan) with immense power and strategic prowess. The first word 'snapping' comes from this movie. The movie's villain, Thanos, cares for the universe and yet he sets out to destroy half of the lives in it to preserve its lost balance and restore what he believes to be a 'just peace' that was long lost due to uncontrollable population growth. In the movie, when he snapped his fingers, wearing the magically designed gauntlet with six powerful infinity stones, half of the lives in the universe were instantly turned into gray dust. Thanos snaps his fingers because he believes that there is no alternative to destroying lives in order to regain the necessary balance of the universe so that the rest can live in peace. This solution rests upon seeing the others as expendables for the good that such destruction brings about. His snapping that kills half of the lives in the universe is done indiscriminately without regard to gender, belief, ethnicity, or class.

The ideal that has driven Thanos to acquire the destructive infinity gauntlet in order to destroy half of the lives in the universe is the belief that the existing resources could no longer sustain the ever-increasing population who depend on them. Once he had successfully wiped out half of the universe to bring back the balance between population and natural resources, he retired to watch the sunrise again 'while an ungrateful universe reaps the rewards'.[10] If popular culture could be seen as reflecting a part of contemporary public imagination, then Thanos' mission simply follows a long line of fictional as well as non-fictional villains/thinkers/revolutionaries who believe in a similar solution.

[10] See a fascinating academic analysis of the film and especially Thanos' role in Peter W. Lee, 'Growing, Growing Gone: The Ecological Thanos Needs a Chill Pill in *Avengers: Infinity War*', https://medium.com/@pwylee/growing-growing-gone-the-ecological-thanos-needs-a-chill-pill-in-avengers-infinity-war-74b9e935ae2f, accessed 26 July 2019.

In Dan Brown's fourth Robert Langdon adventure, *Inferno*,[11] the prime villain is Bertrand Zobrist, a geneticist with a fascination, if not obsession, with Dante's *Inferno*. He designs a plague virus that would make one-third of the world's population infertile in order to solve the imminent doomsday caused by overpopulation. Like Thanos, Zobrist believes that to do nothing is to welcome Dante's hell. Thinking of himself as a global saviour, he decides to proceed with the horrible action convinced that all salvation comes at a price. He also believes that one day the world will grasp the beauty of his sacrifice that will herald 'the posthuman age'.[12]

In the novel *Inferno*, Brown uses one non-fictional graph to illustrate the key environmental issues which have the greatest impact on human health in the world: demand for clean water, global surface temperatures, ozone depletion, consumption of ocean resources, species extinction, carbon dioxide concentration, deforestation, and global sea levels. Not only have these negative indicators been on the rise over the last century, but they are also 'accelerating at terrifying rates'.[13]

Popular culture's dangerous fictional vision of the world and its solution does find its counterparts in the real world. In the past, social Darwinists and eugenicists were proud to defend their civilizations with schemes to control population growth which include sterilization and genocide. To kill off half of the lives in the universe or one-third of the global population in the name of stabilizing the population and existing global resources, or to ensure the holy status of the free market economy—as in Susan George's fierce satire: *The Lugano Report*[14]—depicts a cold and heartless 'rational' calculus.

One could trace the grim vision these villains shared back to the prophetic words of the eighteenth-century philosopher of demography, Thomas Malthus, who writes:

> The power of population is so superior to the power in the earth to produce subsistence for man, that premature death must in some shape or other visit the human race. The vices of mankind are active and able

[11] Dan Brown, *Inferno* (New York and London: Doubleday, 2013).

[12] Brown, *Inferno*, 145.

[13] Ibid., 138.

[14] Susan George, *The Lugano Report* (London: Pluto Press, 1999).

> ministers of depopulation. They are the precursors in the great army of destruction; and often finish the dreadful work themselves. But should they fail in this war of extermination, sickly seasons, epidemics, pestilence, and plague advance in terrific array and sweep off their thousands and ten thousands. Should success be still incomplete, gigantic inevitable famine stalks in the rear, and with one mighty blow levels the population with the food of the world.[15]

Looking back at population growth in the last two centuries, it is understandable why such a grim vision has been prevalent in popular culture. World population reached 1 billion in 1800. Over the next 130 years, it rose to 2 billion. Then in less than 100 years, the world population now is at 7.6 billion. With the planet's limited resources, it is easy to arrive at the harrowing conclusion that, for example, if poor children continue to be saved, then the planet with its limited resources will be killed by overpopulation. In fact, the UN predicts that by 2100, world population will reach 12 billion. But it is important to note that the number of babies per woman, perhaps the single most important contributor to population growth, has been dramatically dropping since 1965. The period of fast population growth is expected to be over soon. True, women in countries such as Somalia and Chad continue to have between five and eight children. But that is because infant mortality rates in these countries are high, and if the children survive, their labour is needed. With better health care and education for women, as well as informed birth control practices, they will have fewer children. This is why some argue that the proven method for curbing population growth is not superimposing control on the birth of children but eradicating extreme poverty.[16]

While Rosling et al. are persuading the world to look at 'facts' with a critical awareness on how 'instincts' have successfully prevented most of us from appreciating the fact that the conditions of the world have improved, I am puzzled at the ease by which the killing solution seems to be prevalent in popular imagination as well as the cruelty shown to

[15] Thomas Malthus, *An Essay on the Principle of Population* (London: Printed for J. Johnson, in St. Paul's Church-Yard, 1798), 44.

[16] Hans Rosling with Ola Rosling and Anna Rosling Ronnlund, *Factfulness: Ten Reasons We're Wrong About the World and Why Things Are Better than You Think* (London: Sceptre, 2019), 80–91.

migrants, refugees and civilians in today's deadly conflict. Thanos' solution to return the universe to ecological balance, and thereby create a more just and peaceful place for all beings who still live is the result of this thinking, shared by many in fictional as well as in real life. It is this understanding of 'thinking' that needs to be called into question.

Thinking

Almost half a century ago, E.F. Schumacher wrote *Small is Beautiful*,[17] an influential book considered 'an eco-bible' by *Time*. Both its subtitle 'Economics as if People Mattered' and the prominently innovative chapter on 'Buddhist Economics' are creative critiques of mainstream economics. Schumacher began the chapter on the subject with the sentence: '"Right Livelihood" is one of the requirements of the Buddha's Noble Eightfold Path. It is clear, therefore, that there must be such a thing as Buddhist economics'.[18] Claiming that the keystone of Buddhist economics is simplicity and nonviolence, he writes: 'From an economist's point of view, the marvel of the Buddhist way of life is the utter rationality of its pattern—amazingly small means leading to extraordinarily satisfactory results'.[19]

Fully aware that he has to deal with dominant ideas of the time, Schumacher begins part two of the book on resources with a chapter on education which he considers 'the greatest resource'. Towards the end of this chapter, he delineates six leading ideas he believes still dominate the modern mind. They are the ideas of evolution, competition, materialism, sexual instinct, relativism, and positivism.[20] These ideas are those of the nineteenth century and early twentieth-century thinkers: Darwin, Marx, Freud, Einstein, and Comte. I believe Schumacher missed a most distinguished seventeenth-century philosopher who continues to influence modern as well as postmodern thought when it comes to epistemology, namely René Descartes.[21]

[17] E.F. Schumacher, *Small Is Beautiful* (London: Blond and Briggs, 1973).

[18] Ibid., 32.

[19] Ibid., 35.

[20] Ibid., 56.

[21] Charles Taylor, *Philosophical Arguments* (Cambridge: Harvard University Press, 1997), vii–viii.

Descartes is known the world over as the prime proponent of the dominant rationalist worldview. As the philosopher Charles Taylor has suggested, this 'dominant view' means 'an outlook which has to some extent colonized the common sense of our civilization'. To understand this worldview, imagine a picture of:

> agents who in perceiving the world take in "bits" of information from their surroundings, and then "process" them in some fashion, in order to emerge with the "picture" of the world they have; who then act on the basis of this picture to fulfill their goals, through a "calculus" of means and ends.[22]

This is the picture of the thinking agent as a 'disengaged' human. Disengagement can therefore be seen as getting free of the perspective of embodied experience which makes the human person give undue importance to the senses and imagination in the attempt to know. Hence from this perspective the thinking activity of the mind should be essentially free from these distorting media influencing the body. As a result, in Descartes' thought, the human mind is categorically separated from his/her body. Descartes writes:

> I, who was thinking them, had to be something; and observing this truth: I am thinking therefore I exist, was so secure and certain that it could not be shaken by any of the most extravagant suppositions of the sceptics, I judged that I could accept it without scruple, as the first principle of the philosophy I was seeking.[23]

He continues:

> .... if I had merely ceased thinking, I would have no reason to believe that I existed, even if everything else I had ever imagined had been true. I thereby concluded that I was a substance whose whole essence or nature resides only in thinking, and which, in order to exist, has no need of place and is not dependent on any material thing. Accordingly this 'I', that is to say, the Soul by which I am what I am, is entirely distinct from the

[22] Ibid., 63.

[23] René Descartes, *A Discourse on the Method of Correctly Conducting One's Reason and Seeking Truth in the Sciences*, translated with an introduction and notes by Ian Maclean (Oxford: Oxford University Press, 2006), 28.

> body and is even easier to know than the body; and would not stop being everything it is, even if the body were not to exist.[24]

It is the thinking activity of the person as a disengaged actor that defines one as being human, elegantly enshrined in the renowned philosophical formulation: '*Cogito ergo sum*' or 'I think therefore I am'. The image of the Cartesian human could very well be August Rodin's statue 'the Thinker' but with a roofed wall blocking him from the world, with all lives effectively left outside.

There are many ways one could ponder the limitations of the Cartesian thinking process. For example, one could rely on philosophical thoughts as explicated by Martin Heidegger's *Being and Time*, or Charles Pierce's brilliant attack of Descartes' idea in 1868.[25] But here I would prefer to elucidate the Cartesian limitation through the problem of alcoholism as discussed in Gregory Bateson's anthropology.

In analyzing the ways alcoholics try to achieve sobriety, Bateson found that the acceptance of the Cartesian dualism works as the 'disabling ontological orientation' for them. Since alcoholics believe that the only way to get rid of their drinking habit is by their force of will, they keep testing themselves by drinking to see if their mind is in control over their body, as evidence of their ability to limit alcohol consumption. Bateson points out that due to the commitment to a radical separation between mind and body, influenced by the Cartesian thinking process, they end up in risk-taking behavior with a self-destructive personal ethic. Bateson maintains that the failure of alcoholics is the result of a 'wrongful local epistemology' and argues that it was the Cartesian thinking process that has produced such disabling beliefs about the nature of the self and its relationship to the world. On the contrary, the success of Alcoholic Anonymous is owed to its rejection of such Cartesian dualism and the isolated thinking mode: instead it allows alcoholics who seek treatment to posit an alternative mode of the self where the human mind is not a domain that is separate from the world around it. Rejecting the Cartesian mode of thinking, the AA session takes place in a circle of other alcoholics, who later share their own stories, usually beginning by stating the name of the person, followed with the statement: 'I am an alcoholic'. It could be

[24] Ibid., 33.

[25] See Richard J. Bernstein, *Philosophical Profiles: Essays in a Pragmatic Mode* (Philadelphia: University of Pennsylvania Press, 1986), 200–201.

argued that the success of AA's method lies in the fact that its members do not try to win over alcoholism with their minds by constantly testing themselves to achieve victory, but by admitting defeat in the context of trusted relationship. Bateson's analysis indicates that the retention of an inadequate and truncated relation between the self and the world would lead to 'the neglect of value commitments, institutional presuppositions, and models of individual and collective responsibility' which are necessary for a successful life outside the influence of alcohol. [26]

The idea that it is the isolated human being doing the thinking, separated from the rest of the world, is in some ways responsible for one's inability to see the life in other living beings, or, if the life is seen, then it becomes a means to the ends of the seeing agent, which consequently makes it easier to indulge oneself in the globalized world of indifference instead of facing and preparing for the ecological crisis as well as other crises that would impact all of us. Instead, there is a need to find an alternative formulation for being human that would be conducive to care for the earth and simultaneously engage other humans nonviolently.

Breathing

The focus on the human being as a thinking agent is not only found in seventeenth-century Europe. From a religious perspective, it could be said that Buddhism's intense attention on the human mind is somewhat similar to Descartes' thinking activity as a defining feature of being human. The major difference lies in the fact that while the Cartesian being is real, in Buddhism it is not but is instead constitutive of withering and impermanent conditions of the world.

It is said that if there is a feature that distinguishes Buddhism from other religious beliefs, it is mindfulness. Some scholars argue that mindfulness is equated with Buddhism itself because it is considered the only way that leads beings to purity, to the transcending of sorrow, the appeasement of pain and sadness, and finally to the realization of *Nirvana* (enlightenment) which is its ideal goal.[27]

[26] Michael J. Shapiro, *Language and Political Understanding: The Politics of Discursive Practices* (New Haven and London: Yale University Press, 1981), 4–5. The phrase in quotes is a paraphrase of Shapiro's.

[27] Edward Conze, *Buddhist Thought in India: Three Phases of Buddhist Philosophy* (Ann Arbor: The University of Michigan Press, 1973), 51–52.

The Buddha discovered a form of meditation which helps one gain insight into the nature of things, leading to the liberation of the mind and the realization of *Nirvana*. The most important discourse on mental development the Buddha gave is called: *Satipatthana–sutta* (the setting of mindfulness). The discourse is divided into four sections: body; feelings and sensation; mind; and moral and intellectual subjects. One of the best known, popular and practical examples of 'meditation' connected with the body is called 'The Mindfulness or Awareness of in-and-out breathing' (*anapanasati*). The focus of this mental exercise is to concentrate one's mind on one's breathing in and out to the point when one's mind is totally concentrated on it so that other things including one's environment are forgotten. At the point when one's mind truly concentrates on breathing so much that one loses oneself completely in the mindfulness of breathing, in that split second the person will experience a moment of ultimate joy.[28]

This is the Buddhist training to attain inner peace, achieved by progressively diminishing the impact of external stimuli. By withdrawing from sensory data and renouncing all interest in them, those who are concentrated regain the inward calm which always dwells in their hearts. 'Subjectively *Samadhi* is marked by a soft, tranquil and pacified passivity, objectively by the abstraction into an unearthly world of experience which lifts us out of this world, and bestows a certainty greater than anything the senses can teach'.[29] Philosophically, Buddhism's teaching on breathing in and out leads one to understand the human ontology of non-self. Meditation is Buddhism's deconditioned epistemological process designed to deal with the egotistic self. Through training in breathing, practitioners regain the ability to see the 'truth' of impermanence, which enables Buddhists to discover that their attachment to the self, a futile effort, is actually a major cause of all human suffering. With this newfound understanding, detached from the egotistical self, the meditator emerges to reconnect with the world in a new way, lighter and liberated from attachment to things in this world.

Echoing the Buddhist understanding of the self as the origin of suffering, the Chinese scientist turned American novelist, Yiyun Li writes:

[28]Walpola Rahula, *What the Buddha Taught* (New York: Grove Press, 1974), 68–70.

[29]Conze, *Buddhist Thought in India*, 53.

> ....I have resisted forming an attachment to any object, or any place. I wished then and I wish now that I had never formed an attachment to anyone in the world either. I would be all kindness. I would not have done anything ruinous. I would never have to ask that question—when will I ever be good enough for you?—because by abolishing you, the opposite of *I*, I could erase that troublesome *I* from my narrative, too.[30]

I would argue, however, that for the human agent who thinks, no matter how isolated or concentrated he/she is, a realistic philosophical assumption will have to also assume that the thinking agent is alive. To be alive means to be able to breathe. But there are different ways to imagine a human being who breathes as a replacement to the unrealistic disengaged self trapped in a Cartesian wall doing the thinking in isolation. From Buddhist exercise in concentrated breathing, the being who emerges will reconnect with the world, care for the world, and act in the world framed by the language of detachment in pursuit of cessation of sufferings born from the illusion of attachment.

It is interesting to note that in Islam too, breathing is what makes us human. But while in the Buddhist lexicon the key word is detachment, in Islam it is through remembering the profound way in which humans are connected to God that makes us human and different from other living beings.

In *Al-Qur'an*, God says: 'Behold, I am about to create mortal man out of sounding clay, out of dark slime transmuted; and when I have formed him fully and breathed into him of My Spirit, fall down before him in prostration' (15: 28–29).[31] In another place, God also says: 'Thus, He begins the creation of man out of clay; then He causes him to be begotten out of the essence of a humble fluid; and then He forms him in accordance with what he is meant to be, and breathes into him His Spirit; and (thus, O men) He endows you with hearing and sight, and feelings as well as minds' (32:7–9).[32]

The idea that the Creator breathes Life/His Spirit into humans created from clay could be understood as saying that the human

[30] Yiyun Li, *Dear Friend, From My Life I Write to You in Your Life* (New York: Random House, 2017), 33.

[31] *The Message of the Qur'an*, translated and explained by Muhammad Asad (Gibraltar: Dar Al-Andalus, 1980), *surah* (chapter) 15: *ayah* (verses) 28–29, 386.

[32] Ibid., *surah* 32: *ayah* 7–9, 634.

being is the connecting point between the lowly clay and the noblest Creator. Muhammad Asad understands this mysterious verse depicting the breathing of God's Spirit into humans as 'obviously a metaphor'.[33] He explains that it is a metaphor of the divine gift of life and consciousness, or of a 'soul' (one of various meanings of the Arabic term 'ruh'). As such, it can be said that 'the soul of every human being is of the spirit of God'.[34]

But the *Qur'an* says that once God breathes life into the human, then hearing, sight, feelings as well as minds follow (32:9). This means that once a human is with life, then he/she could hear, see, and feel the world around them. What then does one see when looking at the world? This Qur'anic verse would invite humans to see the lives of all created living things. But how would one see these lives?

While in the act of creation God breathes life into humans and so they themselves breathe. For a human, breathing is the act of living. But what does he/she breathe? In the physical world, in the company of other humans, is it the dust of lives from the others that he/she is breathing in? We share the same source of lives, therefore when we breathe others in, they can be seen and heard as the precious life that we all are. Hearing and seeing other humans would therefore enable us to feel how they are, and as a result care for them. This would enable us to replace the philosophic '*Cogito ergo sum*' ('I think therefore I am') which has trapped humans in a Cartesian wall, with '*Spiro ergo sum*' ('I breathe therefore I am'). Breaking down the Cartesian wall to breathe in others in the context of the natural world, one emerges with a sense of how it is the wind of life that connects us all.

Conclusion: Hugging

At the end of the second decade of the twenty-first century, the ecological conditions of the world are going from bad to worse. Reviewing Nathaniel Rich's *Losing Earth: A Recent History* (2019), Michelle Nijhuis writes: 'A heat wave in Japan had killed sixty-five people during a single week, and hospitalized tens of thousands more; roads and rooftops were melting in the United Kingdom; in Finland, north of the Arctic Circle,

[33] Ibid., 386, fn. 26.

[34] Ibid., 364, fn. 9.

temperatures had approached ninety degrees Fahrenheit. The real-time effects of climate change were—and are—all around us, and it is highly likely that they are going to get much worse'.[35] Commiserating the deteriorating state of the world, a well-known journalist once remarked: 'You know that someday we'll ditch this journalism crap and become terrorists'.[36] In the midst of such despair, are there rays of hope?

In his latest book on how humans are killing the planet Earth with their reckless use of natural resources without care, the renowned ecologist Bill McKibben suggests that assuming that humans are capable of working together to reverse the damage already done, there are two technologies that could be deployed. The first technology is the photovoltaic solar panel. Interestingly for peace and nonviolence research, McKibben's second technology is 'nonviolent protest and resistance'. His second proposed technology was in fact drawn from his remarkable experience. In 2009, he organized a global fight to stop using carbon-based fuels. Inspired by Gandhi's nonviolence, and perhaps his faith in the Sermon on the Mount, McKibben's organization '350.org' could engender rallies in 181 countries at its first attempt. His commitment and experience led him to believe that nonviolent protest and resistance would ultimately win. The question, however, is whether it will win in time.[37]

Three decades ago, I suggested that nonviolent struggle would shift from struggles against tyranny to the preservation of nature because conflicts about increasing ecological destruction were on the rise.[38] Given the environmental crisis that has ensued, intensification of ecological conflict would likely follow. Recently an organizer working with the US-based Faith Matters Network discusses the important ways religious leaders help shape contemporary nonviolent resistance movements especially in North America. She suggested that religious faiths provide

[35] Michelle Nijhuis, 'Early Warnings', *The New York Review of Books*, Vol. 66, No. 11 (June 27–July 17, 2019), 37.

[36] Alan Weisman, 'Bringing Down the House', *The New York Review of Books*, Vol. 66, No. 13 (August 15–September 25, 2019), 8.

[37] Bill McKibben, *Falter: Has the Human Game Begun to Play Itself Out?* (New York: Henry Halt, 2019), Part Four.

[38] Chaiwat Satha-Anand, 'Nonviolent Practices in Ecological Struggles: Explaining the Success of the Anti-Dam Campaign in Thailand', in Mahendra Kumar and Peter Low (eds), *Legacy and Future of Nonviolence* (New Delhi: Gandhi Peace Foundation, 1996), 199–211.

nonviolent activists with emotional and spiritual support; help reclaim the legacy of ancestral practices as tools in transformational resistance; and initiate spiritual practice that fosters compassionate participation in civil resistance.[39] But a clearer case of nonviolent action has been used to directly and successfully protect the forest occurred in Chamoli, India.

Almost half a century ago, women and girls from Reni village in the Chamoli district in India saved their forest from being felled. The Chipko movement which reflects the connectivity between ecological justice and nonviolent action was born on March 26, 1974 when Gaura Devi stood in front of the trees marked for felling. She told the men: 'Brothers! This forest is the source of our livelihood. If you destroy it, the mountain will come tumbling down onto our village'. She then placed herself between the trees and the men prepared to cut them down. Some of the men were armed. She said: 'This forest nurtures us like a mother, you will only be able to use your axe on it if you shoot me first'. Despite abuse and threats, the women refused to move. The lumbermen, also from Himachal Pradesh, understood the message. After a three-day standoff, the woodcutters withdrew.[40] The Chipko movement is widely known as a movement of courageous activists who protect the forests with their lives by hugging the trees.

Apart from offering us an inspiring story of ordinary people, mostly women, doing extraordinary things in the fight against deforestation, the protesters knew what they were fighting for, who their opponents were, and were prepared to pay for it with their lives in the Gandhian spirit of nonviolence.[41] Hugging can be powerful. In fact, three decades ago, the late Kenneth E. Boulding wrote that power has three faces: the stick or threat power, the carrot or economic power, and the hug or integrative power. These three faces are closely related to the power to

[39] Carinne Luck, 'Three Ways Faith and Spiritual Leaders are Shaping Movements Today', Minds of the Movement website, 18 June 2019, https://www.nonviolent-conflict.org/blog_post/three-ways-faith-and-spiritual-leaders-are-shaping-movements-today/, accessed 28 July 2019.

[40] Jaskiran Chopra, 'Forty Five Years of Chipko Which Brought Uttarakhand Region Into Global Focus', *Daily Pioneer*, March 25, 2019, https://www.dailypioneer.com/2019/state-editions/forty-five-years-of-chipko-which-brought-uttarakhand-region-into-global-focus.html, accessed 30 July 2019.

[41] Mark Shepard, *Gandhi Today: A Report on Mahatma Gandhi's Successors* (Arcata, CA.: Simple Productions, 1987), 63–80.

destroy, the power to produce and exchange, and the power to create relationships based on love, respect, and friendship, among others.[42]

I am curious, however, about why the woodcutters decided to yield to the women's courage. There are no doubt numerous ways to analyze this—that the woodcutters were from the same areas and understood the ecological threats to their livelihood, that they knew the protesters, and that they knew the women were doing the right thing, among others. Many factors may have contributed to the choice they made. But I am wondering what it is they saw when they sighted the women hugging the trees and prepared to lay down their lives?

If we don't grieve, it is because we don't see them, or see them as lifeless others or as far less than 'we'. With some good reasons, like preserving the ways we live comfortably or preventing 'the others' from entering into our 'home'. If we could snap our fingers and make them disappear, many of us might do just that. And that is because in the process of 'thinking'—calculating and then choosing between our well-being and theirs, there is little chance that the others will be chosen. Even to be compassionate towards them is deemed increasingly difficult given fewer and fewer resources at our disposal.

In this case, the woodcutters stopped because they did see the women hugging those trees. Unless the notion of thinking as an act performed in isolation and as used to define the human itself is called into question, and an alternative such as breathing, inspired by religious teachings such as Buddhism and Islam, among others, is considered as the defining feature of being human, then we will not be able to see the others and grieve for those who are lost to us, and hug those who are not yet.

Imagine this: What if what the woodcutters saw was lives protecting lives? The trees were precious because they are sustainers of human lives in the ecology of beings. While standing, they might hold their breath some of the time, but normally they breathed. The air they breathed was suffused with the beings of the protesters and so they breathed in the others who have then become part of themselves. Seeing the lives of others, breathing in the others may have informed the woodcutters' choice not to harm the people who tried to protect the forests. Perhaps, in a profound sense, this is how nonviolent actions work in the continuing difficult struggle to preserve mother earth and all that lives in her.

[42] Kenneth E. Boulding, *Three Faces of Power* (Newbury Park: Sage, 1990).

Bibliography

Bernstein, Richard J. *Philosophical Profiles: Essays in a Pragmatic Mode*. Philadelphia: University of Pennsylvania Press, 1986.

Boulding, Kenneth E. *Three Faces of Power*. Newbury Park: Sage, 1990.

Brown, Dan. *Inferno*. New York and London: Doubleday, 2013.

Cantu, Francisco. 'Has Any One of Us Wept?'. *The New York Review of Books*, Vol. LXVI, No. 1 (January 17–February 6, 2019), 4–6.

Chopra, Jaskiran. 'Forty five years of Chipko which brought Uttarakhand region into global focus'. *Daily Pioneer*, March 25, 2019. https://www.dailypioneer.com/2019/state-editions/forty-five-years-of-chipko-which-brought-uttarakhand-region-into-global-focus.html, accessed 30 July 2019.

Conze, Edward. *Buddhist Thought in India: Three Phases of Buddhist Philosophy*. Ann Arbor: The University of Michigan Press, 1973.

Descartes, René. *A Discourse on the Method of Correctly Conducting One's Reason and Seeking Truth in the Sciences*. Translated with an Introduction and Notes by Ian Maclean. Oxford: Oxford University Press, 2006.

Frederiksen, Mette. 'Realist and Fair Immigration: A Policy to Unite Denmark'. *Inter Press Service*. May 5, 2019. http://www.ipsnews.net/2019/05/realistic-fair-immigration-policy-unite-denmark/, accessed 29 August 2019.

George, Susan. *The Lugano Report*. London: Pluto Press, 1999.

Harari, Yuval Noah. *Sapiens: A Brief History of Humankind*. New York: HarperCollins, 2015.

Herzfeld, Michael. *Anthropology: Theoretical Practice in Culture and Society*. Malden, MA: Blackwell, 2001.

Lee, Peter W. 'Growing, Growing Gone: The Ecological Thanos Needs a Chill Pill in *Avengers: Infinity War*' https://medium.com/@pwylee/growing-growing-gone-the-ecological-thanos-needs-a-chill-pill-in-avengers-infinity-war-74b9e935ae2f.

Li, Yiyun. *Dear Friend, from My Life I Write to You in Your Life*. New York: Random House, 2017.

Luck, Carinne. 'Three Ways Faith and Spiritual Leaders Are Shaping Movements Today'. Minds of the Movement website, June 18, 2019. https://www.nonviolent-conflict.org/blog_post/three-ways-faith-and-spiritual-leaders-are-shaping-movements-today/.

Malthus, Thomas. *An Essay on the Principle of Population*. London: Printed for J.Johnson, in St. Paul's Church-Yard, 1798.

McKibben, Bill. *Falter: Has the Human Game Begun to Play Itself Out?* New York: Henry Halt, 2019.

Nijhuis, Michelle. 'Early Warnings'. *The New York Review of Books*, Vol. 66, No. 11 (June 27–July 17, 2019), 37–39.

Pedrioli, Carlo A. 'Beyond Aristotle: Alternative Rhetorics and the Conflict Over the U.S. Law Professor Persona(e)'. *Ohio Northern University Law Review*, Vol. 28 (2012), 919–955.

Rahula, Walpola. *What the Buddha Taught*. New York: Grove Press, 1974.

Rosling, Hans with Ola Rosling and Anna Rosling Ronnlund. *Factfulness: Ten Reasons We're Wrong About the World and Why Things Are Better Than You Think*. London: Sceptre, 2019.

Sammons, Jack Lee. 'Legal Writing Scholarship, Making Strange, and the Aesthetics of Legal Rhetoric'. *Mercer Law Review*, Vol. 61 (April 13, 2010), 925–942.

Satha-Anand, Chaiwat. 'Nonviolent Practices in Ecological Struggles: Explaining the Success of the Anti-Dam Campaign in Thailand'. In Mahendra Kumar and Peter Low (eds), *Legacy and Future of Nonviolence*. New Delhi: Gandhi Peace Foundation, 1996, 199–211.

Schumacher, E.F. *Small is Beautiful*. London: Blond and Briggs, 1973.

Shapiro, Michael J. *Language and Political Understanding: The Politics of Discursive Practices*. New Haven and London: Yale University Press, 1981.

Shepard, Mark. *Gandhi Today: A Report on Mahatma Gandhi's Successors*. Arcata, CA: Simple Productions, 1987.

Taylor, Charles. *Philosophical Arguments*. Cambridge: Harvard University Press, 1997.

The Message of the Qur'an. Translated and Explained by Muhammad Asad. Gibraltar: Dar Al-Andalus, 1980.

Weisman, Alan. 'Bringing Down the House'. *The New York Review of Books*, Vol. 66, No. 13 (August 15–September 25, 2019), 4–6.

Chaiwat Satha-Anand is Professor of Political Science at Thammasat University, Thailand, and Chairperson of the Strategic Nonviolence Commission, Thailand Research Fund—a think tank that proposes nonviolent policy alternatives to deadly conflicts in Thai society. He has written, edited, or co-edited some forty books published both in Thailand and overseas. Some of his writings have been translated and published in Arabic, Bahasa Indonesia, Chinese, German, Italian, Japanese, and Korean. His most recent publications include: (co-editor with Olivier Urbain) *The Promise of Reconciliation?: Examining Violent and Nonviolent Effects on Asian Conflicts* (New York: Routledge, 2017); (author) *Nonviolence and Islamic Imperatives* (Sparnas: Irene Publishing, 2017). He was named 'Thailand's Best Researcher in Political Science and Public Administration' by the National Research Council in 2006. He also received the National Sri Burapha Distinguished Writer Award (Bangkok) and the International El-Hibri Peace Education Prize (Washington, DC) in 2012.

PART III

Questioning the Colonial Mindset

CHAPTER 12

'We've Seen the End of the World and We Don't Accept It': Protection of Indigenous Country and Climate Justice

Tony Birch

Introduction

The Aboriginal intellectual and activist, Robbie Thorpe, often reminds us that the ethics of 'Land Rights' should never be reduced to a political slogan. For Thorpe, Land Rights must privilege the rights of the land (country) itself. If we are to combat the ravages of environmental destruction and the impacts of climate change, it is vital that we think and act with humility towards the Earth, towards non-human species and towards each other. A key Aboriginal scholar, Professor Irene Watson, comments

Note This chapter was previously published under the same chapter title in Nicole Oke, Christopher Sonn and Alison Baker (Eds.), Places of Privilege: Interdisciplinary Perspectives on Identities, Change and Resistance (Leiden: Brill, 2018), 139–52, with the exception of the first two paragraphs which were exclusively written for the present volume.

T. Birch (✉)
Victoria University, Carlton, VIC, Australia
e-mail: Anthony.Birch@vu.edu.au

J. Camilleri and D. Guess (eds.), *Towards a Just and Ecologically Sustainable Peace*, https://doi.org/10.1007/978-981-15-5021-8_12

on the historical legacy of colonial violence that has contributed to the parlous situation we find ourselves in. She writes, 'universal notions of civility justified the demonisation of our "nativeness" and were used to destroy our relationship to the land and our sovereign Indigenous being'.[1]

In a study of the relationship between racism, colonialism and climate change Razmig Keucheyan writes of the urgency to 'demolish the idea that humanity suffers the consequences of the ecological crisis in a uniform way'. He contends that the wealthy and powerful 'North', overwhelmingly responsible for global carbon emissions, owes a compensatory 'ecological debt' to the less prosperous and more environmentally vulnerable 'South'.[2] Keucheyan is not the first to make such a claim on wealthy 'first world' nations or to direct our attention to the inequalities that the combination of capitalism and colonialism has produced and then reinforced. Rob Nixon's seminal *Slow Violence and the Environmentalism of the Poor* (2011) documents in detail the scale of inequity suffered by Indigenous and poorer communities as a result of western industrial development and the structural connections between colonisation and industrial expansion. The recent Anthropocene thesis argues for a new epoch, the Anthropocene, whose commencement is variously dated to 1610 or 'around' 1780, the birthdate of the Industrial Revolution, or 1945, the start of the nuclear age. Whichever date is accepted, the link between this 'new epoch' and the history of global colonial expansion, industrialisation and the increased reliance on fossil fuels is unambiguous.[3] While wealthier nations and wealthier communities within nations will be shielded by the more severe impacts of climate change in the short to medium term, it is evident that vulnerable nations and communities are faced with the immediate and permanent destruction of life and place as a consequence of environmental and ecological damage

[1] Irene Watson, 'Aboriginality and the Violence of Colonialism', *Borderlands E-Journal* 2009, 8(1), 2. Retrieved from www.borderlands.net.au.

[2] Razmig Keucheyan, *Nature Is a Battlefield: Towards a Political Ecology* (Cambridge: Polity, 2016), 8.

[3] For an analysis of the Anthropocene thesis in relationship to colonisation and climate change see Tony Birch, 'The Lifting of the Sky: Outside the Anthropocene', in *Humanities for the Environment: Integrating Knowledge, Forging New Constellations of Practice*, Joni Adamson and Michael Davis (Eds.) (New York: Routledge Environmental Humanities Series, 2016), 195–209.

involving the loss of human and non-human life and the inevitability of forced physical, social and spiritual dislocation. Such communities suffer from what Henry Shue, in his *Climate Justice: Vulnerability and Protection* refers to as 'compound injustice', which occurs when 'an initial injustice paves the way for further injustices'.[4]

Within Australia climate change poses a direct threat to Indigenous country and this is in line with what we witness within colonial societies globally. The impact of a changing climate and the related extreme and erratic weather events is linked to histories of dispossession and the appropriation of Indigenous land to service agricultural and industrial expansion. Five hundred years of global colonial violence have contributed directly to the environmental crisis that many Indigenous communities face today. For Indigenous nations and communities, the loss and destruction of country and the consequent climate injustice are not only a contemporary manifestation. For Indigenous people, there is nothing new in a cultural, spiritual and physical sense, about the 'mad-made epoch' that has created our current crisis. A reading of the voluminous Anthropocene debate often leaves Indigenous scholars and activists bemused. As Kyle Powys Whyte states, 'climate injustice, for Indigenous peoples is less about the spectre of a new future and more like the experience of déjà vu'.[5] Within the Australian context, climate injustice is a two-hundred-year (and counting) project, shaped by 'the British Empire's exploitative, extractive project, with the concomitant effects of obscuring or effacing Indigenous peoples' autonomy, their connection to "country", their knowledges, and their agency and presence'.[6]

For climate change to be tackled seriously climate justice, predicated on a recognition of the destructive legacies of colonialism, must be a priority. In the Australian context recognition, rather than carrying little more

[4] Henry Shue, *Climate Justice: Vulnerability and Protection* (Oxford: Oxford University Press, 2014), 4.

[5] Kyle Powys Whyte, 'Is It colonial déjà vu?: Indigenous Peoples and Climate Injustice', in *Humanities for the Environment: Integrating Knowledge, Forging New Constellations of Practice*, 88–105, Joni Adamson and Michael Davis (Eds.) (New York: Routledge Environmental Humanities Series, 2017), 88.

[6] M. Davis, 'Walking Together into Knowledge: Aboriginal/European Collaborative Environmental Encounters in Australia's North-East', in *Humanities for the Environment: Integrating Knowledge, Forging New Constellations of Practice*, Joni Adamson and Michael Davis (Eds.) (New York: Routledge Environmental Humanities Series, 2017), 181–194, 184.

than symbolic and reductive weight, must also include responsibility for the overt attempts by colonial authorities to devalue Indigenous country not only through direct violence but also through a fabricated narrative of Indigenous savagery and civil absence, a narrative that legitimated theft of land and destruction of people:

> The forces of British imperialism had the power to construct and determine who was 'truly' primitive and the universal horizon beyond which all humanity would become assimilated. Universal notions of civility justified the demonisation of our 'nativeness' and were used to destroy our relationships to land and our sovereign Indigenous being.[7]

For genuine social and political change to occur in Australia, change that will value and protect country, innovative relationships need to be forged between Indigenous and non-Indigenous communities. Such relationships must privilege genuinely restorative measures that compensate for past acts of dispossession. They must also alleviate the suffering experienced by country itself. As long as government responsibility (at a commonwealth and state level) for the unfinished business of colonial violence committed against people and country remains elusive, climate justice will also remain a tenuous goal. And yet it is an outcome we must strive for. At the core of the collective challenge that we face in relation to climate change is an acceptance of ownership by white Australia for colonial policies and practices that created the agricultural, mining and manufacturing industries. These industries have not only brought about the degradation of local ecologies, but have also assisted the global expansion of industrialisation dependent on increased burning of fossil fuels.

The Legacies of Colonialism

The colonial practice of constructing a narrative of savagery around Indigenous peoples in Australia acted as a device that necessitated violent dispossession and legitimated the extraction of wealth from country. The appropriation of land for wide-acre single crop agricultural practice, alongside the farming of sheep and cattle, led to the rapid destruction of local ecologies and a wasteful use of precious natural resources such as

[7] Irene Watson, 'Aboriginality and the Violence of Colonialism', 2.

water and soil.[8] But before land could be farmed unhindered by opposition from Indigenous owners of country, people were killed, forcibly removed, or at a minimum, pacified with the aid of repressive legislative measures. Jon Altman asserts that 'the brutal colonisation and political marginalisation of Indigenous Australians can be understood as a conflict over land and resource rights'.[9] Colonisation was, and to some extent remains, a conflict dependent on violence for its success. Deborah Bird Rose, among others, has correctly defined this ideology and practice as 'a dual war: a war against Nature and a war against the natives [that] includes both genocide and ecocide'.[10]

In his discussion of the disastrous failures of colonial agricultural practice in Australia Cameron Muir concludes that forms of land usage underpinning imperialist expansion 'sought to remake the continent's interior as a paddock for England' by utilising 'brutality, massacres, corruption, animal cruelty and environmental waste on a scale that threatened to derail the entire settler project'.[11] As in Australia so on a global scale the 'settler project' often failed miserably, leaving behind extensive environmental damage and the destruction of human and non-human life, and producing the 'erosion of environmental justice' that we face today.[12] In the absence of responsibility for the genocidal and ecocidal practices of global colonial expansion which contributed to climate change, narratives of the triumphant colonial settler struggling against country remain vital to the contemporary colonial psyche. My discussion below on coal is a recent example of these narratives. Additionally, in an attempt to obliterate the reality of Indigenous sovereignty and ownership of country, the blank slate *terra nullius* is repeatedly called upon to assist the privileging, production and (re)performance of the colonial 'primal scene'

[8] For an expansive discussion of this history see Cameron Muir, *The Broken Promise of Agricultural Progress: An Environmental History* (New York: Routledge, 2014).

[9] Jon Altman, 'People on Country as Alternate Development', in *People on Country: Vital Landscapes Indigenous Futures*, Jon Altman and Seán Kerins (Eds.) (Alexandria, NSW: The Federation Press, 2012), 1–22, 7.

[10] Deborah Bird Rose, *Reports from a Wild Country: Ethics for Decolonisation* (Sydney: University of New South Wales Press, 2004), 33–34.

[11] Muir, *The Broken Promise of Agricultural Progress: An Environmental History*, 2.

[12] Rob Nixon, *Slow Violence and the Environmentalism of the Poor* (Cambridge: Harvard University Press, 2011), 8.

of immaculate foundation, producing a fiction of the pure and innocent occupation of virgin land. While it may appear counter-intuitive, the colonial-foundational pantomime will occasionally 'acknowledge an original Indigenous presence but obscure the historical fact of violent dispossession, thereby enabling a nostalgic and idealized commemoration of the colonial past'.[13]

In Australia, this foundation stone of settler history requires constant underpinning to legitimate itself, with flag-waving Australia Days reinforcing the barely disguised anxieties that undermine 'the [seeming] might of sovereign power'.[14] This tradition of colonial pageantry is also part of a global tradition, enacted to mask the realities of land theft and deny responsibility for environmental vandalism. As Rob Nixon had also noted:

> Something similar applies to so-called post-conflict societies whose leaders may annually commemorate, as marked on the calendar, the official cessation of hostilities, while ongoing intergenerational slow violence ... may continue hostilities by other means.[15]

'Celebration of the nation' mantras additionally inhibit the potential for a mature collective discussion that may begin the process of accepting responsibility for the extensive damage done to land, air and waterways as a result of unabated colonial expansion. In the absence of reflection the country lingers in a state of delusion, an act of blind or feigned faith in the fantasy of the continued prosperity of the national farm or mining industry. The fantasy at times calls upon a sanitised past of the heroic pioneer, while in contemporary Australia rather than heed scientific warnings about the dangers of fossil fuel burning and its impact on climate change, the commonwealth government itself continues to invest in a narrative of hagiography surrounding fossil fuels, and in particular the little black rock, coal (discussed below). Material wealth for some at the expense of others, which others include ecological systems as well

[13] Simone Bignall, Daryle Rigney, and Robert Hattam, 'Colonial Letters Patent and Excolonialism: Forgetting, Counter-Memory and Mnemonic Potentiality', *Borderlands E-Journal*, 2015, 14(2), 4. Retrieved from www.borderlands.net.au.

[14] Bignall, Rigney, and Hattam, 7.

[15] Nixon, 8.

as human and non-human species, relies on a story that resists interrogation. Decolonised ways of thinking and acting are vitally necessary if country is to be valued not as a saleable commodity or an exploited symbol of national pride, but as an autonomous entity that each of us is connected to, but does not own. 'Land Rights' is not simply a political catch-cry or the basis of a legitimate and important political struggle. In relation to acting on climate change, we must engage with the idea of the inherent rights of land, including the right to reciprocity, recompense for past damages and a future of healing as a vital tenet of climate justice:

> Turning toward place [land] necessitates acknowledgement and reparations based on these histories: of settler colonialism, capitalism ... separations of mind from body, body from land. As humans make our planet increasingly toxic, unlivable, and at the same time increasingly inequitable, at what point might these cleavages be sewn back together, might we account for our pasts and to future generations?[16]

In order to appreciate the connection between colonialism and ecocide, we must produce a more vigorous critique of the relationship and co-dependence between colonialism and capitalism. In his essay, *Extinction: A Radical History*, Ashley Dawson writes that neither colonialism nor extinction can 'be understood in isolation from a critique of capitalism and imperialism'. Capitalism, argues Dawson, 'requires people to be [not only] destructive of the environment'[17] but of nations and communities standing in the way of unhindered expansion. Historically those who have been the most forceful barrier to an exploitative capitalist extraction mentality have been Indigenous nations, who, as a consequence of their acts of sovereignty and self-determination, have suffered levels of violence and dispossession beyond that experienced by other vulnerable and exploited communities.

In his analysis of colonial global expansion and its relevance to Australia's colonial history, Patrick Wolfe notes that the Industrial Revolution 'required colonial land and labour to produce its raw materials just as centrally as it required metropolitan factories and an industrial proletariat

[16] Eve Tuck, and Marcia McKenzie, 'Relational Validity and the "Where" of Inquiry: Place and Land in Qualitative Research', *Qualitative Inquiry*, 2015 21(7), 633–638. https://doi.org/10.1177/1077800414563809.

[17] Ashley Dawson, *Extinction: A Radical History* (New York: Or Books, 2015), 41.

to work in them'.[18] The invasion of and theft of Indigenous country was supported by the sleight-of-hand, *terra nullius*, at a sociopolitical level, thus legitimating a systemic practice of colonial violence that Wolfe refers to as 'the logic of elimination', articulated in Wolfe's (oft-quoted) conclusion that 'invasion is a structure not an event'. The structure of 'frontier homicide' was an institutional requirement in a society underpinned by the compulsion to 'destroy to replace'.[19] People were destroyed, as was land, producing the destruction of balanced ecologies which contributes to the situation we find ourselves in today, confronted by an unpredictable and damaging climate.

Global imperial expansion did not discriminate. Its violence was panoptical. Colonial rule, 'rather than replacing one owner with another' simply 'replace[d] an entire system of ownership with another'.[20] It could not countenance existing sovereign structures, at either a legal or cultural/spiritual level. In Australia over the course of the nineteenth century (flowing into the twentieth century), in addition to direct frontier violence, structural models supporting conquest became increasingly dependent on a bureaucratic regime within which Indigenous people were catalogued, counted and legislated against in order to perversely reinforce a status of nonexistence—non-people in an 'empty land' (empty of Indigenous people, that is)—which in turn refused the vitality of Indigenous ecological knowledge, particularly as it applied to local country. Not surprisingly, having extracted economic value from country, colonial authorities targeted Indigenous family and kinship structures via interventionist dispersal and 'caste' legislation, undermining people's direct connection to land and subsequent attempts to legally claim country. As has been noted elsewhere in relation to global colonial expansion, 'to get to the land, they had to remove the women and children'.[21] The exploitation of country, ecological damage and attempts to destroy Indigenous family and community life are inextricably linked. Institutional strategies

[18] Patrick Wolfe, 'Settler Colonialism and the Elimination of the Native', *Journal of Genocidal Research*, 2006, 8(4), 387–409, 394.

[19] Patrick Wolfe, 'Settler Colonialism and the Elimination of the Native', 387–388.

[20] Patrick Wolfe, *Traces of History: Elementary Structures of Race* (London: Verso, 2016), 34.

[21] Naomi Klein, 'Dancing the World into Being: A Conversation with Idle No More's Leanne Simpson', *Yes! Magazine*. 2013, March 5. Retrieved from http://www.yesmagazine.org.

enacted to eliminate Indigenous life in order that land could be fully exploited were pervasive in approach:

> the continuing operations of the logic of elimination can include officially encouraging miscegenation, the breaking-down of Native title into alienable individual freeholds, Native citizenship, child abduction, religious conversion, resocialisation in total institutions such as missions and boarding schools, and a whole range of cognate biological cultural assimilations. All these strategies, including frontier homicide, are modalities of settler colonialism. All of them come back to the issue of land.[22]

The consequences from the devaluation of, and the failure to see, the deliberate erasure of Indigenous knowledge of country are now being recognised. In Bruce Pascoe's groundbreaking work, *Dark Emu Black Seeds: Agriculture or Accident*, he discusses the first sightings and documentation of Indigenous people by European explorers and squatters. Pascoe notes that the new arrivals, often to their own surprise, did not find an 'untamed and primitive' land, as a literal interpretation of *terra nullius* would demand. In reality, it was 'common [to find] in the first colonial records' detailed evidence and descriptions of Indigenous agricultural practices based on a deep knowledge of country. While land could be poetically described in explorers' journals as resembling 'a gentleman's park' in England, Pascoe notes that the optimistic tone of language masked a darker intention. 'Few were here to marvel at a new civilisation', he writes. 'They were here to replace it'.[23]

The narratives produced by eighteenth-century and early nineteenth-century travellers to Indigenous country continued a global imperial tradition that eulogised and coveted the sovereign lands of those the would-be colonist sought to conquer. The desire for exploitable land, reliant on conquest and obliteration for its 'success', had begun centuries before the British ventured to the southern hemisphere. Ashley Dawson notes that Christopher Columbus, writing about North America in the late fifteenth century, described the land as 'so beautiful and rich for planting and sowing, and rearing cattle of all kinds, and for building

[22] Patrick Wolfe, 33–34.

[23] Bruce Pascoe, *Dark Emu Black Seeds: Agriculture or Accident?* (Broome: Magabala Books, 2016), 15, 33–35.

towns and villages'. The power of imperial writing would have far-reaching consequences. The Columbus narrative 'set the tone for the European imperial expansion in the subsequent five centuries', with the 'greed and lust for power drip[ping] from Columbus's pen', as Dawson describes it,[24] eventuating in not only the genocide of millions of Indigenous people but the transformation of land into regimented agricultural holdings to meet capitalist global demands.

The first wide-scale agricultural practice to exploit and transform Indigenous country in what would become Australia was the wool industry. In particular the farming of merino sheep provided the raw material for textile mills located predominantly in the north of England. The rapid expansion of wool farming and the need for an agricultural labour force had an immediate and devastating impact on Indigenous people and country:

> It took the discovery and development of a key export commodity, Australian merino wool, to provide the impetus for the frontier expansion and accompanying large-scale immigration that culminated in the settler takeover of the continent.[25]

On a global scale, European agricultural practices wilfully ignored the particularities of the Indigenous ecological knowledge systems, labelled by imperials as the 'new world', a telling and sad irony, considering that Indigenous land management was informed by traditions and practices reaching back many thousands of years.[26] The European practices were driven by an extraction mentality whereby cash crops, drawing valuable nutriment from soils and over-harvesting freshwater, were 'grown for export to the imperial metropole'. The impact was devastating:

> In addition to displacing and killing many millions of people, the monocultures of the plantation economy quickly exhausted land in the colonies, destroying soil fertility, and increasing vulnerability to pests.[27]

[24] Ashley Dawson, *Extinction: A Radical History*, 39.

[25] Patrick Wolfe, *Traces of History: Elementary Structures of Race*, 40.

[26] David Wade Chambers and Richard Gillespie, 'Locality in the History of Science: Colonial Science, Technoscience, and Indigenous Knowledge', *Osiris*, 2nd series (15), 2000, 221–240, 228.

[27] Ashley Dawson, *Extinction: A Radical History*, 47.

In Australia, additional strain was put on Indigenous nations with the arrival of a frenetic gold-mining industry from around 1850 onwards. In the colony of Victoria alone the discovery of gold drastically reduced the documented population of Indigenous people to around ten per cent of pre-Invasion numbers by the 1880s.[28] The desire for gold proceeded across Indigenous country uninhibited, absent of any legal or moral obligation to either sovereign Indigenous nations or the manner in which Indigenous nations engaged with and maintained country. It was not necessary to test the legal strength of a concept such as *terra nullius* to acquire a mining lease. The appropriation of Indigenous country was predicated on an acceptance 'that such peoples as did exist had no sovereign rights or proprietary interests'.[29]

The glaring hypocrisy of colonialism is that the coloniser, whether the explorer, squatter or mining prospector, was repeatedly confronted by another reality, an alternative universe, through constant meetings with and observation of people who articulated a sovereign engagement with country which did incorporate legal and cultural systems of ownership. In the historical records, there are many instances in the historical record of colonial officials documenting such encounters and colonial visitors recognising a deep knowledge of country conveyed by Indigenous people.[30] And yet the same diarists, many of them government officials with legal status refused to accept the lawful existence and rights of the very people living with country of which the coloniser had only had the briefest experience. These early encounters offered 'another way' of seeing and acting towards Indigenous nations. The legacy of failure to act in accordance with the rights of people and land is being experienced today. Climate change has a history, one that too few have grasped. We have learned little from it and continue to repeat its failures, as evidenced in the continued extraction of fossil fuels, often at the cost of Indigenous nations and their attempts to protect country.

[28] Patrick Wolfe, *Traces of History: Elementary Structures of Race*, 44.

[29] Bignall, Rigney, and Hattam, 5.

[30] See Bruce Pascoe *Dark Emu Black Seeds* in relation to Indigenous agricultural practice; also, the dairies of the explorer and government official, Sir Thomas Mitchell, as one of many such records.

This Little Black Rock

On a summer afternoon in 2017, the commonwealth Treasurer, Scott Morrison, provided the lower house of the Australian federal parliament with a bizarre piece of theatre, assisted by a lump of coal about the size of a house-brick. Morrison presented himself as a business-suited gladiator displaying a trophy of war, pre-empting victory in a battle against those campaigning against the Adani mega-coal mine proposed for central Queensland. If the mine proceeds, it will be located on country belonging to the Wangan and Jagalingou Traditional owners. They are opposed to the mine because of the threat it poses to the ecological and spiritual health of country.[31] Lest any member of the House, or media viewers, may have been unsure of the object Morrison held aloft he went on to heighten the sense of melodrama. 'This is coal—don't be afraid, don't be scared', he announced. 'It's coal. It was dug up by men and women who live and work in the electorates of those who sit opposite'.[32] Morrison then grinned across the chamber towards the Labor Opposition benches before the piece of coal was passed around the Coalition benches, shared between senior government ministers, including the energy minister, Josh Frydenberg, and the deputy prime minister, Barnaby Joyce.

From mid-2015 onwards Frydenberg has been repeatedly stating that Australia's energy future necessitates the continuation of coal mining. The Adani mine has also been justified partly on moral grounds. According to Frydenberg's position, first articulated in 2015,[33] in India, the country to which the Adani coal would be exported, 'two billion people today are using wood and dung for their cooking, due to the impoverished conditions suffered by many millions of people'. The attempt by

[31] For a detailed summary of the issue see Kristen Lyons, Morgan Brigg, and John Quiggin, *Unfinished Business: Adani, the State and the Indigenous Rights Struggle of the Wangan and Jagalingou Traditional Owners Council* (Brisbane: The University of Queensland, 2017).

[32] J. Butler, 'Scott Morrison Brought a Lump of Coal and Waved It Around in Parliament', *Huffington Post*, 2017. Retrieved from http://www.huffingtonpost.com.au/2017/02/08/scott-morrison-brought-a-lump-of-coal-and-waved-it-around-in-par_a_21710206/.

[33] J. Kelly, 'Strong Moral Case For Adani's Coal Mine: Josh Frydenberg', *Australian*, 18 October 2015. Retrieved from http://www.theaustralian.com.au/business/strong-moral-case-for-adanis-carmichael-coal-mine-josh-frydenberg/.

Frydenberg to tug at the guilt of first-world heartstrings has been subsequently undermined by India's own national energy minister, Piyush Goyal, who stated as recently as June 2017 that 'thanks to the increased productivity of domestic mines, cheaper renewables and lower than expected energy demand', imported coal on the scale projected by the Adani company would not benefit the Indian economy.[34] The Australian government's seeming unconditional support of the Adani company has been questioned further following the Australian Broadcasting Commission's *4 Corners* investigation,[35] highlighting issues of financial corruption, worker and community exploitation and environmental vandalism involving the Adani company within India.

If there is an imperative guiding the Australian government's support for the Adani mine, moral or otherwise, it is one driven by a tradition of settler colonialism's fixation with extraction as a nation-building project. Scott Morrison's 'booster' parliamentary soliloquy on the virtues of coal exemplifies this tradition. The beatification of mining, the enshrined notion that colonial societies must extract 'value' from land, in whatever form and at whatever cost, is vital to white Australia's sense of identity and sovereign legitimacy. Consequently, the mining industry has been a major beneficiary of government policy and legislation framed to ensure that the exploitation of land proceeds unhindered, with companies taking advantage of privileges ranging from generous tax subsidies to the granting of mining leases that disqualify Indigenous peoples' ability to access country.

The Adani coal mine would require the compulsory acquisition of 2750 hectares of Wangan and Jagalingou land. The mine would also result in a vital commodity, water, which is currently harvested for both domestic and farming use, being reallocated to the mine. Six massive open-cut pits would be constructed along with several underground mines. The mine would have a lifespan of approximately fifty years; five decades into an environmental future that cannot afford further fossil fuel poisoning. The coal would need to be transported 400 kilometres across country, much of it already in an ecologically fragile condition, on a yet to be constructed rail-line (for which the Adani company has applied for

[34] M. Safi, 'India Has Enough Coal Without Adani Mine, Yet Must Keep Importing, Minister Says', *Guardian*, 13 June 2017. Retrieved from https://www.theguardian.com/environment/2017/jun/13/.

[35] Australian Broadcasting Commission, Digging Into Adani. *4 Corners*. Initial broadcast, 3 October 2017.

a one-billion-dollar loan from the state government) to the Abbots Point terminal located on the edge of the Great Barrier Reef.[36] The Barrier Reef is the world's largest coral reef system, composed of 2900 individual reefs and 900 islands across an area of approximately 344,400 square kilometres. In recent years, large areas of the reef have suffered extensive coral bleaching as a result of an increase in water temperature. A rise in air temperature linked to climate change is partly responsible for the water temperature rise. A second contributing factor to reef degradation is the run-off into the ocean of water containing chemicals produced by the agricultural, mining and development industries. It is feared that the transportation of the coal would add further stress to the reef as a result of increased run-off.[37]

While the commonwealth's environmental minister made claims for domestic coal as the saviour of 'third world' poverty, within Australia the project would continue a colonial tradition of disenfranchising Indigenous owners of country and culture in favour of mining. To restate the comment of Patrick Wolfe, a colonial 'event' occurs within time, whereas a 'structure', more powerful and pervasive, exists across time. These events are embedded within this continuum and linked to the structural forces of dispossession of the nineteenth and twentieth centuries, which also included the appropriation, supported by favourable legislation, of Indigenous land for agriculture and mining ventures.

Despite the threat to global ecologies posed by climate change and the increasing demands for climate justice being articulated by Indigenous nations and communities, the Australian government support for the proposed Adani mine refuses Indigenous sovereign rights and threatens the environmental health of the planet. As global citizens, we are currently faced with a clear choice between an urgently required shift towards a philosophy and lifestyle that *privileges* country rather than colonial power or continuing to maintain a destructive relationship with country dependent on exploitation and extraction at the cost of both Indigenous rights and ecological maintenance.

[36] Lyons et al.

[37] For a history of the Great Barrier Reef, including the various environmental threats it has faced see Iain McCalman, *The Reef: A Passionate History* (Melbourne: Penguin Books, 2014).

Coal is a major contributing factor to the destabilisation of global atmospheric conditions and the increase in temperatures which underpins climate change.[38] Within Australia coal mining has caused extensive environmental damage and has been responsible for the most unnatural of disasters, including the destructive and extensive fires of February 2014 that devastated communities and country in Victoria's Latrobe Valley.[39] Along with other forms of mining, coal has failed to deliver claimed economic and social benefits to Indigenous nations, who have witnessed the destruction of vital ecologies, while being legally excluded from country they are culturally bound to protect.[40]

While the Adani group has been provided with vocal support by the state and commonwealth governments, in recent years it and other coal mining conglomerates have been losing the public relations war. In response the Minerals Council of Australia (MCA), the peak body representing mining interests in Australia, initiated the advertising campaign, *Coal—It's an amazing thing*, distributed through the mass media and in particular on social media platforms. The advertisement begins with an image of seams of coal. The mineral appears to contain talismanic qualities. The coal glistens, it sparkles, as if alive. Coal is transformed from a major atmospheric polluter into a black diamond. Its beauty is highlighted in the closing image of a lump of coal (perhaps the same one that Scott Morrison displayed in the parliament) rotating slowly for our wonderment. The image is accompanied by a seductive story, delivered in hushed, erotic tones. We are told that coal 'can provide endless possibilities. It can provide light and jobs ... delivering six billion dollars in wages for Australians ... injecting forty billion dollars each year ... isn't it amazing what this little black rock can do'.[41]

It is vital that we fully consider what 'this little black rock can do'. From the commencement of the Industrial Revolution, it has done quite a lot in contributing to climate change. In order that the Adani mine

[38] Andreas Malm, *Fossil Capital: The Rise of Steam Power and the Roots of Global Warming* (London: Verso, 2016).

[39] For a post mortem of the 2014 fire at the Hazelwood coal-powered station in the Latrobe Valley see Tom Doig, *The Coal Face* (Melbourne: Penguin Books, 2015).

[40] For a summary of the relationship between mining companies and Indigenous communities, see Jon Altman.

[41] Australian Mining Corporation, 'This Little Black Rock', 2015. Retrieved from YouTube https://youtu.be/IKp8W1jBuHw.

might proceed, the Australian government, with the full support of the Labor Opposition, rushed an amendment to the Native Title Act through parliament to the disadvantage of Traditional owner groups.[42] The use of parliamentary legislation to disempower Indigenous people is nothing new to the Wangan and Jagalingou nation. From the mid-nineteenth century onward they have fought to overcome barriers imposed by colonial authorities, including being subject to living 'under the Act', the Queensland government legislation framed to dispossess Indigenous people of country. As discussed above, Protection Acts were designed to steal not only land, but family, children, livelihood and human dignity. Within Queensland as a result of being governed by the Act:

> first and foremost, you lost almost every right for yourself and your family, rights which are, for the rest of us so basic as to defy listing … families were destroyed and women and children were kidnapped as sex objects, and it was common practice to capture adults and children and 'break them in' as servants.[43]

The connection between the 'protection' acts and the situation Indigenous communities confront today is direct and unambiguous. Murrawah Johnson, a young Indigenous woman, is a key spokesperson in the Wangan and Jagalingou battle against Adani. Johnson attributes her own fight to protect and sustain country to the political influence of her grandfather, Bowman Johnson. In 1954, fed up with the restrictive and demeaning regulations of the Act, Johnson moved his family off a government mission to the Wondai rubbish tip on the edge of a Queensland country town after being ordered by government officials to move his family to the Woorabinda mission. Instead, he chose freedom for his family, in the full knowledge that he would pay an immediate material cost, in that the family could no longer rely on the government support.[44]

[42] B. Merhab, 'Native Title Changes Pass Parliament', *News.com.au*, 14 June 2017. Retrieved from http://www.news.com.au/national/breaking-news/native-title-changes-pass-the-senate/.

[43] Rosalind Kidd, *Aboriginal Protection in Queensland*. Retrieved from http://www.linksdisc.com/roskidd/tpages/13htm.

[44] A. McGuire, 'Murrahwah Johnson and the Indigenous Fight Against Adani', *The Saturday Paper*, 3–9 June 2017. Retrieved from https://www.thesaturdaypaper.com.au/news/indigenous-affairs/2017/06/03/murrawah-johnson-and-the-indigenous-fight-against-adani.

Such practices, attempting to coerce or force Indigenous communities into submission by way of bureaucratically sanctioned poverty, have been historically utilised by governments. They also provide a grossly inequitable bargaining tool for mining companies who are able to subsequently 'negotiate' the right to mine land with people so impoverished that they struggle to feed their own families.

To fully appreciate the situation the Wangan and Jagalingou people face we must come to terms with the ramifications of this history. We must also remind ourselves that they are not a materially wealthy community. The proposed mining royalty deal being offered by Adani might appear attractive. And yet they have refused money. They are more concerned with the protection of sacred country. Power is a relative concept. Murrawah Johnson articulated a sense of true power in late 2016, when she announced she would refuse to become 'the broken link in the chain' and witness the loss of country to a coal mine. Johnson, on behalf of the Wangan and Jagalingou, has chosen a self-determining and sovereign pathway to the future. 'We've seen the end of the world', she has told us, 'and we've decided not to accept it'.[45]

Conclusion: A Story of Place

Adrian Burragubba is a senior Wangan and Jagalingou elder leading the campaign against the Adani mine. He has taken his nation's struggle to the streets of Australia's major cities, to multinational bankers based in New York City and back home to country. He has made it clear that action to stop the mine from proceeding must be led by the traditional owners of country. He also understands that his people cannot succeed without the help of others:

> I'm going to convince all of our people to stand together as one people, one voice. And then we're going to ask all Australian people and people from all over the world to stand with us, unite with us to fight this fight.[46]

[45] Murrawah Johnson, 'Wangan Jagalingou Traditional Owners: We Will Not Surrender', 2016. Retrieved from https://www.youtube.com/watch?v=xIN8b1MAwvs.

[46] A. Burragubba, 'Stop Adani Destroying Our Land and Culture', 2015. Retrieved from https://www.youtube.com/watch?v=ZB2JC6yKy_E.

Burragubba's statement should not be mistaken as a plea. It is an offering to the Australian nation and people the world over to engage in new relationships that are genuinely equitable, that recognise Indigenous rights and claims to justice and that also have the potential to deliver ecological justice to the planet. Murrawah Johnson recently stated that 'the most forward-thinking people' involved in the Adani fight 'are the ones who are the most disenfranchised and have the least resources', and yet 'we're doing it'.[47] The writer and activist, Naomi Klein, believes settler-societies 'have been colonized by the logic of capitalism, and that has left us uniquely ill-equipped to deal with this particular crisis [climate change]'.[48] Through the expression of their opposition to the Adani mine and through their fight to protect country, Burragubba and Johnson are inviting non-Indigenous people to re-assess their place in this country, and their views 'on country', by raising a series of profound questions that we need to answer urgently. Does the Australian nation possess the level of maturity required to recognise the sovereign authority and leadership of the Wangan and Jagalingou to protect country? Is non-Indigenous Australia capable of realising that country is an autonomous entity with inherent rights rather than a resource to be endless plundered? Is the Australian nation mature enough to accept an invitation to stand with Indigenous people to protect country and, in doing so, break the shackles of the destructive logic of colonialism?

Klein, along with a range of activists and thinkers, believes that it is within Indigenous communities that a pathway forward exists for dealing with environmental instability and ecological maintenance.[49] In moving forward, we will also need to address how knowledge of country and ecology can best be transmitted and shared amongst the non-Indigenous communities without that knowledge being corrupted, commodified or culturally 'stolen'. Zoe Todd has written extensively about this issue and how it impacts on Indigenous/non-Indigenous negotiations and relationships. She has drawn attention to the relationship between Indigenous knowledge systems, climate change and climate justice. Todd is

[47] Murrawah Johnson.

[48] Naomi Klein, 'You and What Army? Indigenous Rights and the Power of Keeping Our Word', *This Changes Everything: Capitalism vs. the Climate* (London: Penguin Books, 2014), 12, 367–387.

[49] See Naomi Klein, 'You and What Army?' for a detailed commentary on Indigenous Rights and leadership with regard to climate change.

critical of forms of western knowledge that 'quickly erase arctic Indigenous peoples and their laws and philosophies from their discourses'.[50] While engagement with Indigenous ecological knowledge by outsiders is increasing, Todd remains cautious. 'Indigenous thought', she writes, 'is not just about social relations and philosophical anecdotes, as many in ethnography would suggest', but something more complex, underpinned by Indigenous people increasingly 'fighting for recognition – fighting to assert their laws, philosophies and stories on their own terms'.[51] Indigenous thought is not simply a body of knowledge that can be opportunistically attached to an environmental cause, nor can it be reduced to the status of quaint folklore or myth. While Indigenous knowledge is sometimes disseminated as 'story', the narrative exists in 'a framework within which more detailed empirically derived knowledge about relationships can be placed and evaluated'.[52]

The history of colonialism in Australia cannot be understood without a critical engagement with the competing forces of Indigenous ownership and protection of country and a colonial mentality driven by an insatiable attachment to extraction—extraction of minerals, energy, nutriment and people opposed to the exploitation of country. For colonialism to unravel and climate justice and protection of country to become a reality vital and, at times, difficult conversations must take place. While the work will be hard, the potential benefits are clear. Facing a future of climate uncertainty and the potential for disastrous knock-on effects, we will require both patience and radical thinking. Joseph Roe, speaking of his relationship with Goolarabooloo Country in north-western Australia, once stated, 'I don't have to be an environmental activist. If I do my job and look after country, protection for the environment will flow from our law and culture'.[53] This is the shift in thinking we require. It involves the recognition of Indigenous sovereignty, law and culture, which will not only reflect an acceptance of Indigenous knowledge by non-Indigenous society, but

[50] Zoe Todd, 'An Indigenous Feminist's Take on the Ontological Turn: "Ontology" Is Just Another Word for Colonialism', *Journal of Historical Sociology*, 2016, 29(1), 4–22, 6, https://doi.org/10.1111/johs.12124.

[51] Todd, 18.

[52] Todd, 12.

[53] S. Muecke, 'Indigenous-Green Knowledge Collaborations and the James Price Point Dispute', in *Unstable Relations: Indigenous People and Environmentalism in Contemporary Australia*, E. Vincent and T. Neale (Eds.). Perth: UWA Publishing, 2016), 252–272, 252.

also prepare the ground for a genuine exchange of knowledge based on reciprocity and trust.

Bibliography

Altman, J. C. 'People on country as alternate development'. In J. C. Altman & S. Kerins (Eds.), *People on country: Vital landscapes indigenous futures*. 1–22. Sydney: The Federation Press, 2012.

Australian Broadcasting Commission. 'Digging into Adani'. *Four Corners*. 2017, October 3.

Australian Mining Corporation. 'This little Black rock'. 2015. Retrieved from https://youtu.be/IKp8W1jBuHw.

Bignall, S., Rigney, D., & Hattam, R. 'Colonial letters patent and excolonialism: Forgetting, counter-memory and mnemonic potentiality'. 2015. Retrieved from http://www.borderlands.net.au.

Birch, T. 'The lifting of the sky: Outside the anthropocene'. In J. Adamson & M. Davis (Eds.), *Humanities for the environment: Integrating knowledge, forging new constellations of practice*. 195–209. New York, NY: Routledge Environmental Humanities Series, 2016.

Bird Rose, D. *Reports from a wild country: Ethics for decolonisation*. Sydney: University of New South Wales Press, 2004.

Burragubba, A. *Stop Adani destroying our land and culture*. 2015. Retrieved from https://www.youtube.com/watch?v=ZB2JC6yKy_E.

Butler, J. 'Scott Morrison brought a lump of coal and waved it around in parliament'. *Huffington Post*. 2017. Retrieved from http://www.huffingtonpost.com.au/2017/02/08/scott-morrison-brought-a-lumpof-coal-and-waved-it-around-in-par_a_21710206/.

Chambers, D. W., & Gillespie, R. 'Locality in the history of science: Colonial science, technoscience, and indigenous knowledge'. *Osiris*, 15, 2000, 221–240.

Clark, N. 'Shock and awe: Trauma as the new colonial frontier'. *Humanities*, 5(14), 2016, 1–22. https://doi.org/10.3390/h5010014.

Clarke, P. A. 'The use and abuse of aboriginal ecological knowledge'. In I. D. Clark & F. Cahir (Eds.), *The Aboriginal story of Burke and Wills: Forgotten narratives*. 61–79. Melbourne: CSIRO Publishing, 2016.

Davis, M. 'Walking together into knowledge: Aboriginal/European collaborative environmental encounters in Australia's north-east'. In J. Adamson & M. Davis (Eds.), *Humanities for the environment: Integrating knowledge, forging new constellations of practice*. 181–194. New York, NY: Routledge Environmental Humanities Series, 2017.

Dawson, A. *Extinction: A radical history*. New York, NY: Or Books, 2015.

Doig, T. *The coal face*. Melbourne: Penguin Books, 2015.
Donaldson, M. 'The end of time? Aboriginal temporality and the British invasion of Australia'. *Time and Society*, 1996, 5(2), 187–207.
Dundas, A. *Towards climate justice: Decolonising adaptation to climate change*. Sydney: Environment Institute, University of Sydney, 2016.
Hage, G. 'Etat de siege: A dying domesticating colonialism?' *Open Anthropology: A Public Journal of the American Anthropological Association*, 2016, 43(1), 38–49. https://doi.org/10.1111/amet.12261.
Holthaus, E. 'Northern hemisphere temperature breaches a terrifying milestone'. *New Scientist*. 2006, March 7. Retrieved from http://www.newscientist.com/article/2079775.
Johnson, M. *Wangan Jagalingou traditional owners: We will not surrender*. 2016. Retrieved from https://www.youtube.com/watch?v=xIN8b1MAwvs.
Kastner, J., & Najafi, S. 'The wall and the eye: An interview with Eyal Weizman'. *Cabinet Magazine*. 2002/2003. Retrieved from http://www.cabinetmagazine.org.
Kelly, J. 'Strong moral case for Adani's coal mine: Josh Frydenberg'. *The Australian*. 2015, October 18. Retrieved from http://www.theaustralian.com.au/business/strong-moral-case-for-adanis-carmichaelcoal-mine-josh-frydenberg/.
Keucheyan, R. *Nature is a battlefield: Towards a political ecology*. Cambridge: Polity Press, 2016.
Kidd, R. *Aboriginal protection in Queensland*, n.d. Retrieved from http://www.linksdisc.com/roskidd/tpages/13htm.
Klein, N. 'Dancing the world into being: A conversation with idle no more's Leanne Simpson'. *Yes! Magazine*. 2013, March 5. Retrieved from http://www.yesmagazine.org.
Klein, N. 'You and what army? Indigenous rights and the power of keeping our word'. In N. Klein (Ed.), *This changes everything: Capitalism vs. the climate*. 367–387. London: Penguin Books, 2014.
Lyons, K., Brigg, M., & Quiggin, J. *Unfinished business: Adani, the state and the indigenous rights struggle of the Wangan and Jagalingou traditional owners council*. Brisbane: The University of Queensland, 2017.
Malm, A. *Fossil capital: The rise of steam power and the roots of global warming*. London: Verso Books, 2016.
McCalman, I. *The reef: A passionate history*. Melbourne: Penguin Books, 2014.
McGuire, A. 'Murrahwah Johnson and the indigenous fight against Adani'. *The Saturday Paper*. 2017, June 3–9. Retrieved from https://www.thesaturdaypaper.com.au/news/indigenous-affairs/2017/06/03/murrawah-johnson-and-the-indigenous-fight-against-adani.

Merhab, B. 'Native title changes pass parliament'. *News.com.au*. 2017, June 14. Retrieved from http://www.news.com.au/national/breaking-news/native-title-changes-pass-the-senate/.
Mitchell, M. T. *Three expeditions into the interior of Eastern Australia; with descriptions of the recently explored region of Australia Felix and the present colony of New South Wales* (2nd ed., Vol. 2). London: T. & W. Boone, Project Gutenberg Australia, 2007. Retrieved from http://gutenberg.net.au.
Muecke, S. 'Indigenous-green knowledge collaborations and the James Price point dispute'. In E. Vincent & T. Neale (Eds.), *Unstable relations: Indigenous people and environmentalism in contemporary Australia*. 252–272. Perth: UWA Publishing, 2016.
Muir, C. *The broken promise of agricultural progress: An environmental history*. New York, NY: Routledge, 2014.
Nixon, R. *Slow violence and the environmentalism of the poor*. Cambridge, MA: Harvard University Press, 2011.
Pascoe, B. *Dark emu Black seeds: Agriculture or accident?* Broome: Magabala Books, 2016.
Powys Whyte, K. 'Is it colonial déjà vu? Indigenous peoples and climate injustice'. In J. Adamson & M. Davis (Eds.), *Humanities for the environment: Integrating knowledge, forging new constellations of practice*. 88–105. New York, NY: Routledge Environmental Humanities Series, 2017.
Pravinchandra, S. 'One species, same difference? Postcolonial critique and concept of life'. *New Literary History*, 2016, 47(1), 27–48. https://doi.org/10.1353/nlh.2016.0002.
Prochnik, G. 'George Prochnik interviews Eyal Weizman: The desert threshold'. *Los Angeles Review of Books*. 2015, October 18. Retrieved from https://lareviewofbooks.org/article/the-desert-threshold/.
Robertson, J. 'Dangerous global warming will happen sooner than thought—Study'. *The Guardian*. 2016, March 10. Retrieved from http://www.theguardian.com/environment/2016/mar/10.
Safi, M. 'India has enough coal without Adani mine, yet must keep importing, minister says'. *The Guardian*. 2017, June 13. Retrieved from https://www.theguardian.com/environment/2017/jun/13/.
Shue, H. *Climate justice: Vulnerability and protection*. Oxford: Oxford University Press, 2014.
Todd, Z. 'An indigenous feminist's take on the ontological turn: 'Ontology' is just another word for colonialism'. *Journal of Historical Sociology*, 2016, 29(1), 4–22. https://doi.org/10.1111/johs.12124.
Tuck, E., & McKenzie, M. 'Relational validity and the "where" of inquiry: Place and land in qualitative research'. *Qualitative Inquiry*, 2015, 21(7), 633–638. https://doi.org/10.1177/1077800414563809.

Watson, I. 'Aboriginality and the violence of colonialism'. *Borderlands E-Journal*, 2009, 8(1), 1–8. Retrieved from http://www.borderlands.net.au.
Wolfe, P. 'Settler colonialism and the elimination of the native'. *Journal of Genocidal Research*, 2006, 8(4), 387–409.
Wolfe, P. *Traces of history: Elementary structures of race*. London: Verso Books, 2016.

Tony Birch is a Professorial Research Fellow in the Moondani Balluk Academic Centre at Victoria University. He is also a novelist and poet.

CHAPTER 13

Reimagining Decolonising Praxis for a Just and Ecologically Sustainable Peace in an Australian Context

Anne Elvey

'Nothing has been the same for Aboriginal People after the "arrival" of Cook on the Australian continent', writes Anne Pattel-Gray in her 1998 volume, *The Great White Flood*, in which she details and analyses racism in Australia.[1] Racism continues to be pervasive in Australian society and its structures, despite the desires of many citizens to see themselves as having outgrown the worst of Australia's colonial history. But this history is not past. Rather, the vestiges of genocidal colonial policies and practices of land theft, dispossession, and assimilation not only echo in the present for First Peoples in experiences of transgenerational trauma, but also persist in contemporary policies and practices of Indigenous child removal and

[1] Anne Pattel-Gray, *The Great White Flood: Racism in Australia: Critically Appraised from an Aboriginal Historico-Theological Viewpoint*, American Academy of Religion Cultural Criticism Series 2 (Atlanta: Scholars Press, 1998), 15.

A. Elvey (✉)
Monash University, Melbourne, VIC, Australia
e-mail: aelvey@tpg.com.au

University of Divinity, Melbourne, VIC, Australia

J. Camilleri and D. Guess (eds.), *Towards a Just and Ecologically Sustainable Peace*, https://doi.org/10.1007/978-981-15-5021-8_13

incarceration of children and adults that continue, under other names, the officially censured procedures of the Stolen Generations and Deaths in Custody.[2] At the same time, mining companies collude with Federal and State Governments to extract wealth from First Nations lands, exploiting divisions within and between Indigenous communities to suit their corporate purposes, while threatening vulnerable ecosystems and potentially exacerbating climate change.[3] As Timothy Neale and Eve Vincent point out, however, Indigenous people should not be painted as passive victims of corporate interests; the situation and the situational alliances with mining companies, for example, are more complex.[4] In contemporary Australia, then, what is at stake when scholars and activists seek to bring together concepts and practices of ecological sustainability, peace and justice? What do we need to consider at the outset when mapping effective principles for a just and ecologically sustainable peace? In this essay, I argue that a renewed engagement with Indigenous sovereignty is an essential beginning point for such ecological peace work.

[2] On the historic Royal Commission investigating Aboriginal Deaths in Custody (final report signed 15 April 1991) and the National Inquiry into the Stolen Generations (final report published 1997), see 'Royal Commission into Aboriginal Deaths in Custody—Fact Sheet 112', National Archives of Australia, https://www.naa.gov.au/sites/default/files/2020-05/fs-112-royal-commission-into-aboriginal-deaths-in-custody.pdf (accessed 9 June 2020) and *Bringing Them Home: Report of the National Inquiry into the Separation of Aboriginal and Torres Strait Islander Children from Their Families* (Commonwealth of Australia 1997), https://www.humanrights.gov.au/sites/default/files/content/pdf/social_justice/bringing_them_home_report.pdf (accessed 5 July 2019). On the contemporary situation, see, for example, 'Deaths in Custody', *The Guardian* Australia, https://www.theguardian.com/australia-news/deaths-in-custody (accessed 5 July 2019); 'The Women Fighting against a Rising Tide of Indigenous Child Removals', NITV, https://www.sbs.com.au/nitv/feature/women-fighting-against-rising-tide-indigenous-child-removals (accessed 5 July 2019). Lorena Allam, Calla Wahlquist and Nick Evershed, 'Aboriginal Deaths in Custody: Black Lives Matter Protests Referred to Our Count of 432 Deaths. It's Now 437', *The Guardian* Australia (9 June 2020), https://www.theguardian.com/australia-news/2020/jun/09/black-lives-matter-protesters-referred-to-our-count-of-432-aboriginal-deaths-in-custody-its-now-437 (accessed 9 June 2020).

[3] See, for example, Timothy Neale and Eve Vincent, 'Instabilities and Inequalities: Relations between Indigenous People and Environmentalism in Australia Today', in Eve Vincent and Timothy Neale (eds), *Unstable Relations: Indigenous People and Environmentalism in Contemporary Australia* (Crawley, WA: UWAP Scholarly, 2016), 1–24 (esp. 6–14); Jillian K. Marsh, 'Decolonising the Interface between Indigenous Peoples and Mining Companies in Australia: Making Space for Cultural Heritage Sites', *Asia Pacific Viewpoint* 54, no. 2 (August 2013): 171–84.

[4] Neale and Vincent, 'Instabilities and Inequalities', 6–14.

Peace, Ecology and First Nations Sovereignty

Writing of the emerging field of 'peace ecology', Úrsula Oswald Spring, Hans Günter Brauch and Keith G Tidball comment that the '*ecology* concept coined by Ernst Haeckel (1834–1919)' was 'developed by many scientists from different disciplines and world regions, based in part upon observations of indigenous cultures in the Americas, in China, India and the Middle East where knowledge on the use and dangers of plants and animals were crucial for human survival and cultural development'.[5] Members of the Earth Bible Team, founded in Adelaide, acknowledge their learning from, as well as collaboration with, the First Peoples of what is now Australia, as do many scholars in the environmental humanities more generally.[6] The indebtedness of ecological thought to First Nations epistemologies underscores the necessity for an Australian-based project toward a just and ecologically sustainable peace to be founded in respect for Indigenous knowledges and to ask on what authority such knowledges are employed by non-Indigenous researchers and activists.

In this chapter, I refer to a concept taken up from English and given new meaning by First Peoples in Australia, namely 'Country', used without the article, and signifying the lands, waters and skies, of a particular Indigenous language group, for example, Boon Wurrung Country, Woi Wurrung Country, Nyoongar Country. Trawloolway man and theologian, Garry Worete Deverell writes:

> Because we believe that our country is still filled with the presence of the ancestor-creators who formed it a long time ago, we also regard all that is alive in that land as our kin, our family. For we are descendants

[5] Úrsula Oswald Spring, Hans Günter Brauch, and Keith G. Tidball, 'Expanding Peace Ecology: Peace, Security, Sustainability, Equity, and Gender', in Úrsula Oswald Spring, Hans Günter Brauch, and Keith G. Tidball (eds), *Expanding Peace Ecology: Peace, Security, Sustainability, Equity and Gender. Perspectives of IPRA's Ecology and Peace Commission*, Springer Briefs in Environment, Security, Development and Peace: Peace and Security Studies 12 (Cham, Switzerland: Springer, 2014), 1–30 (6), emphasis in original.

[6] Norman C. Habel, 'Introducing the Earth Bible', in Norman C. Habel (ed), *Readings from the Perspective of Earth*, Earth Bible 1 (Sheffield: Sheffield Academic Press, 2000), 25–37 (26); The Earth Bible Team, 'Guiding Ecojustice Principles', in Norman C. Habel (ed), *Readings from the Perspective of Earth*, Earth Bible 1 (Sheffield: Sheffield Academic Press, 2000), 38–53 (52); Deborah Bird Rose, 'Country and the Gift', in Joni Adamson and Michael Davis (eds), *Humanities for the Environment: Integrating Knowledge, Forging New Constellations of Practice* (London: Routledge, 2017), 33–44 (33).

> of those ancestor-creators. We are their offspring. That is why, whenever we approach a new place that is not familiar to us, we talk to the spiritual presences in the land, the ancestor-creators, asking for their permission to cross that country and for their protection while we are there.[7]

Proper protocols and permissions are essential to First Nations epistemologies and practices in relation to Country, which—whatever its overlay, for example, as evident especially in highly populated urban areas—remains unceded territory, where the sovereignty of the nation state is unsettled by barely recognised Indigenous sovereignties.[8] I use 'First Nations', 'First Peoples' and 'Indigenous' interchangeably to refer to the first peoples of Australia today; where possible I refer to a specific group such as the Boonwurrung on whose Country I live in Seaford, a suburb of Melbourne, and whose sovereignty I recognise though I am still learning what that means.

Failure to recognise Indigenous sovereignties in ways that are meaningful and life-affirming for the several hundred First Nations of what is now Australia is based on colonial violence that continues in the ongoing moment of invasion and a 'collective memory of imperialism', experienced most keenly and, more than occasionally, in life-threatening ways by Indigenous peoples.[9] Drawing from both peace and environmental studies, Christos Kyrou describes forms of violence, all of which have human as well as environmental costs and include physical, structural, cultural, epistemic, psychological, ecological and time violence.[10] Each of these types of violence is evident in Australian colonial practices which are not entirely past; for example, physical violence occurred not only in

[7] Garry Worete Deverell, *Gondwana Theology: A Trawloolway Man Reflects on Christian Faith* (Reservoir, VIC: Morning Star Publishing, 2018), 14; see also, Deborah Bird Rose, *Nourishing Terrains: Australian Aboriginal Views of Landscape and Wilderness* (Canberra: Australian Heritage Commission, 1996), 6.

[8] Dianne Otto, 'A Question of Law or Politics? Indigenous Claims to Sovereignty in Australia', *Syracuse Journal of International Law and Commerce* 21 (1995): 65–103 (66).

[9] On invasion as a structure rather than an event, see Patrick Wolfe, 'Settler Colonialism and the Elimination of the Native', *Journal of Genocide Research* 8, no. 4 (2006): 387–409 (388). The term 'collective memory of imperialism' is from Linda Tuhiwai Smith, *Decolonizing Methodologies: Research and Indigenous Peoples*, 2nd edn (London: Zed Books, 2012), Kindle edn loc. 237 of 5874.

[10] Christos N. Kyrou, 'Peace Ecology: An Emerging Paradigm in Peace Studies', *International Journal of Peace Studies* 12, no. 1 (Spring–Summer 2007): 73–92 (80).

the underacknowledged contact wars of invasion, subsequent massacres, the wrenching of children from parents under policies of assimilation, but also in contemporary racist violence toward First Nations peoples in custody.[11] The historical suppression and subsequent loss of many Indigenous languages is a form of what Kyrou describes as 'epistemic violence'. Cultural violence occurs in the subordination and outright denial of the value, of First Nations cultures, and continues in failures to recognise and make space for cultural keeping, except as a marketing ploy of nation or a tourist attraction. Psychological violence, and the trauma of invasion, occurs in the displacement of many First Peoples from their traditional lands and the difficulty in maintaining environmentally sensitive practices that are part of living on Country, so contributing to ecological violence endemic to Western consumerist capitalism. 'Structural violence', says Kyrou, 'takes place when laws, formal institutions and cultural or societal structures and processes discriminate against particular groups of people based on traits such as gender, skin color, or ethnic background'.[12] Physical, epistemic, cultural and psychological kinds of violence are interconnected and mutually support each other, leaving deep traces in societal structures, as well as cross-cultural communal and personal encounters, so that the logic of invasion is a daily life-taxing reality for most, if not all, Indigenous peoples in Australia. Racism is institutionalised and internalised and, as Deverell explains, is also encountered in the churches.[13]

'Time violence' adds another dimension. For Kyrou: 'Time violence refers to intergenerational conflict – the victimization of future generations due to today's practices and behaviors'.[14] Anthropogenic climate change—a process in which human agency in conjunction with the material properties and agencies of fossil fuels, oceans, air and so on creates a situation of dangerous global heating—is a prime global example of time violence. As Ellen van Neerven comments, Indigenous peoples will be 'disproportionally' affected by climate change and the 'uncertain futures'

[11] Evidence for the examples of violence described in this paragraph is strong and hardly needs repeating.

[12] Kyrou, 'Peace Ecology', 80.

[13] Deverell, *Gondwana Theology*, 47–50. See also the earlier analysis of Pattel-Gray, *The Great White Flood*.

[14] Kyrou, 'Peace Ecology', 81.

it augurs, so time violence has a particular relevance to First Peoples.[15] Indeed, like their Pacific neighbours, First Peoples of Mer, in what is called colonially the Torres Strait Islands, already report impacts of climate change on their region.[16] Tony Birch describes the impact of the 'slow violence' of environmental damage through climate change and other causes, 'on the disadvantaged, including Indigenous communities'; this violence is perpetrated predominantly by 'wealthier nations and their acts of collective irresponsibility'.[17] Racist stereotyping of Indigenous people, colonial dispossession and climate change are linked, through the violent legitimation of mining on Indigenous Country, as Birch demonstrates.[18]

My description of intersecting types of violence here risks characterising First Nations as victims, with the resonances of passivity such portrayal can imply. That is not my intention. Rather, I am describing a situation, underacknowledged in the national story of Australia, in which founding violence persists in a past that is not past. Traumas of British colonisation impacting Indigenous people from the late eighteenth century to now—invasion, contact wars, massacres, destruction of Indigenous agriculture and culture, dispossession and assimilation, indentured labour, which amounted to wage slavery especially for Indigenous farm and cattle station workers and domestic servants, deaths in custody, intervention, racism, failures to deal justly with claims for land rights and calls for self-determination, and damage to Country through nuclear tests, mining and erroneous agricultural practices including water mismanagement—are a primary context for any discussion of a just and ecologically sustainable

[15] Ellen van Neerven, 'The Country Is Like a Body', *Right Now: Human Rights in Australia* (26 October 2015), http://rightnow.org.au/essay/the-country-is-like-a-body/ (accessed 27 June 2019).

[16] Aaron Smith, 'Torres Strait Islander Musicians Marou and Busby Use Trip Home to Highlight the Real Impact of Climate Crisis on the Region', NITV News (5 July 2019), https://www.sbs.com.au/nitv/article/2019/07/05/climate-change-actually-affecting-torres-strait-islanders-right-now-marou (accessed 5 July 2019).

[17] Tony Birch, 'Climate Change, Mining and Traditional Knowledge in Australia', *Social Inclusion* 4, no. 1 (2016): 92–101 (94).

[18] Tony Birch, '"We've Seen the End of the World and We Don't Accept It": Protection of Indigenous Country and Climate Justice', in Nicole Oke, Christopher Sonn, and Alison Baker (eds), *Places of Privilege: Interdisciplinary Perspectives on Identities, Change and Resistance* (Leiden: Brill, 2018), 139–52 (146–50).

peace.[19] Alongside this narrative of trauma and ongoing disadvantage are strong histories of Indigenous resistance, resilience and recovery. Equally part of this contemporary context, these practices of vigorous survival inform the question of First Nations' sovereignties. Native American, Anishinaabe, writer and professor of literature, Gerald Vizenor coins the term 'survivance' to describe a quality of First Peoples' transformative resilience and 'tribal courage' that resists and counters domination.[20] This quality is evident, too, in contemporary Australian First Nations' literatures, arts, communities and histories.[21]

Important to this discussion is the articulation of the concept of Indigenous sovereignty by First Peoples as being multifaceted and properly a call from Indigenous people to Indigenous people to claim their ground, to assert what they already know as true about their long-standing in, and as, the multiple nations that pre-exist the nation of Australia.[22] On this basis, Indigenous affirmation of sovereignty as First Nations is a challenge to the wider Australian polity to recognise the multiple sovereignties persisting in this nation and to allow the narrative of national sovereignty to be interrupted and told differently, more truthfully, through a deep and transformative acknowledgement, by the majority of Australians, of the violent colonial lies on which Australia is based. These lies include the founding lie of 'terra nullius' and its accompanying myth of 'no sovereignty'.[23] More than recognising these lies, Australian-based advocates for a just and ecologically sustainable peace

[19] My thanks to an anonymous reviewer for clarifying the point about indentured labour and its relation to slavery.

[20] Gerald Vizenor, *Manifest Manners: Narratives on Postindian Survivance* (Lincoln: University of Nebraska Press, 1994); see also, Gerald Vizenor (ed), *Survivance: Narratives of Native Presence* (Lincoln: University of Nebraska Press, 2008); Sharon D. Welch, *After Empire: The Art and Ethics of Enduring Peace* (Minneapolis: Fortress, 2004), 43.

[21] Just one of many examples is the recent novel by Goorie writer, Melissa Lucashenko, *Too Much Lip* (St Lucia: Queensland University Press, 2018), shortlisted for several major awards.

[22] Tony Birch, 'On Sovereignty', *Overland* 229 (2017): 88–89.

[23] Pattel-Gray, *The Great White Flood*, 17–18. The churches were complicit in the dual lie of *terra nullius* and 'no sovereignty', argues Pattel-Gray, as they became enforces of government policies such as protection (127). When she was writing in the late 1990s, the churches remained ambivalent regarding recognising Indigenous sovereignty (158–59). While some progress has been made, this is largely still the case in practice.

need to respond—as they pursue their scholarly and activist vocations—with robust and practical respect to First Nations' sovereignties. What might this mean in practice?

Reimagining Decolonisation

Discussion of decolonisation is decades old, yet not enough has changed, because the question in Australia is not simply linguistic window dressing or intellectual assent. Engaging with a Western philosophical tradition, the late environmental philosopher Val Plumwood, building on years of activist work, described a logic of colonisation, a mode of thinking and action modelled on the relation between master and slave, of unacknowledged dependence of the privileged on those they construct as other, in hyper-separate pairings where the first is seen as superior and opposed to the second: men/women, settler/Indigenous, mind/body, spirit/matter, culture/nature, human/animal, human/nature, God/human.[24] In this othering, those constructed as other tend to be cross-identified, so that Indigenous people, for example, are both naturalised and feminised, and in the process, both people and Country are subject to appropriative violence. Yet, this analysis itself is partial reflecting a failure to listen closely to Indigenous understandings of decolonisation and epistemologies in relation to Country.[25]

Over subsequent years, Australian environmental humanities scholars have reimagined humankind as enmeshed in a more-than-human Earth community, as ecological selves, in ways that Indigenous peoples have

[24] Val Plumwood, *Feminism and the Mastery of Nature* (London: Routledge, 1993), esp. 41–68; see also her *Environmental Culture: The Ecological Crisis of Reason* (London: Routledge, 2002), and 'Decolonizing Relationships with Nature', in William M. Adams and Martin Mulligan (eds), *Decolonizing Nature: Strategies for Conservation in a Post-colonial Era* (London: Earthscan, 2003), 51–78.

[25] Plumwood's critical work stemmed from lived experience of near-death encounter with a crocodile in 1985 and her realisation that a kind of Western settler hubris meant she had not sought advice from local Indigenous people before she canoed in the East Alligator area of Kakadu; Val Plumwood, 'Being Prey', *Terra Nova* 1, no. 3 (Summer 1996): 32–44. While in *Feminism and the Mastery of Nature*, Plumwood cites work by scholar of colour bell hooks and Indigenous elder Bill Neidje, she focuses on critiquing a Western tradition from within.

long known.[26] Informed by their work with Indigenous peoples, Deborah Bird Rose and Mark Brett have explored decolonising ethics, theologies and re-readings of biblical texts.[27] Plumwood has argued for decolonising place names.[28] From a background of activism, Clare Land proposes a decolonising ethic for non-Indigenous solidarity with First Peoples.[29]

In Aotearoa (New Zealand), Linda Tuhiwai Smith writes of decolonising methodologies primarily for an Indigenous audience of writers and researchers, who need to both negotiate and counter Western epistemologies and modes of learning, teaching and storying. She cautions that:

> Taking apart the story, revealing underlying texts, and giving voice to things that are often known intuitively does not help people to improve their current conditions. It provides words, perhaps, an insight that explains certain experiences – but it does not prevent someone from dying.[30]

This distinction is crucial, too, for non-Indigenous proponents of a just and ecologically sustainable peace who seek to act in solidarity with First Nations, in the context of the continuing effects of colonial violence for people and Country.

In the wider region, Kuan-Hsing Chen suggests a renewed understanding of method where cultural, postcolonial, globalisation and Asian studies are addressed through a 'geocolonial historical materialism' in which 'Asia' as a cultural imaginary for Asians, rather than as subject of Western perspectives, becomes the reference point for a decolonising,

[26] See, for example, Freya Mathews, *The Ecological Self* (London: Routledge, 1991); Norman C. Habel (ed.), *Readings from the Perspective of Earth* (Sheffield: Sheffield Academic Press, 2000); Norman C. Habel and Peter Trudinger (eds), *Exploring Ecological Hermeneutics* (Atlanta: SBL, 2008).

[27] Deborah Bird Rose, *Reports from a Wild Country: Ethics for Decolonisation* (Sydney: UNSW Press, 2004); Mark Brett, *Decolonizing God: The Bible in the Tides of Empire* (Sheffield: Sheffield Phoenix, 2008). See also, Graham Paulson and Mark Brett, 'Five Smooth Stones: Reading the Bible through Aboriginal Eyes', *Colloquium: The Australian and New Zealand Theological Review* 45, no. 2 (November 2013): 199–214.

[28] Plumwood, 'Decolonizing Relationships with Nature', 67–75.

[29] Clare Land, *Decolonizing Solidarity: Dilemmas and Directions for Supporters of Indigenous Struggles* (London: Zed Books, 2015).

[30] Smith, *Decolonizing Methodologies*, loc. 287 of 5874.

deimperialising approach across multiple fields.[31] In this practice, contextualised experience becomes method, and 'The aim is not simply to rediscover the suppressed voices of the multiple subjects within the social formation, but to generate a system of multiple reference points that can break away from the self-reproducing neo-colonial framework that structures the trajectories and flow of desire'.[32] This method is a deliberate, self-determining shift away from the framework of the coloniser, and so from critique alone of colonial praxis, toward positively focused deimperialism.

Colonisers, too, Chen argues, remain in need of deimperialisation, through interrogation of the complex workings of imperial desire and the way these desires play out in experience. This deimperialisation of the coloniser is particularly necessary for a nation like Australia where more recent migrants as well as contemporary descendants of the settler-colonisers—both, to greater or lesser extents, beneficiaries of the violence on which the nation is founded—greatly outnumber Indigenous people. Several things are needed for a decolonising praxis for a just and ecologically sustainable peace. These include: truth-telling; contextualising ecological action in relation to Country; recognising Indigenous activist ancestries; and submitting to Indigenous sovereignty both with and beyond what is offered by processes toward formal constitutional recognition.

Truth-Telling

The baseline for truth-telling in Australia is to find ways to bring to public consciousness the underacknowledged, suppressed and sometimes denied violence, lies and theft on which the nation was founded. For just and ecologically sustainable peace, however, more is needed. The impact of past and present colonising practices, locally and globally, on the current situation of ecological devastation needs to be acknowledged. In relation to climate change, for non-Indigenous Australians, particularly those

[31] Kuan-Hsing Chen, *Asia as Method: Toward Deimperialization* (Durham: Duke University Press, 2010), Kindle eBook loc 186 and 275 of 5681. Chen distinguishes between colonialism and imperialism, arguing that imperialism is broader than colonialism, that 'colonialism is a deepening of imperialism' (loc 359 of 5681).

[32] Chen, *Asia as Method*, ch. 3 (loc. 2035 of 5681).

of British settler ancestry, Birch writes of a challenge to accept ownership 'for the colonising practices that created the agricultural, mining and manufacturing industries. These industries have not only brought about the degradation of local ecologies but have also assisted the global expansion of industrialisation dependent on increased burning of fossil fuels'.[33] Many have been involved in informal truth-telling processes but this will need to be formalised, and become an effective turn (or in biblical terms, *metanoia*, conversion) in Australia's story about itself as a nation. For a just and ecologically sustainable peace, such truth-telling should include, alongside recognition and redress for violence toward Indigenous people from contact to now, acknowledgement of colonial contributions to ongoing ecological trauma, including climate change.

Contextualising Ecological Action in Relation to Country

Eve Vincent and Timothy Neale describe relations between Indigenous people and environmentalism in Australia as 'unstable'.[34] While some Indigenous people are environmental activists, Neale and Vincent 'argue that it is untenable for analysts to proceed as though non-Indigenous political agendas and intentions are the determining aspect of Indigenous-environmentalists engagements'.[35] They add that it is also flawed 'to proceed as if Indigenous actors are either naturally conservationist in their orientation, or if they are not, that their cultural difference has been somehow corrupted or diluted'.[36] Both of these approaches they argue 'deny the presence and agency of Indigenous people in shaping these relations with environmentalists'.[37] Recognition of Indigenous presence and agency is central to a renewed engagement with Indigenous sovereignties. This includes attention to Indigenous perspectives on ecological crisis and environmental action on Country.

Writing on 'protection of Indigenous Country and climate justice', Birch borrows the title, 'We've Seen the End of the World and We Don't

[33] Birch, '"We've Seen the End of the World and We Don't Accept It"', 142.

[34] Vincent and Neale, eds, *Unstable Relations*.

[35] Neale and Vincent, 'Instabilities and Inequalities', 19.

[36] Neale and Vincent, 'Instabilities and Inequalities', 19.

[37] Neale and Vincent, 'Instabilities and Inequalities', 19.

Accept It' from Murrawah Johnson a young Indigenous spokesperson for the Wangan and Jagalingou people, who have been resisting the Adani mega coal mine in the Galilee Basin in far north Queensland.[38] As Birch points out, eschatological apocalyptic 'end time' stories about climate catastrophe are not new to Indigenous Australians, who have been living their own invasion catastrophe for Country and its peoples for over two centuries.[39] Birch quotes Kyle Powys Whyte, 'climate injustice for Indigenous peoples is less about the spectre of a new future and more like the experience of déjà vu'.[40] Invasion is an ongoing process he says, in which 'colonisation was and to some extent remains a conflict dependent on violence for its success'.[41] He cites Rose who describes this ongoing invasive colonial ideology as a dual war against Earth/nature/Country and Indigenous people, 'that includes both genocide and ecocide'.[42] Turning to the history of British invasion of Australia, Birch sees the current climate situation as built in Australia on successive capitalist extractive colonisations of land—deemed *terra nullius* precisely so that it could be invaded and utilised—first for the wool industry, then gold, and later agribusiness and mining; he describes the 'beatification of mining'.[43] This extractive praxis is of a piece with the contemporary invasion of Wangan and Jagalingou Country for mega coal mines like Adani's, in the process of which the Queensland Government has resisted the people's sovereignty by undermining that weakest of land rights in law: their Native Title.[44]

[38] Birch, '"We've Seen the End of the World and We Don't Accept It"', 148; Wangan and Jagalingou Family Council, https://wanganjagalingou.com.au/ (accessed 9 July 2019).

[39] Birch, '"We've Seen the End of the World and We Don't Accept It"', 140.

[40] Kyle Powys Whyte, 'Is It Colonial *Déjà vu*? Indigenous Peoples and Climate Injustice', in Adamson and Davis (eds), *Humanities for the Environment*, 88–105, cited in Birch, '"We've Seen the End of the World and We Don't Accept It"', 140.

[41] Birch, '"We've Seen the End of the World and We Don't Accept It"', 141.

[42] Rose, *Reports from a Wild Country*, 34, cited in Birch, '"We've Seen the End of the World and We Don't Accept It"', 141.

[43] Birch, '"We've Seen the End of the World and We Don't Accept It"', 144–46, esp. 146.

[44] 'Traditional Owners Continue to Resist Adani's "Invasion"', Wangan and Jagalingou Family Council, Latest News (Friday, 25 January 2019), https://wanganjagalingou.com.au/traditional-owners-continue-to-resist-adanis-invasion/ (accessed 9 July 2019).

Naming racism as endemic across multiple sites of Australian society, Pattel-Gray highlights how corporate practices—she cites the Ranger uranium mine opened in 1980—often ignore Indigenous health and safety, contaminating water used by local people.[45] Similar concerns have been raised in relation to the proposed Adani coal mine.[46] Keynotes of the announcement—'We've Seen the End of the World and We Don't Accept It'—from a young Indigenous woman whose lands are being usurped by corporations and governments are not despair or resignation, but resilience, recovery, reconnection and resistance. It is not corporate and government practices alone that constitute risk for First Nations as understood from this perspective of what, as I noted above, Vizenor terms 'survivance'. Birch comments on 'the risks at stake for Indigenous people entering into […] conversations' with environmentalists and environmental humanities scholars.[47] He writes: 'Not only are Indigenous communities subject to knowledge appropriation, the concept of inclusion itself, however meaningful, can be debilitating and disempowering for Indigenous people'.[48] In this regard, as Land argues, the onus is on non-Indigenous scholars and activists to decolonise themselves and this includes recognising when we 'inadvertently display [our] privilege and power', and educating each other in such decolonising praxis.[49] She writes: 'Solidarity should be directed to decolonization; and the way solidarity is undertaken needs to be decolonized'.[50] In relation to ecological action, the primary decolonising imperative is to give precedence to Indigenous agency. This is not easy or simple. For example, in conversation with Timothy Neale, environmental activist Anthony Esposito describes a process of personal change that led him to prioritise Indigenous agency in relation to the Wangan and Jagalingou defence of country

[45] Pattel-Gray, *The Great White Flood*, 61.

[46] 'Qld Govt's Adani Water Plan Approval Imperils Sacred Doongmabulla Springs', Wangan and Jagalingou Family Council, Latest News (Friday, 14 June 2019), https://wanganjagalingou.com.au/adani-water-plan-approval-imperils-sacred-doongmabulla-springs%e2%80%8b/ (accessed 9 July 2019).

[47] Tony Birch, 'Climate Change, Recognition and Social Place-Making', in Vincent and Neale (eds), *Unstable Relations*, 356–83 (367).

[48] Birch, 'Climate Change, Recognition and Social Place-Making', 367.

[49] Land, *Decolonizing Solidarity*, 6.

[50] Land, *Decolonizing Solidarity*, 5.

campaign, but his responses show a complicated relation to Indigenous sovereignty that is not a straightforward submission.[51]

Giving precedence to Indigenous agency in relation to environmental action has another key aspect that is often forgotten, namely the capacity of First Nations to contribute to a just and ecologically sustainable peace that is Indigenous-led. Bruce Pascoe argues for, and models this, in his work to re-establish Indigenous agriculture more suited to Australian conditions.[52] In relation to climate change, Van Neerven states: 'If Australia does respond to climate change, but does so without seeking the input of its Indigenous people, this response will be perpetuating this country's colonial history'.[53] Birch makes a similar point, arguing that:

> Greater recognition of the knowledge maintained within Indigenous communities relative to localized ecologies and the effects of climate change would go some way to addressing injustice by configuring Indigenous people globally as valuable arbiters of changes rather than helpless victims of the First World.[54]

It is all too easy for non-Indigenous advocates for a just and ecologically sustainable peace to see First Nations knowledges solely as resource and to avoid making space for Indigenous leadership.

Recognising First Nations Activist Ancestries in the Work for a Just and Ecologically Sustainable Peace

Another challenge is found in the way non-Indigenous activists imagine, recollect and narrate the ancestries of movements for social justice, peace and ecological sustainability, ignoring the agency of Indigenous peoples in these fields. In a North American context, Sharon Welch writes of how aspects of the social structure of First Nations, for example, the Iroquois

[51] Anthony Esposito (with Timothy Neale), 'Never Squib the Rights Issues in Favour of Conservation', in Vincent and Neale (eds), *Unstable Relations*, 336–55 (348, 355).

[52] Bruce Pascoe, *Dark Emu: Black Seeds: Agriculture of Accident?* (Broome: Magabala Books, 2014).

[53] Van Neerven, 'The Country Is Like a Body'.

[54] Birch, 'Climate Change, Mining and Traditional Indigenous Knowledge in Australia', 92.

were taken up by Thomas Jefferson and Benjamin Franklin to inform the foundation of the United States as a democracy; later, observation of Iroquois women's status helped inspire the women's suffrage movement.[55] In Australia, a substantial history exists of Indigenous protest for justice, not only against racist structures and land theft, but for other people suffering violence and oppression. William Cooper's leadership extended from Yorta Yorta struggles for their rights to advocacy for many others; in 1938, after Kristallnacht he led a protest to Melbourne's German Consulate against the German Government's treatment of Jewish people, when most across the Western world were silent.[56]

Clare Land narrates a history of Indigenous protest in the south-east of Australia as part of 'the broader history of Indigenous resistance to colonialism across the continent from the time of colonization onwards'.[57] Without co-opting the history of Indigenous protest to broader movements of social justice, peace and environmentalism, non-Indigenous activists need to recognise Indigenous movements for self-determination, sovereignty and basic human rights not only as post-contact protest movements parallel to other protest movements, but also as a central part of the ongoing story of activism for a just and ecologically sustainable peace in what is now Australia. We will need to acknowledge Indigenous resistance and activism since 1788 as ancestral to contemporary social justice movements in Australia, not in order to own Indigenous protest or to appropriate Indigenous resistance to shore up our own protest histories, but to recognise that non-Indigenous struggles for justice, peace and ecological wholeness are not completely separate from Indigenous-led movements, and may well be in debt to them. This debt accrues not only in the benefit of being privileged through the theft of Indigenous Country and all the violence underlying that theft, but also in the resilience—the 'survivance'—that marks Indigenous resistance movements. To acknowledge such Indigenous histories as living ancestors of our own (sometimes meagre) activist endeavours is to open to the ongoing work of affirming Indigenous leadership and sovereignty.

[55] Welch, *After Empire*, 34.

[56] Adam Manovic, 'William Cooper: A Koorie's Protest against the Nazis', NITV (31 May 2018), https://www.sbs.com.au/nitv/article/2018/05/31/william-cooper-kooories-protest-against-nazis (accessed 9 July 2019).

[57] Land, *Decolonizing Solidarity*, 38–50, esp. 41.

Conclusion: Submitting to Indigenous Sovereignty

Aileen Moreton-Robinson describes the way 'race' marks the politics of possession in a 'patriarchal white sovereignty' in the nation of Australia and argues that this is rarely recognised because 'Indigenous sovereignty is never positioned as central to the shaping of the terms and conditions of the very making of the nation'.[58] Moreover, the 'continuing refusal' of Indigenous sovereignty shapes 'a politics based on white anxiety of dispossession'.[59] Part of my argument in this chapter is that non-Indigenous activists and scholars building toward a just and ecologically sustainable peace are not immune to this anxiety of dispossession in a nation built on what Moreton-Robinson describes as 'the white possessive'. I wonder, and I expect I am by no means the first to wonder, if our repression of Indigenous sovereignty has the force in the Western psyche of the repression of the maternal and if this is part of the underlying coincidence of the *other* as nature, Indigenous people, women, Earth. This repression has the potency of an open secret.[60] As often as we tell the truth of Indigenous sovereignty, we effectively forget it at the same time, a kind of forgetting that both contains a suppressed anger and enacts violence. For some the anger is not suppressed, the violence enacted in hate speech and worse. We cannot assume that the problem lies with those who are overtly racist, not us. Acknowledgement of Country, while essential, can itself become part of this convention of remembering in order to forget.

In 2017, a group of First Nations leaders and representatives gathered from around Australia for the National Constitutional Convention and made what is known as the Uluru Statement from the Heart.[61] This statement offered an open hand to the nation of Australia for a way forward together, recognising sovereignty, the founding and ongoing violence of invasion, and inviting a Makarrata, a commission 'to supervise a process

[58] Aileen Moreton-Robinson, *The White Possessive: Property, Power, and Indigenous Sovereignty* (Minneapolis: University of Minnesota Press, 2015), 152.

[59] Moreton-Robinson, *The White Possessive*, 152.

[60] See Michael Farrell, *Writing Australian Unsettlement: Modes of Poetic Intervention 1796–1945* (New York: Palgrave Macmillan, 2015), 86, referring to D. A. Miller on the structure of 'the secret'. D. A. Miller, *The Novel and the Police* (Berkeley: University of California Press, 1988).

[61] 'Uluru Statement from the Heart', https://www.clc.org.au/files/pdf/Uluru-Statement-3.pdf (accessed 9 July 2019).

of agreement-making between governments and First Nations and truth-telling about our history'.[62] Effectively, the Federal Government rejected the proffered olive branch and the Recognition movement stalled. In its wake, Indigenous-led calls for sovereignty and treaties gained renewed focus.[63] For many, Yorta Yorta elder Monica Morgan's response to Eve Vincent's question 'What are your priorities right now?' (published in 2016 before the Statement from the Heart) continues to ring true: 'Ultimately, it has to be respect for the sovereign nature of our Peoples' rights, and to get to a point that there can be a proper treaty put in place for each First Nation'.[64] If we, non-Indigenous activists and scholars, think that the conversion to submitting to Indigenous sovereignty is slow and difficult, Morgan's analysis is a stark reminder; she speaks of the way partial (token) acknowledgements of Indigenous sovereignty impact on people's agency (and energy), scattering action across multiple demands to attend to this waterway agreement, this species need, this regulator and so on, and thus treating water, species and Country as separate things in government and corporate accounting, contrary to Indigenous practice.[65] The challenge for non-Indigenous advocates of a just and ecologically sustainable peace is to respect Indigenous epistemologies and to learn and relearn what it means to submit to First Nations sovereignty as a priority in relation to our national life, our scholarship and our activism.

[62] 'Uluru Statement from the Heart'.

[63] See, for example, Birch, 'On Sovereignty'; cf. Asmi Wood, 'Constitutional Recognition or a Treaty: Horns of a Dilemma or False Dichotomy', *Woroni* (7 May 2017), https://www.woroni.com.au/words/constitutional-recognition-or-a-treaty-horns-of-a-dilemma-or-false-dichotomy/ (accessed 9 July 2019).

[64] Monica Morgan (with Eve Vincent), 'We're Always Pushing the Boundaries for First Nations' Rights', in Vincent and Neale (eds), *Unstable Relations*, 301–16 (316).

[65] Morgan, 'We're Always Pushing the Boundaries', 316.

Bibliography

Adamson, Joni, and Michael Davis, eds. *Humanities for the Environment: Integrating Knowledge, Forging New Constellations of Practice*. London: Routledge, 2017.

Allam, Lorena, Calla Wahlquist, and Nick Evershed. 'Aboriginal Deaths in Custody: Black Lives Matter Protests Referred to Our Count of 432 Deaths. It's Now 437'. *The Guardian* Australia (9 June 2020). https://www.theguardian.com/australia-news/2020/jun/09/black-lives-matter-protesters-referred-to-our-count-of-432-aboriginal-deaths-in-custody-its-now-437. Accessed 9 June 2020.

Birch, Tony. 'Climate Change, Mining and Traditional Knowledge in Australia'. *Social Inclusion* 4, no. 1 (2016): 92–101.

Birch, Tony. 'Climate Change, Recognition and Social Place-Making'. In *Unstable Relations*, edited by Eve Vincent and Timothy Neale, 356–83. Crawley, WA: UWAP Scholarly, 2016.

Birch, Tony. 'On Sovereignty'. *Overland* 229 (2017): 88–89.

Birch, Tony. '"We've Seen the End of the World and We Don't Accept It": Protection of Indigenous Country and Climate Justice'. In *Places of Privilege: Interdisciplinary Perspectives on Identities, Change and Resistance*, edited by Nicole Oke, Christopher Sonn, and Alison Baker, 139–52. Leiden: Brill, 2018.

Brett, Mark. *Decolonizing God: The Bible in the Tides of Empire*. Sheffield: Sheffield Phoenix, 2008.

Bringing Them Home: Report of the National Inquiry into the Separation of Aboriginal and Torres Strait Islander Children from Their Families. Commonwealth of Australia 1997. https://www.humanrights.gov.au/sites/default/files/content/pdf/social_justice/bringing_them_home_report.pdf. Accessed 5 July 2019.

Chen, Kuan-Hsing. *Asia as Method: Toward Deimperialization*. Durham: Duke University Press, 2010. Kindle eBook.

'Deaths in Custody'. *The Guardian* Australia. https://www.theguardian.com/australia-news/deaths-in-custody. Accessed 5 July 2019.

Deverell, Garry Worete. *Gondwana Theology: A Trawloolway Man Reflects on Christian Faith*. Reservoir, VIC: Morning Star Publishing, 2018.

The Earth Bible Team. 'Guiding Ecojustice Principles'. In *Readings from the Perspective of Earth*, edited by Norman C. Habel, 38–53. Earth Bible 1. Sheffield: Sheffield Academic Press, 2000.

Esposito, Anthony, with Timothy Neale. 'Never Squib the Rights Issues in Favour of Conservation'. In *Unstable Relations*, edited by Eve Vincent and Timothy Neale, 336–55. Crawley, WA: UWAP Scholarly, 2016.

Farrell, Michael. *Writing Australian Unsettlement: Modes of Poetic Intervention 1796–1945*. New York: Palgrave Macmillan, 2015.

Habel, Norman C. 'Introducing the Earth Bible'. In *Readings from the Perspective of Earth*, edited by Norman C. Habel, 25–37. Earth Bible 1. Sheffield: Sheffield Academic Press, 2000.

Habel, Norman C., ed. *Readings from the Perspective of Earth*. Sheffield: Sheffield Academic Press, 2000.

Habel, Norman C., and Peter Trudinger, eds. *Exploring Ecological Hermeneutics*. Atlanta: SBL, 2008.

Kyrou, Christos N. 'Peace Ecology: An Emerging Paradigm in Peace Studies'. *International Journal of Peace Studies* 12, no. 1 (Spring–Summer 2007): 73–92.

Land, Clare. *Decolonizing Solidarity: Dilemmas and Directions for Supporters of Indigenous Struggles*. London: Zed Books, 2015.

Lucashenko, Melissa. *Too Much Lip*. St Lucia: Queensland University Press, 2018.

Manovic, Adam. 'William Cooper: A Koorie's Protest against the Nazis'. NITV (31 May 2018). https://www.sbs.com.au/nitv/article/2018/05/31/william-cooper-koories-protest-against-nazis. Accessed 9 July 2019.

Marsh, Jillian K. 'Decolonising the Interface between Indigenous Peoples and Mining Companies in Australia: Making Space for Cultural Heritage Sites'. *Asia Pacific Viewpoint* 54, no. 2 (August 2013): 171–84.

Mathews, Freya. *The Ecological Self*. London: Routledge, 1991.

Miller, D. A. *The Novel and the Police*. Berkeley: University of California Press, 1988.

Moreton-Robinson, Aileen. *The White Possessive: Property, Power, and Indigenous Sovereignty*. Minneapolis: University of Minnesota Press, 2015.

Morgan, Monica, with Eve Vincent. 'We're Always Pushing the Boundaries for First Nations' Rights'. In *Unstable Relations*, edited by Eve Vincent and Timothy Neale, 301–16. Crawley, WA: UWAP Scholarly, 2016.

Neale, Timothy, and Eve Vincent. 'Instabilities and Inequalities: Relations between Indigenous People and Environmentalism in Australia Today'. In *Unstable Relations*, edited by Eve Vincent and Timothy Neale, 1–24. Crawley, WA: UWAP Scholarly, 2016.

Otto, Dianne. 'A Question of Law or Politics? Indigenous Claims to Sovereignty in Australia'. *Syracuse Journal of International Law and Commerce* 21 (1995): 65–103.

Pascoe, Bruce. *Dark Emu: Black Seeds: Agriculture of Accident?* Broome: Magabala Books, 2014.

Pattel-Gray, Anne. *The Great White Flood: Racism in Australia: Critically Appraised from an Aboriginal Historico-Theological Viewpoint*. American Academy of Religion Cultural Criticism Series 2. Atlanta: Scholars Press, 1998.

Paulson, Graham, and Mark Brett. 'Five Smooth Stones: Reading the Bible through Aboriginal Eyes'. *Colloquium: The Australian and New Zealand Theological Review* 45, no. 2 (November 2013): 199–214.

Plumwood, Val. 'Being Prey'. *Terra Nova* 1, no. 3 (Summer 1996): 32–44.

Plumwood, Val. 'Decolonizing Relationships with Nature'. In *Decolonizing Nature: Strategies for Conservation in a Post-colonial Era*, edited by William M. Adams and Martin Mulligan, 51–78. London: Earthscan, 2003.

Plumwood, Val. *Environmental Culture: The Ecological Crisis of Reason*. London: Routledge, 2002.

Plumwood, Val. *Feminism and the Mastery of Nature*. London: Routledge, 1993.

Rose, Deborah Bird. 'Country and the Gift'. In *Humanities for the Environment*, edited by Joni Adamson and Michael Davis, 33–44. London: Routledge, 2017.

Rose, Deborah Bird. *Nourishing Terrains: Australian Aboriginal Views of Landscape and Wilderness*. Canberra: Australian Heritage Commission, 1996.

Rose, Deborah Bird. *Reports from a Wild Country: Ethics for Decolonisation*. Sydney: UNSW Press, 2004.

'Royal Commission into Aboriginal Deaths in Custody—Fact Sheet 112'. National Archives of Australia. https://www.naa.gov.au/sites/default/files/2020-05/fs-112-royal-commission-into-aboriginal-deaths-in-custody.pdf. Accessed 9 June 2020.

Smith, Aaron. 'Torres Strait Islander Musicians Marou and Busby Use Trip Home to Highlight the Real Impact of Climate Crisis on the Region'. NITV News (5 July 2019). https://www.sbs.com.au/nitv/article/2019/07/05/climate-change-actually-affecting-torres-strait-islanders-right-now-marou. Accessed 5 July 2019.

Smith, Linda Tuhiwai. *Decolonizing Methodologies: Research and Indigenous Peoples*, 2nd edn. London: Zed Books, 2012. Kindle edn.

Spring, Úrsula Oswald, Hans Günter Brauch, and Keith G. Tidball. 'Expanding Peace Ecology: Peace, Security, Sustainability, Equity, and Gender'. In *Expanding Peace Ecology: Peace, Security, Sustainability, Equity and Gender. Perspectives of IPRA's Ecology and Peace Commission*, edited by Úrsula Oswald Spring, Hans Günter Brauch, and Keith G. Tidball, 1–30. Springer Briefs in Environment, Security, Development and Peace. Peace and Security Studies 12. Cham, Switzerland: Springer, 2014.

'Uluru Statement from the Heart'. https://www.clc.org.au/files/pdf/Uluru-Statement-3.pdf. Accessed 9 July 2019.

van Neerven, Ellen. 'The Country Is Like a Body'. *Right Now: Human Rights in Australia* (26 October 2015). http://rightnow.org.au/essay/the-country-is-like-a-body/. Accessed 27 June 2019.

Vincent, Eve, and Timothy Neale, eds. *Unstable Relations: Indigenous People and Environmentalism in Contemporary Australia*. Crawley, WA: UWAP Scholarly, 2016.

Vizenor, Gerald. *Manifest Manners: Narratives on Postindian Survivance*. Lincoln: University of Nebraska Press, 1994.

Vizenor, Gerald, ed. *Survivance: Narratives of Native Presence*. Lincoln: University of Nebraska Press, 2008.

Wangan and Jagalingou Family Council. https://wanganjagalingou.com.au/. Accessed 9 July 2019.

Welch, Sharon D. *After Empire: The Art and Ethics of Enduring Peace*. Minneapolis: Fortress, 2004.

Whyte, Kyle Powys. 'Is It Colonial Déjà vu? Indigenous Peoples and Climate Injustice'. In *Humanities for the Environment*, edited by Joni Adamson and Michael Davis, 88–105. London: Routledge, 2017.

Wolfe, Patrick. 'Settler Colonialism and the Elimination of the Native'. *Journal of Genocide Research* 8, no. 4 (2006): 387–409.

'The Women Fighting against a Rising Tide of Indigenous Child Removals'. NITV. https://www.sbs.com.au/nitv/feature/women-fighting-against-rising-tide-indigenous-child-removals. Accessed 5 July 2019.

Wood, Asmi. 'Constitutional Recognition or a Treaty: Horns of a Dilemma or False Dichotomy'. *Woroni* (7 May 2017). https://www.woroni.com.au/words/constitutional-recognition-or-a-treaty-horns-of-a-dilemma-or-false-dichotomy/. Accessed 9 July 2019.

Anne Elvey is a poet, editor and researcher living and working on Boonwurrung Country in Seaford, Victoria. Her recent scholarly books include *Ecological Aspects of War: Engagements with Biblical Texts* (coedited with Keith Dyer and Deborah Guess, Bloomsbury T&T Clark, 2017), *Reinterpreting the Eucharist: Explorations in Feminist Theology and Ethics* (coedited with Carol Hogan, Kim Power and Claire Renkin, Equinox, 2013), and *The Matter of the Text: Material Engagements between Luke and the Five Senses* (Sheffield Phoenix, 2011). Her most recent poetry books are *On Arrivals of Breath* (Poetica Christi, 2019) and *White on White* (Cordite Books, 2018). She is an Adjunct Research Fellow, School of Languages, Literatures, Cultures and Linguistics, Monash University, and an Honorary Research Associate, Trinity College Theological School, University of Divinity, Melbourne, Australia.

CHAPTER 14

From Mendicant Nation to Global Citizen: Towards a New Australian Foreign Policy for the Twenty-First Century

Allan Patience

Introduction

This chapter challenges the assumed 'middle power' status that has been central to Australian foreign policy-making since the closing months of World War II. It argues that the country's assumption of what amounts to a counterfeit middle power façade is no longer tenable in the face of the profound foreign policy and security challenges confronting Australia in the twenty-first century. Australia's current middle power posturing masks the fact that it is a mendicant nation, a credulous supplicant to the United States for its security, while expecting China to underwrite its economic growth. This amounts to policy schizophrenia. It is no longer possible for the country to have its American cake while eating its Chinese cake too. Australia should now be aiming to earn respect and influence as an independent global citizen proposing effective niche diplomatic initiatives

A. Patience (✉)
School of Social and Political Sciences,
University of Melbourne, Melbourne, VIC, Australia
e-mail: allan.patience@unimelb.edu.au

J. Camilleri and D. Guess (eds.), *Towards a Just and Ecologically Sustainable Peace*, https://doi.org/10.1007/978-981-15-5021-8_14

in collaboration with a range of like-minded states. These initiatives will need to embrace policies that contribute positively to peace-building and to the making of a just and ecologically sustainable world. To advance in these directions, Australia's policy makers will need to draw on lessons of the post-World War II Non-Aligned Movement.

To Be or Not to Be a Middle Power

Since the 1980s, there has been a growing interest in 'powers great and powers small and some in between trying to speak louder than the latter and to exert some influence on the former'.[1] Despite a sharpening empirical and theoretical focus on those 'in between' states, much of the relevant literature is in furious agreement that conceptualising them as 'middle powers' is an opaque, even implausible, way of trying to understand them. As Mark Beeson has noted: '"Middle power" is not an entirely happy formulation. Like "globalisation" it can obscure as much as it reveals'.[2] John Ravenhill has concluded that 'the concept of middle power is so lacking in analytical utility that it should be abandoned'.[3] Nonetheless, amid the turmoil of contemporary international politics, some states are exercising levels of influence that do not always equate with their economic or military resources. This suggests that much of the academic handwringing about the difficulties of *defining* middle powers is missing the point. A better way of understanding what may be characterised as middle powers will be gained by focusing on their *actions*. What are those states actually *doing*, what are they *achieving*, in regional and global affairs when allegedly they 'punch above their weight'?

Previously, it has been argued that there are two kinds of action that provide certain states with the *appearance*, but not the *substance* of

[1] John C. Campbell, 'Review of Carsten Holbraad, *Middle Powers in International Relations*,' *Foreign Affairs*, 1984, 62 (5): 1247–1248. See also Carsten Holbraad, *Middle Powers in International Politics* (New York: St. Martin's Press, 1984).

[2] Mark Beeson, 'Can Australia Save the World? The Limits and Possibilities of Middle Power Diplomacy', *Australian Journal of International Affairs*, 2011, 65 (5): 564.

[3] John Ravenhill, 'Entrepreneurial States: A Conceptual Overview', *International Journal*, 2018, 73 (4): 503.

middle power status.[4] The first action relates to alliance-making between a smaller state and a big power, in the belief that the alliance will guarantee the smaller state's security and amplify its influence in regional and global forums. In this instance, the smaller state's claim to middle power status is based on a *dependency* relationship it connects with its big power ally. Given its substantial security dependency on the United States, via the ANZUS treaty, Australia is a precise example of such a state. The second action is evident where a smaller state becomes a member of a formally constituted regional organisation, such as the European Union (EU) or the Association of Southeast Asian Nations (ASEAN). Membership of a regional organisation can boost the profile of the smaller state, providing it with recognition in regional and global affairs that it would otherwise be unlikely to attract to itself. Regional associations can also assist the smaller state to brush off international criticisms, or even bypass sanctions, for example, for human rights violations and related abuses. A standout example is Myanmar/Burma in ASEAN.

In both cases—that is, by allying with a big state, or by joining a regional grouping of states—smaller states can cloak themselves with a middle power *façade* which has little to do with their domestic policy achievements or their international diplomacy. The old adage about not judging a book by its cover is relevant to both of these *apparent* (or counterfeit) middle powers. In each case, their claim to middle power status lacks substance. To adopt a Trumpian trope, we could even say they are fake middle powers.

A third action by a state that is capable of 'punching above its weight' relates to its capacity to engage in effective niche diplomacy (e.g. peace-building negotiations, initiating and/or sustaining arms-limitation and nuclear non-proliferation agreements, human rights advocacy, promoting international climate change accords). Andrew Cooper defines niche diplomacy as 'the ability of countries, like biological species or firms,

[4] Allan Patience, 'Imagining Middle Powers', *Australian Journal of International Affairs*, 2014, 68 (2): 210–224; Allan Patience, *Australian Foreign Policy in Asia: Middle Power or Awkward Partner?* (London: Palgrave Macmillan, 2018); Allan Patience and Chiraag Roy, 'Interlocutors for Peace? Bringing Middle Powers in from the Theoretical Cold', in Tanguy de Swielande et al. (eds.), *Rethinking Middle Powers in the Asian Century: New Theories, New Cases* (London: Routledge, 2019), 62–73.

to identify and fill niche space on a selective basis through policy ingenuity and execution'.[5] When successfully deployed in regional and global contexts, this level of diplomacy can provide a small (or an 'in between') state with international recognition as a 'good global citizen'—that is, a state that speaks louder than smaller states and which is listened to by bigger states on issues judged to be regionally and/or globally auspicious. This assumes that the middle power as global citizen is characterised by attributes that include a well-educated population, access to resources, a sound economic base, diplomatic adeptness and associated cultural and intellectual assets. In short, to be a middle power in this sense is to be a global citizen, that is the fact of middle power action—to exercise influence, even power, regionally and/or globally, over and above what its hard power resources would normally allow.

In addition to their astuteness in identifying opportunities for niche diplomacy—and providing leadership accordingly—it is important to note that these states (which are likely to be states with sustained records of social democratic policy-making) are also characterised by their *governance integrity*. This refers to their sustained records of domestic public policy that enhances the security and well-being of their citizens. Norway, for example, has an enviable reputation for exercising authority and leadership in international forums through its record in human rights advocacy; its role in reducing deforestation activities in Colombia; its leadership in promoting peace talks between the Israelis and the Palestinians via the Oslo accords; its peace-making efforts in Myanmar/Burma; and its international activism in movements to limit the spread of nuclear weapons. As James Taulbee explains: 'Norway has exhibited sustained involvement in foreign policy action that goes well beyond narrow definitions of immediate interest. This involves actively committing resources, not just rhetorical support' (Taulbee et al. 2014: 14). John Langmore highlights a survey conducted by the Norwegian Ministry of Foreign Affairs which found that the Norwegian people view their country's foreign policy activism as their country's 'international hallmark'.[6]

[5] Andrew F. Cooper, 'Introduction,' in Andrew F. Cooper (ed.), *Niche Diplomacy: Middle Powers After the Cold War* (Basingstoke: Macmillan, 1997), 5. See also Alan K. Henriksen, 'Niche Diplomacy in the World Public Arena: The Global "Corners" of Canada and Norway', in Jan Melissen (ed.), *The New Public Diplomacy: Soft Power in International Relations* (London: Palgrave Macmillan, 2005), 67–87.

[6] John Langmore, 'Learning from Norway,' *Griffith Review*, 2011, 32: 99.

Norwegians, in short, are proud of their international achievements. At the same time, the country's economic, education, health and welfare policies are widely respected. Of course, these positive attributes are qualified by Norway's membership of NATO which led it to its declining to sign up to the United Nations 2017 Treaty on the Prohibition of Nuclear Weapons. Nonetheless, and to its credit, it has tried to play a moderating role in fostering debates on the humanitarian impacts of nuclear weapons. These very debates culminated in the 2017 Treaty.

Similar positive policy achievements are evident among the other Nordic states—Finland, Denmark, Iceland and Sweden—lending weight to the idea of the 'Nordic model of social democracy'. As Mary Hilson explains:

> In contrast to states such as the USA and the UK, where political debate is more adversarial and there is more open conflict, the Nordic countries have been described as 'consensual democracies', where the political systems have a high degree of legitimacy and support, where social divisions have been relatively mild; and where political deliberations aim to neutralize conflict and achieve compromise. Other important features of the Nordic political model have included the electoral dominance of social democratic labour parties within multi-party systems, and the high level of participation and organization among citizens, with a tendency to favour neo-corporatist methods of decision making.[7]

The Nordic achievements in governance integrity constitute the springboard for the kinds of niche diplomacy that are characteristic of a genuine middle power—a state that is respected as a 'good global citizen'. Those states demonstrate that, even with relatively limited resources, they still have sufficient capability and willingness to act with varying degrees of independence on issues of regional and global importance. Moreover, they do not seek to posture as 'middle powers' per se. Claiming that label is not an immediate diplomatic priority for them. Rather, acting as what John Ravenhill describes as 'entrepreneurial states', they are effectively contributing to a wide range of initiatives that offer them influence that they have hitherto lacked. Their entrepreneurialism 'involves a proactive "selling" of policy proposals to others and gaining their support'.

[7] Mary Hilson, *The Nordic Model: Scandinavia since 1945* (London: Reaktion Books, 2008). See also Nik Brandel, Øivind Bratberg and Dag Einar Thorsen, *The Nordic Model of Social Democracy* (London: Palgrave Macmillan, 2013).

Moreover, they generally do not act alone: 'Being an entrepreneurial state is all about attracting followers'. They do so by proposing 'constructive policy proposals that at least give credible appearance that they will materially benefit their potential partners as well as the entrepreneurial state itself'.[8] This is Professor Ravenhill's reason for wanting to abandon the concept of 'middle power'. His argument has merit. However, it is also a reason for identifying some of the Nordic states as 'global citizens'. Their activism is early evidence of a potentially new development in global politics that could help reshape international politics. Small states' actions in their national interests that are of benefit not simply to themselves but to regional and global partners, are providing models of global citizenship not hitherto well understood, let alone valued in orthodox International Relations theorising.

This paper focuses on the first and third kinds of actions by states claiming middle power status: the former arrogating middle power standing to itself on the basis of its alliance-making with a big power; the latter being accorded 'good global citizen' recognition by other states.

Counterfeit Middle Power Thinking

Among states assuming middle power recognition on the basis of their big power alliances, there is a commonly held belief that they benefit from the reflected prestige and influence (or even the power) of their big power ally. Confident in this belief, they rarely acknowledge the fact that there are also real costs associated with such an alliance. These costs amount to a significant curtailment of sovereignty. Moreover, as one of the high priests of American realism reminds us: 'alliances are only temporary marriages of convenience: today's alliance partners might be tomorrow's enemy, and today's enemy might be tomorrow's alliance partner'.[9] A (if not *the*) defining feature of big powers is their focus—above all else—on their own economic and security concerns. If (or when) the big power decides that its national interests are best served by limiting the alliance's sphere of activity—or by backing away from it altogether—there is little the junior partner can do about it. Meanwhile, the junior partner may be pressured

[8] Ravenhill, 'Entrepreneurial States', 510.

[9] John Mearsheimer, *The Tragedy of Great Power Politics* (New York: W. W. Norton, 2001).

by the big power into allowing the construction of strategic bases on the junior partner's territory, with the junior partner having limited (or no) control over the use of those facilities. These bases could well become a potential target for attack by an enemy of the big power. Moreover, the junior partner will generally be expected to support the big power's foreign and defence strategies, including going to war with it, without there necessarily being a quid pro quo. The asymmetry of the alliance means that the junior partner's claim to middle power status is built on a foundation of sand. Inevitably, it becomes a ***mendicant nation*** constantly obliged to supplicate its 'great and powerful friend' to shore up its security and any other benefits associated with the alliance arrangements. What is overlooked by these states is that those benefits are often less real than imagined. All middle power dreaming by a mendicant nation is simply dreaming.

Australia as a Mendicant Nation

In the 1990s, Gareth Evans and Bruce Grant expressed a view that has become the orthodox consensus among Australia's foreign and defence policy establishment: 'Australia is a middle power. We are manifestly not a great or even a major power; nor, however, are we small or insignificant'.[10] That assumption has been at the core of Australian foreign policy-making from the closing months of World War II until today. However, the substance of the claim that Australia is recognised (indeed *respected)* as a middle power, by allies and contenders alike, is rarely interrogated. This amounts to an astonishing act of policy complacency on the part of the country's politicians, bureaucrats, academics and commentators—a complacency that overshadows nearly every aspect of the country's security and defence policy today.

That act of policy complacency has convinced many Australians that the ANZUS alliance guarantees their country's protection, absolutely, under the United States' security umbrella, across the Asia Pacific and

[10] Gareth Evans and Bruce Grant, *Australia's Foreign Relations in the World of the 1990s*, second edition (Melbourne: Melbourne University Press, 1995), 344.

across the globe.[11] They overlook the fact that the alliance has resulted in Australian military personnel and materiel participating in all of the United States' wars since 1945. In strategic or national interest terms, Henry Reynolds is correct in labelling these as 'unnecessary wars'.[12] The exorbitant budget outlays that participation in America's wars really entails needs to be clearly spelt out to Australia's citizens. Showing how these substantial outlays could otherwise have been used to fund more beneficial international and domestic purposes will be the measure of wise political leadership.

Additionally, the American communications facilities at Pine Gap and the North-West Cape raise serious questions about the integrity of Australia's sovereignty.[13] Nonetheless, these factors are rarely taken into account when Australia's orthodox foreign policy aficionados formulate policy based on what is effectively a purblind act of faith, proclaiming not only that Australia is recognised as a middle power (because of its alliance with the United States), but that it is respected as such by all its allies and contenders alike. The alliance, in short, remains the foundation of this faith, as attested by former Prime Minister John Howard:

> The relationship we have with the United States is the most important we have with any single country. This is not only because of the strategic, economic, and diplomatic power of the United States. But of equal, if not more significance, are the values and aspirations we share.[14]

[11] Coral Bell, *Dependent Ally: A Study in Australian Foreign Policy* (Melbourne: Oxford University Press, 1988); Alison Broinowski, *Allied and Addicted* (Melbourne: Scribe, 2007); Andrew Carr, 'ANZUS and Australia's Role in World Affairs,' in Peter J. Dean, Stephan Frühling and Brendan Taylor (eds.), *Australia's American Alliance* (Melbourne: Melbourne University Press, 2017), 66–85; Clinton Fernandes, *What Uncle Sam Wants: U.S. Foreign Policy Objectives in Australia and Beyond* (London: Palgrave Macmillan, 2019); John Langmore, *Dealing with America: The UN, the US and Australia* (Sydney: Newsouth, 2005); Thomas K. Robb and David James Gill, 'The ANZUS Treaty During the Cold War: A Reinterpretation of U.S. Diplomacy in the Southwest Pacific),' *Journal of Cold War Studies*, 2015, 17 (4), 109–157.

[12] Henry Reynolds, *Unnecessary Wars* (Sydney: Newsouth, 2016).

[13] Desmond Ball, *A Suitable Piece of Real Estate: American Installations in Australia* (Sydney: Hale and Iremonger, 1980); Desmond Ball, Bill Robinson and Richard Tanter, 'Australia's Participation in the Pine Gap Enterprise,' The Nautilus Institute, 2016, https://nautilus.org/napsnet/napsnet-special-reports/australias-participation-in-the-pine-gap-enterprise/, accessed 12 February 2019.

[14] Howard quoted in Langmore, *Dealing with America*, 68–69.

In stark contrast to Howard's view, another former Prime Minister, Malcolm Fraser, argued in his 2014 book, *Dangerous Allies*, that it's time for the alliance to be abandoned. He explained:

> We have significantly diminished our capacity to act as a separate sovereign nation by the way we have committed ourselves to American purposes [...] the most likely reason Australia would need to confront an aggressive foe is our strong alliance with the United States. We need America for defence from an attacker who is likely to attack us because we use America for defence. It's not a sustainable policy.[15]

Between the Scylla of Howard's passive-submissive position and the Charybdis of Fraser's clarion call for a robustly independent stance towards the United States, there is today a gradual awakening among thoughtful observers that a comprehensive and nuanced rethinking of the alliance is urgently needed. Nonetheless, probably the majority of Australians pay little attention to this central issue, leaving the country in an increasingly perilous security bind. Stephan Frühling explains why leaving the matter in the too-hard basket is causing confusion about ANZUS and its negative impacts on contemporary Australian foreign and defence policy-making:

> Emphasising the rejection of US 'bases' while agreeing, in principle, to the increasing presence of long-range US strike forces; insisting on Australia's 'full knowledge and concurrency' of the activities of US forces in and from Australia in the absence of any command system that might give this practical effects; and emphasising interoperability and integration with the United States while maintaining the importance of 'independent operations' all highlight the different directions in which Australia's strategic policy is pulled by the benefits and limits of the alliance.[16]

[15] Malcolm Fraser, *Dangerous Allies* (Melbourne: Melbourne University Press, 2014), 257–258.

[16] Stephan Frühling (2016), 'Wrestling with Commitment: Geography, Alliance Institutions and the ANZUS Treaty,' in Peter J. Dean, Stephan Frühling and Brendon Taylor (eds.), *Australia's American Alliance* (Melbourne: Melbourne University Press, 2014), 29.

This is especially concerning because, as Clinton Fernandes points out: 'Australia remains relatively unimportant in U.S. thinking. Only 1 per cent of the U.S. diplomatic cables published by WikiLeaks related to Australia. Australia's share of the world economy is also only one per cent. Its population is about 0.33 per cent of the world's total'.[17] And, as James O'Neill has warned: 'one would have to ask whether the United States is in fact really ready, willing and most importantly, able to fulfill the role Australian planners devoutly hope for'.[18] The confusion created by the Trump administration's security policies highlights the relevance of this question.

Australia's Poor Niche Diplomacy and Governance Integrity Record

Does the ANZUS alliance enable Australia to engage in niche diplomacy to win for the country the reputation of being a 'good global citizen'—for example, by playing a leadership role in peace-making within its region and globally; or negotiations for achieving a global consensus on climate change? Mostly not, with such exceptions as the country's leadership role in the 1991 conference in Paris to bring about peace in post-Khmer Rouge Cambodia,[19] and its sometimes controversial participation in the RAMSI intervention in the Solomon Islands, 2003–2017.[20] Meanwhile, Australia's mendicant gesturing towards the United States has won for it a reputation as a belligerent state because of its unhesitating commitments to its big power ally's conflicts—even to conflicts

[17] Clinton Fernandes, *Island Off the Coast of Asia: Instruments of Statecraft in Australian Foreign Policy* (Clayton: Monash University Publishing, 2018), 49.

[18] James O'Neill, 'When Better Than an Election to Have a Serious and Overdue Debate About Defence and Foreign Policy Objectives?' *Pearls and Irritations*, 24 April 2019, https://johnmenadue.com/james-oneill-when-better-than-at-an-election-to-have-a-serious-and-overdue-debate-about-defence-and-foreign-policy-objectives/, accessed 24 April 2019.

[19] Gareth Evans and Bruce Grant, *Australia's Foreign Relations in the World of the 1990s*, second edition (Melbourne: Melbourne University Press, 1995), 221–237.

[20] Jon Fraenkel, 'Reassessing the 2003–2017 Regional Assistance Mission to the Solomon Island,' *RUSI Journal*, 2019, 164 (1), 52–61.

unrelated to Australia's obligations stipulated by the ANZUS treaty.[21] Moreover, Australia's history as a 'laggard state' on climate change negatively affects its international credibility. This was especially highlighted at the 2019 Pacific Islands Forum meeting in Tuvalu where Prime Minister Scott Morrison was strongly criticised by the other members of the Forum for his government's reluctance to respond to their calls for a curtailment of coal mining and the country's CO_2 emissions. Meanwhile, Lorraine Elliott has shown that Australia is ranked 'towards the bottom of leading industrialised country commitments' in limiting its per capita carbon emissions.[22] Fiji's Prime Minister, Frank Bainimarama, forcefully emphasised the criticism of Australia by his fellow PIF members when visiting Canberra in mid-September 2019. He stated that Australia 'needs to be more than good mates to the Pacific'.[23]

This negativism about Australia's governance integrity in regional and global forums is intensified by the country's policies on asylum seekers which are marked by a 'preoccupation with a threat-focused engagement over a protection-focused agenda for those who are seeking asylum'.[24] Simultaneously, a prolonged period of counterproductive neoliberal economic policy-making has aggravated deep socio-economic cleavages across the country.[25] Underpinning the damage to Australia's foreign relations is its shameful human rights record on Indigenous Australians that continues to haunt the national consciousness while seriously tarnishing the international reputation of the country.[26]

[21] Alison Broinowski, 'Australia's Tug of War: Militarism Versus Independence', in David Stephens and Alison Broinowski (eds.), *The Honest History Book* (Sydney: Newsouth, 2017), 271–286.

[22] Lorraine Elliott, 'The Environment in Australian Foreign Policy,' in Mark Beeson and Shahar Hamieri (eds.), *Navigating the New International Disorder: Australia in World Affairs, 2011–2015* (Melbourne: Oxford University Press, 2017), 176–191.

[23] *Papua New Guinea Post Courier*, 17 September 2019, https://postcourier.com.pg/aust-needs-to-be-more-than-good-mates-to-the-pacific/, accessed 17 September 2019.

[24] Sara Davies, 'Asylum Seekers and Australia's Security,' in Beeson and Hamieri (eds.), *Navigating the New International Disorder: Australia in World Affairs, 2011–2015* (Oxford: Oxford University Press, 2017), 109–122.

[25] Richard Dennis, 'Dead Right: How Neoliberalism Ate Itself and What Comes Next,' *Quarterly Essay*, 2018, Issue 70.

[26] James Curran and Stuart Ward, *The Unknown Nation: Australia After Empire* (Melbourne: Melbourne University Press, 2010), 78–81. See also Lyndall Ryan, 'The

These large policy failures point to a serious lack of governance integrity within the Australian political system.

Meanwhile, Australia orbits around America's security arrangements like a small and lonely satellite, resulting in the country being viewed by a number of countries as an inward-looking nation complacently doing less than it could to make for a peaceful and ecologically sustainable world. Can this be changed? Yes, it can.

The case needs urgently to be elaborated that by rejecting its felt-obligation (under the ANZUS treaty) to go to war every time its American ally instructs it to, Australia could achieve regional and international recognition as a good global citizen in such areas as, for example, well-targeted overseas aid programmes, peace-building initiatives at both regional and global levels, public investment to make it a world leader in the invention and production of clean energy technology. Simultaneously, and linked closely to those global policies, domestic policy developments like rehabilitating the Great Barrier Reef, funding better schools and hospitals, and upgrading infrastructure should be enacted. Educating Australians about these domestic and international benefits to their country presents a challenge to statesmen-like political leaders who appear presently in short supply. However, a new generation of potential leaders could aim to gain international respect for Australia for its governance integrity, enabling the country to join like-minded states in checking big states that threaten a just and ecologically sustainable global order. Identifying this vital nexus between domestic and foreign policy is fundamental to Australia emerging from its mendicant status and becoming a global citizen.

From Mendicant Nation to Global Citizen

What does Australia need to do to transition from being a mendicant nation into a respected global citizen? There are two big steps that the country could take in order to achieve this desirable outcome. The first has to do with the ANZUS alliance. The second has to do with negotiating multi-layered cooperative arrangements with like-minded states with shared interests. These multi-layered arrangements will need to address

Myall Creek Massacre: Was It Typical of the Time?' in Jane Lydon and Lyndall Ryan (eds.), *Remembering the Myall Creek Massacre* (Sydney: Newsouth, 2018): 85–99.

a variety of regional and international issues, via adroitly planned and executed programmes of niche diplomacy.

1. *Rethinking ANZUS*

Most Australians will be unnerved by the proposition that the ANZUS alliance has become an albatross around the country's security neck. The idea of the United States as the country's security backstop, its 'great and powerful friend', is deeply embedded into the collective Australian psyche, at least as strongly as the jingoistic nationalism surrounding the contemporary celebrations of ANZAC Day.[27] For many Australians, the very thought that the alliance should be discarded is unthinkable—even treasonous. Its consideration would trigger a collective anxiety which has very deep roots in the country's historical culture. As Anthony Burke has explained, Australia 'is imagined on the basis of a bounded and vulnerable identity in perpetual opposition to an outside—an *Other*—whose character and claims threaten its integrity and safety'.[28]

In *Dangerous Allies*, Malcolm Fraser made plain that the United States with which Australia signed the ANZUS treaty way back in 1951 is a very changed United States today. What seemed to be a stable 'great and powerful friend' in the 1950s has become a fractious and increasingly isolated country. Its behaviour is characteristic of a great power in decline.[29] While this trend has been escalating over several decades, under President Trump it is likely to gather pace as he seeks to realise his promise to 'make America great again'. To achieve this, he is using a rhetoric imbued with blustering threats, childish complaints, angry outbursts and disdain for any who doubt his foreign policy aims. He has placed NATO on edge, withdrawn from the Trans-Pacific Partnership (TPP), withdrawn the United States from the nuclear agreement with Iran, established bizarre 'bromances' with the likes of Vladimir Putin and Kim Jong-Un, praised authoritarian leaders like Duterte and Orbán, and provided gratuitous advice to leaders of some of America's closest allies. This is not the

[27] David Stephens, 'ANZAC and Anzackery: Useful Future or Sentimental Dream?' in David Stephens and Alison Broinowski (eds.), *The Honest History Book* (Sydney: Newsouth, 2017), 136–150.

[28] Anthony Burke, *In Fear of Security: Australia's Invasion Anxiety* (Cambridge: Cambridge University Press, 2008), 4. See also David Walker, *Anxious Nation: Australia and the Rise of Asia, 1850–1939* (Brisbane: University of Queensland Press, 1999).

[29] Paul Kennedy, *The Rise and Fall of the Great Powers* (New York: Vintage, 1989).

'steady as she goes' United States that Australia has dealt with in the past. And, importantly, Trump is a *symptom*, not the *cause*, of the changed America about which Malcolm Fraser was rightly warning Australia.[30] (It is noteworthy that Mr Fraser's warnings preceded Donald Trump's election to the Presidency.) This changing America shows many of the symptoms of a great power in decline. This very fact is the reality that requires Australians to reassess their security relationship with the United States, to forge an independent foreign and security policy.

Australia, too, is a changed (and changing) country compared to the anxiously provincial British outpost that it was back in the 1950s when it signed the ANZUS treaty. At that time, many Australians still imagined themselves as being safely cocooned within the last vestiges of the British Empire. However, since the negotiation of the ANZUS treaty, developments have been taking the country in a very different direction. In 1947, the Chifley Government had already introduced a large-scale immigration programme which has since significantly increased economic growth and contributed to the development of a cosmopolitan outlook in the country.[31] In 1973, the egregious White Australia policy was finally repealed, opening up the country to non-white immigrants. Today, Australia's most important international trading partners are in the Asia-Pacific (with China being the country's foremost trading partner). While nostalgia for things British still lurks in the interstices of Australia's political culture, realists are well-aware that Australia's future prosperity and security are inexorably tied into the Asia-Pacific region.[32] Since the 1960s (if not well before), as David Walker has argued, it is no longer possible for Australia to 'persist with the comforting evasion that it was a newly minted nation, innocent of history, but [is] a settler colonial society

[30] John Callaghan, Brendon O'Connor and Mark Phythion, *Ideologies of American Foreign Policy* (London: Routledge, 2019).

[31] Christina Ho, 'From Social Justice to Social Cohesion: A History of Australian Multicultural Policy,' in Andrew Jakubowicz and Christine Ho (eds.), *For Those Who've Come Across the Seas: Australian Multicultural Theory, Policy and Practice* (Melbourne: Australian Scholarly Publishing, 2013), 31–41.

[32] See for example: Australian Government, *Australia and the Asian Century* (the Henry Report) (Canberra: Department of Prime Minister and Cabinet, 2012); East Asia Analytical Unit, Department of Foreign Affairs and Trade, *Australia and North-East Asia in the 1990s: Accelerating Change* (Canberra: Australian Government Publishing Service, 1992); Ross Garnaut, *Australia and the Northeast Asian Ascendancy* (the Garnaut Report) (Canberra: Australian Government Publishing Service, 1989).

located in Asia and facing profound questions about its own racial history and regional identity'.[33]

A changed America and a changed Australia mean that ANZUS has become outdated and tangential to the major security concerns of both countries. Yet Australia's mainstream political leaders are extremely nervous about proposing revisions to the original treaty, fearing a massive voter backlash against any such move. However, given the rapidly changing realities surrounding ANZUS today, such faintheartedness amounts to a massive failure of leadership. As Malcolm Fraser warned, the alliance with America has become dangerous for Australia; it constitutes the most likely cause of an attack—for example, from either North Korea or China targeting the American communications bases at Pine Gap and/or the North-West Cape, and/or the military facilities for the American marines rotating through Darwin.

What Australia, as the 'anxious nation', needs to remember is that the country has been through all this before. After World War I, the decline of Britain as an imperial power had become palpable, even for the most dedicated Anglophile. In February 1942, Australians witnessed the catastrophic collapse of Britain's defence arrangements in Singapore. The fall of Singapore to Japan's imperial forces was swift and comprehensive. Thousands of lives were lost (many Australians among them) and thousands became prisoners of war of the Japanese. The bungling of British political and military officials was at the root of this humiliating defeat, demonstrating very clearly to the traumatised Australians that Britain was no longer the reliable security partner they had always assumed it to be.[34] Throughout the 1950s and 1960s nostalgia kept ties with the UK more or less intact. However, in 1971 a declining Britain announced its withdrawal from all its military commitments 'East of Suez', making it plain to its former colonies in Asia-Pacific that it had become a mere shadow of its imperial past. Then, in 1973, after several false starts, Britain joined what was to become the European Union. The sense of betrayal felt in Australia

[33] David Walker, *Stranded Nation: White Australia in an Asian Region* (Perth: University of Western Australia Publishing, 2019), 457.

[34] Christopher F. Shores, Brian Cull and Yasuho Izawa, *Bloody Shambles: The Drift to War to the Fall of Singapore* (London: Grub Street, 1992); David Horner, 'Australia in 1942: A Pivotal Year,' in Peter J. Dean (ed.), *Australia 1942: In the Shadow of War* (Cambridge: Cambridge University Press, 2013), 11–29.

at this time was intense. But the country adjusted, opening up trade relationships with Asia which have been more lucrative for its economy than any attenuated neo-imperial relationship with the UK. Even as the shambles of Brexit unfold, countries like China, Japan, India and the ASEAN states will remain at the centre of Australia's international trade. It is most unlikely that trade with a post-Brexit Britain would ever exceed the present levels of around six per cent.

It is clear, therefore, that Australia has successfully survived the shrinking of Britain into the dependent middle power that it is today (and soon to become a mendicant nation if/when Brexit goes ahead). This is evidence that Australia has the cultural and material resources to deal with the decline of its second 'great and powerful friend', the United States. Equally, Australians need also to recall that New Zealand's withdrawal or expulsion from ANZUS (both interpretations are plausible),[35] way back in 1985, has not resulted in catastrophic consequences for that country's security. It's time for the country to become an independent state actively contributing to peace-making and the making of a post-fossil fuels world. How can this be achieved?

2. *From mendicant nation to global citizen*

In April of 1955, in the new Republic of Indonesia's city of Bandung, a conference of leaders from independent African and Asian states caught the attention of the West. The list of attendees included Zhou Enlai from Communist China who conducted an effective charm offensive throughout the conference's proceedings, much to the chagrin of both Washington and Canberra. As David Walker explains:

> It was the largest gathering of Afro-Asian nations the world had seen, brought together to promote goodwill and cooperation, discuss shared cultural and economic problems, address racism and colonialism and advance world peace. Of even more significance, the Bandung conference expressed a determination that the voices of formerly colonised peoples would at last be heard.[36]

[35] Amy L. Catalinac, 'Why New Zealand Took Itself Out of ANZUS: Observing "Opposition for Autonomy" in A-Symmetric Alliances,' *Foreign Affairs Analysis*, 2010, 6, 317.

[36] Walker, *Stranded Nation*, 193.

The Non-Aligned Movement (NAM) that emerged from Bandung has been unnecessarily maligned over the years, especially by its big power critics who see it as either a nuisance, an irrelevance, or—more likely—a threat to their hegemonic ambitions. However, as the world comes to grips with the post-Cold War era, NAM's guiding principles need to be revisited. This revisiting must begin with the determined persistence of NAM's ideals in the face of concerted attempts to crush the Movement. In the words of Govind Narain Srivastava:

> NAM continues to demonstrate the refusal of the small and the weak nations to be mere observers of international events and followers or satellites of contending power blocs. The Movement stands for the principles of self-determination for all peoples and for peaceful-coexistence between states with different ideologies'.[37]

Over the years, the most salient principles driving NAM have been decolonization, disarmament, development and advocating for détente between conflicting big powers.[38] While these principles have never been realised in full measure—or, perhaps, even in half measure—they are most certainly worthy of re-evaluation as world leaders struggle to deal with the mounting threats of nuclear conflict and environmental destruction on a terrifying scale. Jürgen Dinkel observes that over the years of NAM's existence, Western critics sought to highlight the divergent interests of its members while assuming it would fade away in the wake of the Cold War. He notes:

> Confounding the assumptions of Western observers, the movement neither broke up in the early 1990s nor did it fade into obscurity in the non-aligned world. In Jakarta [1992] the non-aligned governments called for the movement's reform, and they have continued to get together regularly for summit conferences into the present.[39]

[37] Govind Narain Srivastava, 'The Future of the NAM: How to Make the Movement Effective,' in Bantarto Bandoro (ed.), *Non-Aligned Movement: Its Future and Action Programme* (Jakarta: Centre for Strategic and International Studies, 1992), 36.

[38] K.P. Misra, 'Ideological Bases of Non-Alignment (An Overview),' in Hans Köchler (ed.), *The Principles of Non-Alignment* (London: Third World Centre for Research and Publishing, 1982), 62–74.

[39] Jürgen Dinkel, *The Non-Aligned Movement: Genesis, Organization and Politics (1927–1992)* (Leiden: Brill, 2019), 258.

A post-ANZUS Australia could learn a great deal from NAM despite the fact that the removal (or collapse) of the always ambiguous reassurances of America's security guarantees would be a massive challenge for the country to absorb.

There are two broad options facing a post-ANZUS Australia that need to be considered. The first option would mean pursuing a hard power approach—that is, substantially increasing Australia's defence budget, aiming for a version of the armed neutrality strategy that David Martin recommended back in the 1980s.[40] However, there are four immediately obvious disadvantages that make this a less than desirable—or practical—option. (i) It would be extremely costly. Malcolm Fraser anticipated it would mean at least doubling the current defence budget. That is a conservative estimate. Trebling or even quadrupling the country's defence expenditure would be closer to the truth. (ii) It would lead to an arms race in the region resulting in Australia entering a spiral of ever-increasing defence spending to ensure a balance of power of sorts in the region. (iii) Lurking behind this option would be the danger that Australia's defence aficionados would press the Australian government to build a nuclear weapons stockpile, adding to the doomsday scenarios looming over a nuclear weapons proliferation strategy. Moreover, (iv) at the end of the day it is unlikely that Australia could ever be able to develop the military capability to defend itself from a great power with ill intent in the region.

The idea of armed neutrality (itself a contradiction in terms) is today even more problematic than when David Martin was advocating the policy for Australia in the 1980s. The contemporary technologies of war are even more hideous than all the previous centuries of humankind's warring activities put together. First, we are confronted by the grim realities of the latest advances in nuclear weapons production. The capacity of these weapons to completely annihilate the entire globe is a terrifying reality. There are five signatories to the Nuclear Non-Proliferation Treaty (NPT)—the United States, Russia, China, the UK and France—that nonetheless continue to stockpile these weapons. There are three states outside the NPT that openly declare their possession of nuclear weapons—North Korea, Pakistan and India. And there are some states

[40] David Martin, *Armed Neutrality for Australia* (Melbourne: Dove Publications, 1984).

that are known to covertly possess nuclear weapons of varying capabilities—Israel, South Africa and some of the former Soviet Republics.

In addition to the nuclear threat posed by these states (and possibly by other clandestine organisations or equally nefarious non-state actors), the vast array of contemporary warfare technologies is threatening human civilisation ever more each day. They include unmanned bombing drones with a record of civilian fatalities (grotesquely dismissed—or spuriously justified—by military officials as 'collateral damage'). There are guided missiles capable of reaching across whole continents, pinpointing targets with terrifying accuracy. There is the evil of biological warfare that can wipe out whole populations and devastate entire landscapes. There are bombs that can kill entire populations without destroying material infrastructure (buildings, bridges, roads). There are technologies that make night warfare possible. This list goes on. The point is that warfare today is now potentially so horrific that it can no longer be an option in any civilised security planning across the globe. At the same time, the whole propaganda industry glorifying war—jingoistic military parades, D-Day 'celebrations', ANZAC Day remembrances—needs to be critically interrogated. The sheer madness of war needs to be made clear—as Peter Hitchens has pointed out in *The Phoney Victory: The World War II illusion*, in which he argues that while Nazism needed to be defeated, the means used to defeat it by the allies were often as evil as the means used by the Nazis themselves.[41] The old adage that there are never any victors in war is truer today than at any time in human history.

This makes the second option the best strategy for a post-ANZUS Australia to pursue, requiring the country to engage in wide-ranging and sophisticated soft power niche diplomacy. The country cannot act effectively in this regard on its own. Instead of resorting to an isolationist Fortress Australia or Forward Defence strategy, Australia needs to develop a range of multi-layered diplomatic strategies, identifying states with similar interests and like-minded approaches on specific issues, or a range of issues, that it can help to resolve in collaboration with states that are ready to collaborate with Australia. The primary interest that the kinds of cooperation among states envisaged in this strategy would, in particular, seek to replicate is the NAM principle of persuading or forcing big powers (or superpowers) to exercise restraint in their dealings with each

[41] Peter Hitchens, *The Phoney Victory: The World War II Illusion* (London: Tauris, 2018).

other and with all other regional or global interests. Other interests could include regional and global cooperation to address the ravages of climate change, dealing justly with the world's refugee crisis, engaging in peace-making and peace-keeping operations—in short, being a friend of all and enemy of none.

For any of this to succeed however, Australia must also address the country's declining levels of governance integrity. At the top of this list must be the recognition of the rights of Indigenous Australians, the rights of asylum seekers, and transitioning swiftly to a carbon-free economy.

Conclusion

In her book, *The Good University*, Raewyn Connell writes:

> I belong to the generation that grew up with The Bomb, aware that we had an excellent chance of exterminating humanity by high-technology war. The current generation knows we are poisoning the planet and may exterminate humanity with climate change. We desperately need ways to generate better futures.[42]

The Bomb threat is still present, casting a terrible gloom across the globe. A post-ANZUS Australian foreign policy could, for example, collaborate with existing NAM countries in signing and ratifying the 2017 United Nations Treaty on the Prohibition of Nuclear Weapons. The International Campaign Against Nuclear Weapons (ICAN) has been energetically lobbying federal politicians to this effect, and signing and ratifying it has been adopted as policy by the most recent ALP national conference. The destruction of the environment, species extinction, and climate change are equally threatening to the future of the planet. Blind-sided by an alliance with a power that is a major cause of the global dilemmas outlined by Professor Connell, Australian governments over the years have been wilfully and stupidly complicit in their making. It's time for the country to rid itself of its mendicant gesturing towards the United States. It's time for it to step up to the challenges of peace-making in an ecologically sustainable world. It's time for Australia to adopt foreign and security policies that enable it to eschew mendicant posturing to any state, great

[42] Raewyn Connell, *The Good University: What Universities Actually Do and Why It's Time for Radical Change* (Clayton: Monash University Publishing, 2019), 10.

or small, and that will win for it regional and global respect as a good global citizen.

Bibliography

Australian Government (2012), *Australia and the Asian Century* (the Henry Report) (Canberra: Department of Prime Minister and Cabinet).

Ball, Desmond (1980), *A Suitable Piece of Real Estate: American Installations in Australia* (Sydney: Hale and Iremonger).

Ball, Desmond, Bill Robinson and Richard Tanter (2016), 'Australia's Participation in the Pine Gap Enterprise,' The Nautilus Institute, available at: https://nautilus.org/napsnet/napsnet-special-reports/australias-participation-in-the-pine-gap-enterprise/ (accessed 12 February 2019).

Bell, Coral (1988), *Dependent Ally: A study in Australian Foreign Policy* (Melbourne: Oxford University Press).

Beeson, Mark (2011), 'Can Australia Save the World? The Limits and Possibilities of Middle Power Diplomacy,' *Australian Journal of International Affairs*, 65 (5): 563–577.

Brandel, Nik, Øivind Bratberg and Dag Einar Thorsen (2013), *The Nordic Model of Social Democracy* (London: Palgrave Macmillan).

Broinowski, Alison (2007), *Allied and Addicted* (Melbourne: Scribe).

Broinowski, Alison (2017), 'Australia's Tug of War: Militarism Versus Independence,' in David Stephens and Alison Broinowski (eds.), *The Honest History Book* (Sydney: Newsouth): 271–286.

Burke, Anthony (2008), *In Fear of Security: Australia's Invasion Anxiety* (Cambridge: Cambridge University Press).

Callaghan, John, Brendon O'Connor and Mark Phythion (2019), *Ideologies of American Foreign Policy* (London: Routledge).

Campbell, John C. (1984), 'Review of Carsten Holbraad, *Middle Powers in International Relations*,' *Foreign Affairs*, 62 (5): 1247–1248.

Carr, Andrew (2017), 'ANZUS and Australia's Role in World Affairs,' in Peter J. Dean, Stephan Frühling and Brendan Taylor (eds.), *Australia's American Alliance* (Melbourne: Melbourne University Press): 66–85.

Catalinac, Amy L. (2010), 'Why New Zealand Took Itself Out of ANZUS: Observing "Opposition for Autonomy" in A-Symmetric Alliances,' *Foreign Affairs Analysis*, 6: 317–338.

Connell, Raewyn (2019), *The Good University: What Universities Actually Do and Why It's Time for Radical Change* (Clayton: Monash University Publishing).

Cooper, Andrew F. (1997), 'Introduction,' in Andrew F. Cooper (ed.), *Niche Diplomacy: Middle Powers After the Cold War* (Basingstoke: Macmillan): 1–24.

Curran, James and Stuart Ward (2010), *The Unknown Nation: Australia After Empire* (Melbourne: Melbourne University Press): 78–81.
Davies, Sara (2017), 'Asylum Seekers and Australia's Security,' in Beeson and Hamieri (eds.), *Navigating the New International Disorder: Australia in World Affairs, 2011–2015* (Oxford: Oxford University Press): 109–122.
Dennis, Richard (2018), 'Dead Right: How Neoliberalism Ate Itself and What Comes Next,' *Quarterly Essay*, Issue 70.
Dinkel, Jürgen (2019), *The Non-Aligned Movement: Genesis, Organization and Politics (1927–1992)* (Leiden: Brill).
East Asia Analytical Unit, Department of Foreign Affairs and Trade (1992), *Australia and North-East Asia in the 1990s: Accelerating Change* (Canberra: Australian Government Publishing Service).
Elliott, Lorraine (2017), 'The Environment in Australian Foreign Policy,' in Mark Beeson and Shahar Hamieri (eds.), *Navigating the New International Disorder: Australia in World Affairs 2011–2015* (Melbourne: Oxford University Press): 176–191.
Evans, Gareth and Bruce Grant (1995), *Australia's Foreign Relations in the World of the 1990s*, second edition (Melbourne: Melbourne University Press).
Fernandes, Clinton (2018), *Island Off the Coast of Asia: Instruments of Statecraft in Australian Foreign Policy* (Clayton: Monash University Publishing).
Fernandes, Clinton (2019), *What Uncle Sam Wants: U.S. Foreign Policy Objectives in Australia and Beyond* (London: Palgrave Macmillan).
Fraenkel, Jon (2019), 'Reassessing the 2003–2017 Regional Assistance Mission to the Solomon Island,' *RUSI Journal*, 164 (1): 52–61.
Fraser, Malcolm (with Cain Roberts) (2014), *Dangerous Allies* (Melbourne: Melbourne University Press).
Frühling, Stephan (2016), 'Wrestling with Commitment: Geography, Alliance Institutions and the ANZUS Treaty,' in Peter J. Dean, Stephan Frühling and Brendon Taylor (eds.), *Australia's American Alliance* (Melbourne: Melbourne University Press).
Garnaut, Ross (1989), *Australia and the Northeast Asian Ascendancy* (the Garnaut Report) (Canberra: Australian Government Publishing Service).
Henriksen, Alan K. (2005), 'Niche Diplomacy in the World Public Arena: The Global "Corners" of Canada and Norway,' in Jan Melissen (ed.), *The New Public Diplomacy: Soft Power in International Relations* (London: Palgrave Macmillan): 67–87.
Hilson, Mary (2008), *The Nordic Model: Scandinavia Since 1945* (London: Reaktion Books).
Hitchens, Peter (2018), *The Phoney Victory: The World War II Illusion* (London: Tauris).
Ho, Christina (2013), 'From Social Justice to Social Cohesion: A History of Australian Multicultural Policy,' in Andrew Jakubowicz and Christine Ho

(eds.), *'For Those Who've Come Across the Seas': Australian Multicultural Theory, Policy and Practice* (Melbourne: Australian Scholarly Publishing): 31–41.

Holbraad, Carsten (1984), *Middle Powers in International Politics* (New York: St. Martin's Press).

Horner, David (2013), 'Australia in 1942: A Pivotal Year,' in Peter J. Dean (ed.), *Australia 1942: In the Shadow of War* (Cambridge: Cambridge University Press): 11–29.

Kennedy, Paul (1989), *The Rise and Fall of the Great Powers* (New York: Vintage).

Langmore, John (2005), *Dealing with America: The UN, the US and Australia* (Sydney: Newsouth).

Langmore, John (2011), 'Learning from Norway,' *Griffith Review*, 32: 97–110.

Martin, David (1984), *Armed Neutrality for Australia* (Melbourne: Dove Publications).

Mearsheimer, John (2001), *The Tragedy of Great Power Politics* (New York: W. W. Norton).

Misra, K.P. (1982), 'Ideological Bases of Non-Alignment (An Overview),' in Hans Köchler (ed.), *The Principles of Non-Alignment* (London: Third World Centre for Research and Publishing): 62–74.

O'Neill, James (2019), 'When Better Than an Election to Have a Serious and Overdue Debate About Defence and Foreign Policy Objectives?' *Pearls and Irritations*, 24 April, at: https://johnmenadue.com/james-oneill-when-better-than-at-an-election-to-have-a-serious-and-overdue-debate-about-defence-and-foreign-policy-objectives/ (accessed 24 April 2019).

Papua New Guinea Post Courier, 17 September 2019. https://postcourier.com.pg/aust-needs-to-be-more-than-good-mates-to-the-pacific/ (accessed 17 September 2019).

Patience, Allan (2014), 'Imagining Middle Powers,' *Australian Journal of International Affairs*, 68 (2): 210–224.

Patience, Allan (2018), *Australian Foreign Policy in Asia: Middle Power or Awkward Partner?* (London: Palgrave Macmillan).

Patience, Allan and Chiraag Roy (2019), 'Interlocutors for Peace? Bringing Middle Powers in from the Theoretical Cold,' in Tanguy de Swielande et al. (eds.), *Rethinking Middle Powers in the Asian Century: New Theories, New Cases* (London: Routledge): 62–73.

Ravenhill, John (2018), 'Entrepreneurial States: A Conceptual Overview,' *International Journal*, 73 (4): 501–517.

Reynolds, Henry (2016), *Unnecessary Wars* (Sydney: Newsouth).

Robb, Thomas K. and David James Gill (2015), 'The ANZUS Treaty During the Cold War: A Reinterpretation of U.S. Diplomacy in the Southwest Pacific,' *Journal of Cold War Studies*, 17 (4): 109–157.

Ryan, Lyndall (2018), 'The Myall Creek Massacre: Was It Typical of the Time?' in Jane Lydon and Lyndall Ryan (eds.), *Remembering the Myall Creek Massacre* (Sydney: Newsouth): 85–99.
Shores, Christopher F., Brian Cull and Yasuho Izawa (1992), *Bloody Shambles: The Drift to War to the Fall of Singapore* (London: Grub Street).
Stephens, David (2017), 'ANZAC and Anzackery: Useful Future or Sentimental Dream?' in David Stephens and Alison Broinowski (eds.), *The Honest History Book* (Sydney: Newsouth): 136–150.
Srivastava, Govind Narain (1992), 'The Future of the NAM: How to Make the Movement Effective,' in Bantarto Bandoro (ed.), *Non-Aligned Movement: Its Future and Action Programme* (Jakarta: Centre for Strategic and International Studies).
Taulbee, James Larry, Ann Kelleher, and Peter C. Grovenor (2014), *Norway's Peace Policy* (London: Palgrave Macmillan).
Walker, David (1999), *Anxious Nation: Australia and the Rise of Asia, 1850–1939* (Brisbane: University of Queensland Press).
Walker, David (2019), *Stranded Nation: White Australia in an Asian Region* (Perth: University of Western Australia Publishing).

Allan Patience was educated at Monash University and the LSE. He has a Ph.D. from the University of Melbourne where he is a Principal Fellow in the School of Social and Political Sciences. He has held chairs in Political Science and Asian Studies in universities in Australia and the Asia Pacific. Recent publications include *Australian Foreign Policy in Asia: Middle Power or Awkward Partner?* (London: Palgrave Macmillan, 2018); (with Chiraag Roy) "Interlocutors for Peace? Bringing middle powers in from the theoretical cold," in T.S. de Swielande et al. (eds), *Rethinking Middle Powers in the Asian Century* (London: Routledge, 2019).

PART IV

Concluding Reflections

CHAPTER 15

Response: Utopian Versus Prophetic Visions

Mark G. Brett

The essays collected in this volume embrace an impossibly large set of questions, and each essay presents very different arguments. I cannot respond to them all, but this short reflection will observe some of the intersections between the various contributions and reflect on questions of academic strategy. If there is a unifying thread to this discussion it would be that all the contributors are articulating a 'prophetic' hope—the kind of hope that can call people to repentance via strong critique, without giving in to apocalyptic scenarios, whether an evangelical Christian utopia[1] or a Hollywood-style dystopia.

Many readers of this volume will no doubt be asking why it has taken so long for an understanding our planet's ecological crisis to reach the captains of industry and the seats of government. If anything, the economic and political developments of the past decade have dragged

[1]On evangelical apocalyptic thought, see especially: Wylie Carr et al., 'The Faithful Skeptics: Evangelical Religious Beliefs and Perceptions of Climate Change', *Journal for the Study of Religion, Nature and Culture* 6, 2012, 276–299; Matthew H. Goldberg et al., 'Social Identity Approach to Engaging Christians in the Issue of Climate Change', *Science Communication* 41, 2019, 442–463.

M. G. Brett (✉)
Whitley College, University of Divinity, Kew, VIC, Australia
e-mail: mbrett@whitley.edu.au

J. Camilleri and D. Guess (eds.), *Towards a Just and Ecologically Sustainable Peace*, https://doi.org/10.1007/978-981-15-5021-8_15

our ecological indicators backwards. How can this be the case when the clear majority of the scientific community reached disturbing conclusions regarding anthropogenic climate change so long ago? One significant factor was the rise of neo-liberalism in the 1980s, which has managed to incorporate much of the middle class, and at the same time, to undermine climate science.[2] Decades of illuminated ecological argument has not effectively connected with the range of polities within civil societies, nor even with the working classes in Western democracies, if recent voting patterns are anything to go by.

Explanations of the social and economic conditions that have led to the current climate emergency are of course contested, but Joseph Camilleri touches on one of the key indicators of our planetary problems when he describes the subordination of politics to the ideology of economic growth. Shelini Harris goes a step further to suggest that even the UN Sustainable Development Goals seem to be indebted in self-contradictory ways to ideas of growth.[3] The key question is how to decouple the connection between living standards and growth and to adopt, especially in the Global North, an economy of 'degrowth'.[4] Timothy Jackson prefers the terminology of *Prosperity without Growth*.[5] The utopian Left demands the eradication of capitalism, a goal that would take decades to achieve even if all the policymakers in the world agreed on the basic idea. Communist regimes have, however, been no less addicted to fossil fuels, and the question of how to achieve a carbon-neutral global economy will obviously need to be solved in a way that allows for a multiplicity of political systems. Given this political complexity, and acknowledging that sudden economic changes are very likely to be traumatic for the majority of an affected population, what would be the most effective first steps?

[2] Naomi Klein, *On Fire: The Burning Case for a Green New Deal* (New York: Simon and Schuster, 2019).

[3] See, e.g., Viktoria Spaiser, Shyam Ranganathan, Ranjula Bali Swain, and David J. T. Sumpter, 'The Sustainable Development Oxymoron: Quantifying and Modelling the Incompatibility of Sustainable Development Goals', *International Journal of Sustainable Development and World Ecology*, Vol. 24, No. 6, 2017, 457–470.

[4] Giacomo D'Alisa, Federico Demaria, and Giorgios Kallis, *Degrowth: A Vocabulary for a New Era* (London: Routledge, 2014).

[5] Timothy Jackson, *Prosperity without Growth: Economics for a Finite Planet* (London: Routledge, 2011).

Which are the key institutions that need reform, and what might be the catalysts of reform?

Chaiwat Satha-Anand draws attention to the historic resistance of the women of the Chipko movement in the 1970s, inspired by the Gandhian spirit of nonviolence and beginning at the local level. Alternatively, Bruce Duncan emphasizes the extraordinary international advocacy of Pope Francis, who commands the attention of vast numbers on the planet. In some respects, it is no surprise that a Catholic theologian should extol the virtues of *Laudato Si'*; more notable is to find the journalist Naomi Klein acknowledging the significance of such a papal encyclical,[6] as does the influential Australian economist, Ross Garnaut.[7] If there is a kind of spectrum between local and global actions, Allen Patience helpfully explores some of the options for 'middle' or 'non-aligned' state powers. But there is no time left to evaluate the effectiveness of each attempt to address the global predicament; we need action at every level from grass-roots community protest to the international collaboration of complex institutions. And all parties will need to be looking for opportunities to develop an overlapping consensus—such as the unlikely convergence of Pope Francis, Naomi Klein and Ross Garnaut.

At the micro- and the macro-levels, Anne Elvey is right to argue that we need to deconstruct the ideologies that have obscured the realities of cultural and economic violence.[8] Tony Birch calls this intergenerational 'slow violence'. The prevailing patterns of production and consumption are, as Ariel Salleh also points out, patterns of destruction. The implications of economic violence have too easily been neutralized by the imagery of 'growth', which is regularly affirmed as an unquestionable good by industry and governments. But growth can also be cancerous, of course. This point has always been obvious to Aboriginal and Torres Strait Islanders in Australia who have consistently resisted the ideology of industrial capitalism, and have provided enduring testimonies to the most sustainable economies imaginable, only to be regularly overruled

[6] Naomi Klein, 'A Radical Vatican', *The New Yorker*, 10 July 2015.

[7] Ross Garnaut, *Super-Power: Australia's Low-Carbon Opportunity* (Carlton: Back Inc./La Trobe University Press, 2019), 23–24, 48–49, 166, 173.

[8] Elvey refers in particular to Christos N. Kyrou, 'Peace Ecology: An Emerging Paradigm in Peace Studies', *International Journal of Peace Studies* 12, 2007, 73–92, but see also Johan Galtung, 'Cultural Violence', *Journal of Peace Research* 27, 1990, 291–305.

with cultural and economic violence.[9] We also need to acknowledge that the enduring testimonies of Indigenous peoples have derived their power from deep networks of culture and tradition, extending back before the recent inventions of neoliberal theory and even before the half-millennium of colonial indifference to the fate of the earth.

Secular theorists often rush to diminish the significance of 'religious' versions of ecology, imagining that public discourse can fabricate its own motivations without the aid of tradition. This version of Enlightenment rationalism is only marginally displaced by postmodern approaches that can sift the range of traditions on offer, looking for the right kind of ecological sensibility. Thus, for example, Freya Mathews suggests from a seemingly Olympian height that 'Abrahamic traditions' all demonstrate fatal flaws and looks elsewhere. A little closer to the ground, Norman Habel can dismiss whole blocks of biblical tradition attributed to 'the jealous God of Moses or the warrior God of Joshua' in order to find the God who discerns Wisdom in the earth. Intellectuals such as myself may be titillated by such critical insights, but sweeping away the deep convictions of religious communities can itself be understood as a kind of ontological violence that simply creates new forms of resistance to change.[10] Each tradition needs to explore its own ways of thinking about prosperity without growth, or of restoring relationships with the created order. This is, for example, what Zuleyha Keskin and Mehmet Ozalp have done in their contribution to this volume by showing the ecological dimensions of Islam.

Each of the long-standing traditions has their part to play, but several contributors propose new levels of collaboration *between* the major religious traditions.[11] This might be a difficult pill for white secularists to swallow, who are still in the minority when it comes to global statistics. The declaration signed in March 2019 by Pope Francis and Ahmad Al-Tayyeb is certainly one of the most outstanding examples of collaboration

[9] In the Australian context, see especially Anne Pattel-Gray, *The Great White Flood: Racism in Australia* (Atlanta: Scholars Press, 1998); Aileen Moreton-Robinson, *The White Possessive: Property, Power, and Indigenous Sovereignty* (Minneapolis: University of Minnesota Press, 2015).

[10] Erin K. Wilson, '"Power Differences" and "the Power of Difference": The Dominance of Secularism as Ontological Injustice', *Globalizations* 14, 2017, 1076–1093.

[11] See the work of the Australian Religious Response to Climate Change and the resources gathered on their website: www.arrcc.org.au.

to date.[12] Particularly in the sphere of economics, Protestants might need to acknowledge that Catholic Social Thought has been more resilient in holding theology and economics together, as has Islam, so this recent collaboration between Catholic and Islamic leaders could be understood to be building on some historic strengths. Especially in Protestant hands, Christian theologies of economics have been withering away over the past two centuries, and we need strategic efforts to lift the effective participation of the Christian churches within economic life more generally. Even the work over several decades of the World Council of Churches under the heading of 'Justice, Peace and the Integrity of Creation' has yielded mainly very general communiqués rather than detailed interdisciplinary dialogue with economists. Beyond the slogans, where are the concrete proposals?

Even if the Christian traditions are as redeemable as imagined by Pope Francis, and by the World Council of Churches, we would still need comparable developments in India and China, along with the other powerful economies of East Asia, most of which embody quite different cultural and religious traditions. The spirituality of the natural world (its meaning, purpose, connectedness and transcendence) needs to be recovered via all the relevant traditions, whether religious or not. Pope Francis has rightly recognized not only the importance of dialogue with Islam, but also the need to learn from Indigenous peoples: their voices 'should be at the centre of the implementation of the 2030 Agenda and at the heart of the search for new paths for a sustainable future'.[13] Theologians therefore need to push further into the implications of decolonial theologies, within which human and non-human creatures are equally dependent on the larger web of creation, connected to lands, waters and atmosphere. In the conversation with Indigenous peoples, I would want to emphasize three points very briefly.

First, settler colonial constructions of national sovereignty will need to be eclipsed by—to use the words of the Canticle of St. Francis—'Mother

[12] 'A Document on Human Fraternity for World Peace and Living Together', Abu Dhabi, 4 February 2019, http://w2.vatican.va/content/francesco/en/travels/2019/outside/documents/papa-francesco_20190204_documento-fratellanza-umana.html.

[13] Pope Francis, 'Conference on Religions and the Sustainable Development Goals (SDGs): Listening to the Cry of the Earth and of the Poor', Vatican City, 8 March 2019, http://w2.vatican.va/content/francesco/en/speeches/2019/march/documents/papa-francesco_20190308_religioni-sviluppo sostenibile.html.

Earth, who sustains and governs us'.[14] This wording, cited in the very first paragraph of *Laudato Si'*, has special resonance in Australia, most recently in the alternative understanding of sovereignty articulated in the Uluru Statement from the Heart:

> This sovereignty is a spiritual notion: the ancestral tie between the land, or "mother nature" and the Aboriginal and Torres Strait Islander peoples who were born therefrom, remain attached thereto, and must one day return thither to be united with our ancestors. This link is the basis of the ownership of the soil, or better, of sovereignty. It has never been ceded or extinguished, and co-exists with the sovereignty of the Crown.[15]

The extraordinary generosity embodied in the word 'co-exists' should not weaken any resolve on the part of Christian theology to repent of the dreadful legacies of the Doctrine of Discovery and the inventions of settler sovereignty.[16]

Secondly, the Indigenous spirituality that connects humans to the non-human creation extends back in time for hundreds of generations, but also forwards to future generations. Accordingly, an ecological repentance will need to reflect not only on past injustices but also on the injustices that are now being committed against future generations. Our Australian legal system, for example, can confess that native title rights and interests have been extinguished in the past, and even imagine some ways in which current generations may exercise their residual rights, but the rights of future generations seem to defeat even the most imaginative judicial activism. The idolatry of economic growth is considered common sense, while the suffering of our grandchildren can still be swept aside as a matter

[14] Pope Francis, *Laudato Si': On Care for Our Common Home*, 24 May 2015, par. 1, citing 'Canticle of the Creatures', in *Francis of Assisi: Early Documents*, Vol. 1, ed. Regis J. Armstrong, J. A. Wayne Hellman, and William J. Short (New York: New City Press, 1999), 113–114.

[15] The wording of the quotation comes mainly from African lawyer, Nicolas Bayona-ba-Meya, in the Western Sahara case (1975). Western Sahara, Advisory Opinion, ICJ Reports 1975, 85–86.

[16] See, further, Mark G. Brett, *Political Trauma and Healing: Biblical Ethics for a Post-colonial World* (Grand Rapids: Eerdmans, 2016), 145–162, 179–193; Mark Charles and Soong-Chan Rah, *Unsettling Truths: The Ongoing, Dehumanizing Legacy of the Doctrine of Discovery* (Downers Grove: InterVarsity, 2019).

of economic externality—even as the climate scientists relentlessly report their terrifying statistics.[17]

Thirdly, how are the spiritual relationships with the natural world to be rewoven among non-Indigenous people when habits of mind over half a millennium have systematically severed the connections? The vagaries of democracy have introduced a multiplicity of traditions, religions and communities, each with their own worldviews. The multiculturalism of our institutions allows for a seemingly endless debate about the very idea of a common good. And these social conditions also impact on Aboriginal and Torres Strait Islander groups in Australia in complex ways.

This is exemplified in the debates among the Wangan and Jagalingou people who are doing battle with the Adani Mining company in Queensland. A key controversy turns on the validity of an Indigenous Land Use Agreement (ILUA) that was secured under vexed circumstances in April 2016 with some traditional owners of the land. It seems that the intra-Indigenous conflicts have been exploited by the mining industry, and also by the federal government, who amended the *Native Title Act* in 2017 in a way that supported Adani's legal strategy.[18] In this and many other cases, external parties can drive a wedge through Aboriginal groups by offering incentives that divide the community. As Tony McAvoy has rightly argued, the current system puts traditional owners under duress to reach an agreement with a mining company, or risk losing any compensation at all. So even among Aboriginal and Torres Strait Islander groups, people are often torn between their highest spiritual ideals and a pressure to negotiate agreements that represent the lesser of evils. This is an intra-Indigenous debate about what counts as the common good, but the example has many analogies in non-Indigenous society.

When developing effective strategies for dealing with the fossil fuel industries, it will not be possible to stay entirely within one's own tradition and culture. Even the most brilliant works of ecocriticism or legal theory are not enough when it comes to changing the world. We need

[17] Regarding future generations, see the exceptional work of John Quiggin, 'Equity between Overlapping Generations', *Journal of Public Economic Theory* 14, 2012, 273–283; idem., *Economics in Two Lessons: Why Markets Work So Well, and Why They Can Fail So Badly* (Princeton: Princeton University Press, 2019), 196–213.

[18] See the interview with Aboriginal silk Tony McAvoy, himself a Wangan and Jagalingou traditional owner, at https://www.sydneycriminallawyers.com.au/blog/the-bias-in-the-native-title-system-an-interview-with-barrister-tony-mcavoy-sc/. The legislation is available at https://www.legislation.gov.au/Details/C2017A00053.

to develop more multidisciplinary collaborations with a broad range of professionals and activists (including lawyers and economists), each of whom can engage in their very particular idiom with the overlapping complexities of state and federal governments, the courts, the media and the variety of institutions within civil society.

If, for example, people are concerned about jobs in north Queensland and support the Adani mine on those grounds, then creative thinking about employment needs to be part of any effective strategy. Those of us who gravitate towards moral and theological arguments will need to concede that complex policy problems need to be considered through several different lenses—reflecting multiple cultures, traditions and disciplines of thought, including disciplines we are not well disposed towards. Fortunately, the case for a carbon-neutral future in Australia can now be constructed in very convincing ways purely on the basis of economics, as illustrated in Ross Garnaut's book, *Super-Power*.[19] But the necessary changes to policy and practice will only be effected via detailed debates and disputes, and when it comes to debates, there is no virtue in abstract dismissals of an opponent (e.g., in dismissing Garnaut's work because it is 'capitalist'). People of peace will be those who become known for attributing the strongest possible version of an argument to their opponents. Some of us might simply call this love, while others will see it as a philosophical principle of charity that is characteristic of the best forms of argument. Either way, the habit of demonizing our opponents, or actively misrepresenting them, will not lead to a just and economically sustainable peace.

This principle of charity does not, of course, imply a commitment to relativism. Relativism in democratic contexts commonly has the effect of conceding power to those who have the numbers. Every tradition has its own limits, and every religion needs to articulate its own conceptions of reconciliation. Thus, Salim Farrar suggests that truth commissions in the Islamic world need to be framed in dialogue with traditional Islamic ethics, and in Australia, we would need to give weight to Aboriginal and Torres Strait Islander conceptions of truth-telling, rather than impose a conception of reconciliation that derives from the dominant culture. One aspect of the awe and wonder that Heather Eaton speaks about, I would suggest, is an unwillingness to subject others to our own conceptual

[19] See above, fn 7.

schemes. This does not mean that we can live entirely without conceptual schemes, but it does imply a need to develop practices of political conversation that are ready to consider the rights of others—others in the past, present and future, human and non-human.

The current models of democratic argument are in deep trouble. It is often pointless to ask a politician a serious question about the common good, because they will almost inevitably use their 30 second sound bite to demonize their opponents, and to answer in ways that have only tangential relevance to the original question. But demonizing politicians (as I have done in the previous sentence) is actually part of our problem; righteous indignation can stoke the fires just as easily as the power of the coal lobby. It is easy to take down politicians, lawyers, economic policy-makers, church leaders, and so on, and rewards for clever denunciations can have marvellous effects in social media. But an ecologically sustainable future cannot be built on a social media campaign or an academic article that scores well on a citation index. Our grandchildren will not thank us for righteous indignation that simply leaves our politics broken.

Bibliography

Brett, Mark G. *Political Trauma and Healing: Biblical Ethics for a Postcolonial World*. Grand Rapids: Eerdmans, 2016.

Carr, Wylie, Michael Patterson, Laurie Yung, and Daniel Spencer. 'The Faithful Skeptics: Evangelical Religious Beliefs and Perceptions of Climate Change'. *Journal for the Study of Religion, Nature and Culture* 6 (2012): 276–299.

Charles, Mark and Soong-Chan Rah. *Unsettling Truths: The Ongoing, Dehumanizing Legacy of the Doctrine of Discovery*. Downers Grove: InterVarsity, 2019.

D'Alisa, Giacomo, Federico Demaria, and Giorgios Kallis. *Degrowth: A Vocabulary for a New Era*. London: Routlege, 2014.

Francis, Saint. 'Canticle of the Creatures'. In *Francis of Assisi: Early Documents*, Vol. 1, edited by Regis J. Armstrong, Wayne Hellman, and William J. Short, 113–114. New York: New City Press, 1999.

Galtung, Johan. 'Cultural Violence'. *Journal of Peace Research* 27 (1990): 291–305.

Garnaut, Ross. *Super-Power: Australia's Low-Carbon Opportunity*. Carlton: Back Inc./La Trobe University Press, 2019.

Goldberg, Matthew H., Abel Gustafson, Matthew T. Ballew, Seth A. Rosenthal, and Anthony Leiserowitz. 'Social Identity Approach to Engaging Christians in the Issue of Climate Change'. *Science Communication* 41 (2019): 442–463.

Jackson, Timothy. *Prosperity without Growth: Economics for a Finite Planet.* London: Routledge, 2011.

Klein, Naomi. 'A Radical Vatican'. *The New Yorker*, 10 July 2015.

———. *On Fire: The Burning Case for a Green New Deal.* New York: Simon and Schuster, 2019.

Kyrou, Christos N. 'Peace Ecology: An Emerging Paradigm in Peace Studies'. *International Journal of Peace Studies* 12 (2007): 73–92.

McAvoy, Tony. 'The Bias in the Native Title System', 4 August 2018. Available at: https://www.sydneycriminallawyers.com.au/blog/the-bias-in-the-native-title-system-an-interview-with-barrister-tony-mcavoy-sc/. Accessed 8 January 2020.

Moreton-Robinson, Aileen. *The White Possessive: Property, Power, and Indigenous Sovereignty.* Minneapolis: University of Minnesota Press, 2015.

Pattel-Gray, Anne. *The Great White Flood: Racism in Australia.* Atlanta: Scholars Press, 1998.

Pope Francis. *Laudato Si': On Care for Our Common Home*, 24 May 2015. Available at: https://w2.vatican.va/content/dam/francesco/pdf/encyclicals/documents/papa-francesco_20150524_enciclica-laudato-si_en.pdf. Accessed 8 January 2020.

———. 'Conference on Religions and the Sustainable Development Goals (SDGs): Listening to the Cry of the Earth and of the Poor', Vatican City, 8 March 2019. Available at: http://w2.vatican.va/content/francesco/en/speeches/2019/march/documents/papa-francesco_20190308_religioni-svilupposostenibile.html. Accessed 8 January 2020.

Quiggin, John. 'Equity between Overlapping Generations'. *Journal of Public Economic Theory* 14 (2012): 273–283.

———. *Economics in Two Lessons: Why Markets Work So Well, and Why They Can Fail So Badly.* Princeton: Princeton University Press, 2019.

Spaiser, Viktoria, Shyam Ranganathan, Ranjula Bali Swain, and David J. T. Sumpter. 'The Sustainable Development Oxymoron: Quantifying and Modelling the Incompatibility of Sustainable Development Goals'. *International Journal of Sustainable Development and World Ecology*, Vol. 24, No. 6 (2017): 457–470.

Wilson, Erin K. "'Power Differences' and 'the Power of Difference': The Dominance of Secularism as Ontological Injustice". *Globalizations* 14 (2017): 1076–1093.

Professor Mark Brett FAHA, teaches Hebrew Bible and ethics at Whitley College in Melbourne, within the University of Divinity. He was raised in Papua New Guinea, which produced a lifelong interest in the cultural contexts of education and theology. His research has focused on ethnicity and postcolonial studies, and among many other works, he is the author of *Decolonizing God: The Bible in the Tides of Empire* (Phoenix Press, 2008), *Political Trauma and Healing: Biblical Ethics for a Postcolonial World* (Eerdmans, 2016) and *Locations of God: Political Theology in the Hebrew Bible* (Oxford University Press, 2019). During 2005–2008, he also worked with Traditional Aboriginal Owners to develop new frameworks for the negotiation of native title claims within the state of Victoria.

CHAPTER 16

The Winter of Our Discontent and the Promise of Spring

Joseph Camilleri

We appear to have reached one of those extraordinary moments in history when people everywhere feel vulnerable, anxious and resentful. The same may be said of the planet as a whole.

That we live in a deeply troubled world is to state the obvious. We need only think of the succession of financial, environmental and humanitarian crises, epidemics, small and larger conflicts, genocides, war crimes, military interventions and terrorist attacks, and social and political institutions that seem powerless to prevent any of it.

Stress and anxiety are experienced differently in different places. For some, it is the loss of job security, for others grinding poverty in the midst of affluence and for others still the frightening loss of identity in the wake of dramatic demographic and cultural change. For many individuals, communities and even entire societies, it is the ravages of dispossession, humiliation and marginalisation. And for a younger generation, it is the likely trail of devastation climate change will leave in its wake. As

J. Camilleri (✉)
La Trobe University, Melbourne, VIC, Australia
e-mail: j.camilleri@latrobe.edu.au

J. Camilleri and D. Guess (eds.), *Towards a Just and Ecologically Sustainable Peace*, https://doi.org/10.1007/978-981-15-5021-8_16

much of this book is at pains to show, the pervasive violence that afflicts women, indigenous communities, workers and the unemployed, minorities, combatants and civilians in war, refugees and other living species has become the defining feature of life in the twentieth and early twenty-first centuries. Violence assumes many forms—social and ecological as much as physical—but it is ubiquitous as is the experience of stress and anxiety which the use or threat of violence inevitably provokes.

Many of the chapters in this book dwell on the victim's experience of violence, while others focus on the underlying structure and modalities of violence. Some draw our attention to gender violence and its breeding grounds, the persistent mindsets and institutions of patriarchy and the world's war-making machines. Others highlight the habits and processes of 'othering' and 'objectifying', which manifest themselves in diverse social settings. A case in point is Indigenous dispossession, which is largely the product of the colonialist mindset and practice that have been a feature of European settler societies. Others still place the spotlight on the plundering of the environment. Exploitative resource extraction, polluting industries and consumerist lifestyles are all part of the same addictive and incalculably destructive dynamic that has routinely excluded nature from personal and institutional decision-making. Another instance of the exclusivist mentality cited by several authors is the callous response of many Western electorates and their governments to the plight of refugees and asylum seekers. Their disregard of both international law and basic moral norms points to a striking inability to identify with human suffering. We are dealing here with a deeply embedded societal outlook that has been aptly labelled the 'globalisation of indifference'.

The ground covered in the preceding pages is not confined to analysis or diagnosis of the contemporary ailment. We are offered a rich collection of insights, values and principles that can inform the vision of a just, peaceful and ecologically sustainable future. An inclusivist ethic is proposed as the necessary antidote to the practices and structures of exclusion. The inclusion of nature in our decision-making, a recurring theme of the book, requires a new consciousness, a shift to an anthropocosmic or Earth-centric worldview which Freya Mathews suggests would have much more in common with Indigenous cosmologies than with Abrahamic faiths. As a stepping stone to this still distant goal, she proposes practical forms of conservation that are attentive to the needs and rhythms of different forms of life and their habitats, and enable an immersion into 'life-worlds outside the blind bubble of modern civilization'. For Heather Eaton, the key to an alternative vision lies in nonviolence, understood as disengagement from domination and the various forms of oppression which accompany it. But an equally important ingredient is the sense of wonder and awe, which expands awareness and educates, awakens us to the beauty and complexity of the natural world, and instils a deeper appreciation of our responsibility to other humans and to nature.

Applied to Australia's circumstances, this reimagining of the future entails for Anne Elvey a commitment to 'decolonisation', a new approach to Country that acknowledges the truth, namely the unjust dispossession of Indigenous people but also the devastation of land inflicted by European agricultural, mining and manufacturing practices. Importantly, we are reminded that such acknowledgement requires institutional and constitutional change—public apologies and eloquent words are not enough. Tony Birch expands on this theme by his emphasis on Country understood as an 'an autonomous entity with inherent rights', which tells us that a new relationship with the Indigenous peoples necessarily entails a new relationship to the land and all the living species that inhabit it.

Decolonisation as both idea and practice is equally relevant to Australia's engagement with the world, a reflection which has relevance for other allied countries that have also become dependent on one or other great power. For Allan Patience, the time has come for Australia to rethink its global position. If it is to fashion a responsible course as a middle power consistent with the principles of a just and ecologically sustainable peace, it must reject the imperial connection and the implicit illusion that protection lies in dependence on the imperial power.

In somewhat different vein, though pointing very much in the same direction, other contributors to the book concerned with the role of world religions offer a number of evocative and thought-provoking possibilities. Norman Habel proposes the 'God of wisdom' as a guiding concept which he likens to eco-wisdom, 'a driving force IN nature', a dynamic dimension of the universe that humans are called upon to observe in themselves and so give expression to an 'innate wisdom', which he describes as 'a deep awareness of our pivotal relationships with the cosmos, the planet, the natural world, the community and our own psyche'. The interdependence and interconnectedness of things are the centrepiece of Shelini Harris's chapter. She argues that, in contrast to the mechanistic and dualistic conceptions of reality which have become prevalent in modern rationalistic culture, eastern religions, notably Buddhism and Hinduism, point to a deeper spirituality based on 'a sense of self that is continuous with the rest of nature'. This spirituality enables humans to experience and express the interrelatedness and mutuality of all that is, and is therefore coterminous with ecological integrity.

The cosmologies, values and concepts we have just briefly reviewed are to a greater or lesser degree suggestive of pathways that can be followed in pursuit of a peace that is just and ecologically sustainable. Though much of this is implicit, some chapters explicitly outline and evaluate from different standpoints practical initiatives, some of which have already seen the light of day and yielded positive results. Based on a rich exploration of Islamic ethics, Salim Farrar makes the case that within the Islamic frame of reference peace is attainable at least in part, so long as it is pursued with knowledge and wisdom and a generous dose of humility, acknowledging the role of divine providence and grace and making sure that the striving for peace is consistent with the Qur'anic injunction to seek justice, goodness and decency. These precepts are said to have particular significance for societies that have recently experienced armed conflict and human rights violations, and to be applicable to the task of transitional justice, hence the potential role that third parties can play, in particular with the setting up of truth and reconciliation commissions. Bruce Duncan presents a different case study drawn from the Christian tradition, namely Pope Francis's influential encyclical *Laudato Si'*. The result of extensive consultation with scientists and religious scholars, the encyclical has provided the opportunity for intensive dialogue with political and business leaders and a wide range of other organisations, not least within UN forums. The initiative has also created fertile terrain for interreligious dialogue and collaboration, not just with respect to climate change but to the wider and deeper issues at the heart of the relationship between humanity and the larger Earth community.

This collection of essays represents a diverse yet complementary and mutually reinforcing array of observations and reflections, which offer useful signposts along the way to a just and ecologically sustainable peace. But the journey ahead is likely to be long and hazardous, for which we need a roadmap that adequately illuminates the way and resonates with a wide enough public, a roadmap which tells us where we have come from, where we are now, and where we are heading. This is no easy task, and for several reasons. First, it is not enough to revisit the past. To do it well requires us to rethink the normative and cognitive assumptions which guide our understanding of the past. Second, if they are to have traction, our notions of justice, peace and ecology must be recast in theory and practice to take account of the radically altered circumstances of the present age. Third, the response to violence must somehow take cognisance of the normative and institutional changes already under way, some positive, others less so. Finally, the effort to create a more habitable world must explore the many obstacles that stand in the way of a value-centred, dialogue-based exploration of a convivial future. Only another book can do justice to such a large and complex undertaking. What follows are a few preliminary thoughts on how we might constructively address these four interlinked considerations. The underlying argument is that justice, peace and ecology must be approached holistically. Only by viewing them as an integrated whole can we hope to fashion a response equal to the challenges of epochal change in our time.

A closer look at three interlocking and far-reaching trends might go some way towards developing a helpful roadmap. The first relates to the consequences of unfettered globalisation, or to be more precise the spiralling turbulence that has become a feature of the neoliberal order. The second involves the global power and civilisational shift that is now in full swing and the associated danger of new and destructive rivalries. And the third holds the promise, as yet unfulfilled, of a global dialogue that brings to the table the historical grievances, aspirations and insights of the world's religious and ethical traditions.

The first trend, which is central to our current predicament, is the incapacity, bordering on paralysis, of our social, political and legal institutions to regulate ethically and effectively the unprecedented volume, speed and intensity of cross-border flows—flows of goods and services, technology, money, arms, pathogens, greenhouse gases, people, images and information. These flows affect virtually every facet of human activity, including finance, commerce and trade, diplomatic relations, human rights, environment, scientific and cultural relations, information and communication, health and education, technological change, operations at sea and in space, and much else. Guided by unrelenting profit seeking within a largely unfettered market, they continuously transform the way we trade, work, produce, consume, travel, learn and communicate, often in ways that are highly injurious to the environment, human dignity, health and education, social equity and cultural literacy.

Under the neoliberal order, the lowering of barriers to international trade became an article of faith for most national governments. To free up, international trade unions had to be cut down to size, at times simply crushed, for union muscle kept wages high and made it harder to hire and fire as profit margins required. The task of governments everywhere was to make the economy as hospitable as possible to corporate interests generally, and foreign investors in particular. This in turn meant keeping wages law and doing away with or least severely curtailing labour and environmental standards.

And then came a succession of financial crises: the Latin American sovereign debt crisis in 1982, the stock market crash of 1987, the Asian financial crisis of 1997–1998 and the global financial crisis in 2007–2008, considered by many economists to be the worst crisis since the Great Depression. The effects of this last crisis, which are still with us, include the collapse of large financial institutions, slowing economic growth, the current wave of European sovereign debt crises, and rising levels of wealth and income inequality in many advanced economies.

These developments reflect a marked debasement of the quality of governance. National institutions are struggling in advanced industrial states as much as in failed states. Parliaments, governments and political parties, buffeted by volatile transnational markets and poorly planned technical innovation, lack the competence, resources, legitimacy and in many cases the will to manage complex cross-border flows. To hide their irrelevance, they resort to short-term fixes, political spin, security hysteria and fake nationalism.

Despite their promise, global multilateral institutions, including the United Nations, have not fared much better. Powerful states and corporate interests, often working in tandem, have actively resisted the advent of well-functioning cross-border regulatory institutions and processes. The fiasco of international climate change conferences is but the most glaring example of this trend. These days, UN reform, though urgently needed, barely rates a mention in diplomatic circles. As for civil society organisations, they are often lacking in resources, organisational capacity and even imagination. Very few have grasped the significance of the governance deficit of our times, let alone sought to collaborate across their respective areas of engagement to develop creative responses to the deficit.

Compounding the travails of what Richard Falk has described as 'predatory globalisation' is the second trend, the far-reaching geopolitical, geoeconomic and geocultural shift that has been under way for some time. Yet, it remains poorly understood in the West and nowhere more conspicuously than in the United States and some of its English-speaking allies. The failure to appreciate this shift, let alone deal constructively with it, carries immense risks for the future.

Experts and commentators—some rather belatedly—have rightly focused on China's rise. Yet, important as it is, China's renewed capacity to flex economic muscle does not fully explain the profound changes now sweeping across the globe and reshaping the social and political fabric of many countries.

Just as momentous has been the steady decline of US hegemonic power, to which we should add the reassertion of Russian strategic interests under Putin, and the steady but uneven rise of other centres of economic and political influence in Europe and Asia. As we shall see, these geopolitical trends assume their full significance only when placed in a wider cultural and civilisational, indeed planetary context.

First, a few words about the China factor. The obsessive concern with the 'Chinese threat', especially in the English-speaking world, risks causing dangerous and unnecessary tensions. The scale of the Chinese economic miracle is nothing short of remarkable, which is not to say that its effects are fully benign. Since the introduction of market reforms in 1979, China's real annual gross domestic product (GDP) has averaged an annual growth rate of 9.5 per cent; its GDP has doubled every eight years; its manufacturing output overtook that of Japan in 2007 and that of the United States in 2010. All of this amounts to what the World Bank has described as 'the fastest sustained expansion by a major economy in history'.

True enough, China's GDP growth rate has slowed substantially, from 14.2 per cent in 2007 to little more than 6 per cent in 2017, and greater attention is now given to the social and environmental costs of growth. But the comparison with the performance of the US economy is nevertheless revealing. For some time now, even in good years, US GDP has not risen above 3 per cent. Moreover, the US economy faces a perennial budget deficit which, exacerbated by recent tax cuts, was expected to exceed $1 trillion by 2020 and, according to the Congressional Budget Office, reach a staggering $1.5 trillion by 2028.

In the meantime, the US trade gap has continued to widen despite the Trump administration's 'America First' policies, which have led to tit-for-tat tariffs with the European Union, Canada and Mexico and an escalating trade war with China.

US recourse to punitive measures is at least as much about restricting imports as enhancing exports. It highlights the inability of several key US industries to compete against Asian or European imports in their own domestic markets. Simply put, the Trump strategy is an indicator of weakness rather than strength. It is doubtful that Trump's strong-arm tactics will exact from China, Japan, South Korea or the EU the kind of concessions that will lead to a substantial long-term reduction in the US trade deficit

But the paradox of US weakness becomes even starker when it comes to the projection of military power. Taken at face value, US actions and pronouncements since the end of the Cold War have radiated power, yet in many respects this is what may be called 'abstracted' or 'cosmetic' power, the kind of power that can inflict wholesale destruction and mesmerises but delivers few positive results.

Following a decline in the US military budget in the aftermath of the Cold War, military spending soon resumed its growth pattern, exceeding for the first time $700 billion in 2011, and after a temporary dip during the Obama years has risen again, with Congress approving a military budget of $738 billion for FY2020. Yet technological prowess and the flexing of military muscle have not translated into military victory on the ground or effective political influence. The power of the American state is more fragile than appears to the naked eye, as the war on terror, the disastrous war in Iraq, the protracted and punishing conflict in Afghanistan, and the unholy mess in Libya and Syria attest.

The US dilemma of seeking to project power on a global scale while expecting friends and allies to carry a substantial part of the cost, euphemistically described as burden sharing, has simply led to heightened tensions within American alliances. Nowhere is the blowback effect of this clumsy strategy proving more costly than in Europe.

As for NATO's unrelenting expansion to Russia's doorstep and the deployment of anti-ballistic missile systems in Romania and Poland, they have simply led Putin's Russia to confront US power, whether it be in Georgia, Ukraine, Syria or in the upgrading of its own nuclear arsenal.

And in Asia-Pacific, relentless US efforts to contain the expansion of Chinese influence, often justified by reference to China's militarisation activities in the South China Sea, have strengthened Beijing's determination to break through the US containment perimeter. This it is doing through an extensive and strategically applied mix of investment and aid incentives, of which its 'One Belt, One Road' (OBOR) initiative is by far the most ambitious.

Some Asian governments are no doubt wary of China's expanding influence, though less so than is often intimated in Western media. For many, China is their largest trading customer and a source of much needed infrastructure support. Some 78 countries and international organisations across Asia, Europe, Africa and the Middle East have already joined OBOR, conservatively estimated to be a $1 trillion global infrastructure programme. Complementing China's rise is the strong economic performance of other East Asian economies, including South Korea, Taiwan and Singapore, and more recently India, Indonesia and Vietnam. Asia as a whole now accounts for nearly two-thirds of global growth and is the world's most dynamic region by a considerable margin. The World Economic Forum expects Asia's GDP to overtake that of the rest of the world combined in 2020, to contribute roughly 60 per cent of global growth by 2030, and account for the majority of the world's middle class. Economics and geopolitics are at the heart of the seismic shift now under way. In this sense, a multi-centric world is rapidly emerging in which several major centres of wealth and diplomatic and organisational clout—some rising, others declining—are furiously competing to continue their ascent or arrest their decline.

There is more, however, to this epochal shift than economics and geopolitics. Yet, this other dimension, often overlooked, is crucial to the shape of things to come. We are inexorably moving towards a multi-civilisational world, in which the centre of gravity is shifting from the Occident to the Orient. The West-centric world, in which first Europe and then the United States held sway, is slowly but surely giving way to a new world in which other civilisational centres are emerging or re-emerging. Three such centres, the Sinic, Indian and Islamic cultural spheres, each with its uniquely rich and long history, have already made a dramatic appearance on the world stage. This is in no way to foreshadow Samuel Huntington's 'clash of civilizations'. It is simply to highlight the intense intersection of civilisational currents that is fast becoming a reality and brings with it obvious dangers but also new opportunities for mutually beneficial interaction.

Managing the cross-civilisational traffic of intellectual and cultural currents, not to mention competing economic expectations, will be one of the more demanding but potentially fruitful tasks of the next few decades. And dialogue that draws on the wisdom of different civilisations and traditions will be critical to the success of this enterprise. Only a wide-ranging inter-civilisational dialogue can provide the conceptual and practical foundation on which to build a just and ecologically sustainable peace.

In this context, dialogue is best understood as the common search for truth. Participants in dialogue will no doubt differ strongly on what the truth might be when it comes to this or that current or historical circumstance, but they accept the value of ethically based discourse and the need to conduct it in a spirit of truthfulness. They understand that in dialogue there is a place for both difference and commonality.

The world's major ethical traditions share a deep sense of the dignity of human life, a commitment to human fulfilment and a concern for standards of 'rightness' in human conduct. Here, we can include not only Hinduism, Buddhism, Judaism, Islam, Christianity and other world religions but also Confucianism, Western secular humanism and importantly the traditions of indigenous peoples, to which several contributors draw attention, not least Mark Brett in his incisive reflections towards the end of the book. There is enough common ground between these ethical worldviews to make possible an ongoing conversation about ethics in general and social and ecological ethics in particular.

At the same time, dialogue acknowledges the importance of difference. Each tradition has its distinctive ethos and symbolism, linguistic structure and customs, artistic and intellectual achievements, its own perspectives on what constitutes ethical conduct, its own understanding of personal and social relationships, and crucially its own understanding of the natural world and of humanity's place within it. Such differences need not stand in the way of meaningful dialogue either within or between cultural, religious or political traditions. Dialogue is enriched as much by diversity as by commonality. In a profound sense, diversity and commonality are not antithetical but complementary. Diversity is an integral part of the human inheritance. All human beings, regardless of their religious, cultural, philosophical or political loyalties, share the same civilisational inheritance. Regardless of background or viewpoint, we all share something of the priceless achievements and perspectives of different cultures and civilisations.

Dialogue, understood as an encounter across cultural, religious, philosophical, ethical and civilisational boundaries, enables participants to listen to the other, become open, sensitive, even vulnerable to the other's perspectives, concerns and grievances. In this sense, the participants in cross-cultural dialogue may be said to embark on a journey as much of 'self-discovery' as of 'discovery of the other'. This is another way of saying that to engage in dialogue is to hold one's own tradition and political preferences up to critical examination, to rediscover the fundamental ethical impulse which sustains one's worldview and to consider ways of adapting it to the new circumstances of our epoch. Dialogue works best when it fosters profound soul-searching, however painful it may be, within and between communities, faiths, cultures and civilisations. Applying this prescription to the dialogue between the West and Asia, or to be more accurate between Occident and Orient, will be especially taxing. This dialogue will require participants to share their often conflicting narratives, listen to one another's experience of pain, acknowledge past wrongs and accept responsibility for righting the wrongs of the past.

Such dialogue can prove immensely valuable in diverse settings, but three bear directly on the prospect of a just and ecologically sustainable peace: situations where communities have had to endure oppression and marginalisation at the hands of dominant majorities, as in the case of Indigenous minorities that have been subjected to settler colonialism; troubled relationships between the West and the non-West, whose far-reaching and often poorly understood consequences are still with us; and policy debates in relation to the multifaceted ecological crisis that now exercises every level of governance. It is in these highly contested and intersecting contexts that the dialogical project can make its most telling contribution. The formal and informal dialogues that are set in train enable existing entitlements and obligations to be reviewed and a healthier relationship to be cultivated between humans and the rest of the Earth community, and a more prudent balance to be struck between risk taking and risk avoidance. The dialogue of analysis is complemented by the dialogue of action.

The dialogical enterprise is especially suited to our increasingly interdependent, interconnected, yet multicentred world. At a time when governments, parliaments and political parties are in disarray, the dialogue of religions, cultures and civilisations provides civil society with a new focus and new forms of local, national and transnational engagement. The nation state can no longer be considered the exclusive ordering principle in human governance or for that matter the exclusive form of cultural identification and political allegiance. As Mark Brett intimates, if a just and ecologically sustainable peace is the objective, both governance and citizenship have to function within a multi-spatial, multidimensional frame of reference, and this opens up new arenas and possibilities for dialogue.

It enables different voices to be heard and rising levels of discontent, marginalisation and homelessness to receive the attention they deserve. Religious, tribal, ethnic, racial, civilisational and increasingly ecological solidarities are looking for more convivial avenues to express themselves and negotiate differences. In dialogue that escapes the control of the state, they can do so in ways that are mutually beneficial and conducive to the public good understood in its true anthropocosmic sense.

All this will not materialise overnight. No technical fix or social engineering can bring it to fruition. A major cultural shift will be necessary. The signs of such a shift are now in clear view, but time is of the essence.

Appendices

Appendix A: Earth Consciousness

Norman Habel

At the conference in April 2019 in Melbourne entitled *Earth@Peace: A Justand Ecologically Sustainable Peace*, in addition to delivering the gist of the article on eco-wisdom (which appears in this volume as Chapter 8), I was also involved in a workshop on exploring ways to develop an Earth consciousness that would facilitate peace with Earth.

In that context I proposed we consider a covenant or treaty with Earth taking into account biblical precedents where Earth and domains of Earth are third parties in such agreements. It was agreed, however, that a treaty is probably not the best term for a public proclamation of our regenerative intentions towards Earth. As a result, I have prepared a revised document entitled 'The Terra Australis Statement of the Heart' with a subtitle 'Principles for Promoting Peace with Earth' (see Appendix B).

The title clearly reflects the document 'The Uluru Statement of the Heart' which has had a powerful influence in pressing for improved relations with Aboriginal Peoples and the corresponding treaty process. The contents of the 'Terra Australis Statement' reflect the essentials of my paper and the many contributions of the working group which was committed to developing a new personal relationship with Earth as a necessary step towards experiencing peace with Earth.

J. Camilleri and D. Guess (eds.), *Towards a Just and Ecologically Sustainable Peace*, https://doi.org/10.1007/978-981-15-5021-8

In the background also is the 'Serafino Declaration' developed at the Serafino Conference in March 2015 whose aim was to explore the spiritual in creation. The challenges included in this Declaration included the healing of Earth, the spirituality of country, Wisdom in creation, disasters in nature, the rights of nature and the consciousness of Earth beings (Buxton and Habel 2016, xiii–xvii). Also in the background is the Earth-Care Charter which I prepared to facilitate an earth-care consciousness in Lutheran Schools (Habel 2017).

Our hope is that this new *Statement of the Heart* will also be circulated and challenge individuals and organisations in Australia to rethink their relationship with Earth and publicly endorse the principles for promoting peace with Earth enunciated in the *Statement.* These principles include recognising that

- Humans are Earth Beings;
- Earth is a Living Planet;
- Earth is an Eco-Wonder;
- Australia is a Unique Domain;
- Australia is a Wounded Mother;
- Australia has a Relevant Voice.

A detailed analysis of these principles is outlined in the *2019 Charles Strong Trust Lecture* which can be found in the Charles Strong Trust website (www.charlesstrongtrust.org.au).

Appendix B: The Terra Australis Statement of the Heart: Making Peace with the Land of Australia

Based on Principles for Promoting Peace with Planet Earth

Norman Habel

As Earth Beings in Australia, we make the following Statement as a step toward creating peace with Planet Earth, and especially with the Land of Australia:

Principle 1: The Land of Australia is the Mother of Earth Beings

We confirm our identity as Earth Beings who live in the Land of Australia, a mother who has nurtured us and now calls us to make peace with her.

Principle 2: Australia is a Land Sustained in a Cosmic Sanctuary
We declare our intention to gain a deeper appreciation of how the Land of Australia is embraced by a living planet called Earth who also functions as a cosmic sanctuary that protects the Land of Australia.
Principle 3: The Land of Australia is an Eco-Wonder of Planet Earth
We need to discern afresh the deep web of ecological impulses that animate Planet Earth and learn to cooperate with them to create a harmonious community in the Land of Australia.
Principle 4: The Land of Australia is a Wounded Child of Mother Earth
We regret the numerous acts of violence that the Land of Australia has experienced at the hands of European settlers and we seek means of regeneration and restoration of the land previously preserved by her Aboriginal custodians.
Principle 5: Australia is a Land with a Voice
We ask that the voice of the Land of Australia and her Aboriginal custodians be heard so that the government of this land becomes acutely aware not only of the environmental crisis but also of the sovereignty and rights of the original custodians of the Land of Australia.
Principle 6: The Land of Australia has a Sacred Site at its Centre
The Land of Australia is connected by many sacred sites, Uluru being especially significant. Uluru could be viewed as a common spiritual symbol of peace for all Australians, no matter what their faith or orientation. Thereby, all Australians can express peace between themselves and Planet Earth.

Bibliography

Buxton, Graham and Habel, Norman. *The Nature of Things: Rediscovering the Spiritual in God's Creation*. Eugene, OR: Pickwick, 2016.

Habel, Norman C. *The Earth-Care Charter and 95 Eco-theses*. Adelaide: Lutheran Education, Australia, 2017.

Index

J. Camilleri and D. Guess (eds.), *Towards a Just and Ecologically Sustainable Peace*, https://doi.org/10.1007/978-981-15-5021-8

D

The manufacturer's authorised representative in the EU is Springer
Nature Customer Service Centre GmbH, Europaplatz 3, 69115 Heidelberg,
Germany. If you have any concerns regarding our products, please
contact ProductSafety@springernature.com

Printed and bound by CPI Group (UK) Ltd, Croydon, CR0 4YY
30/09/2026
02225502-0013